ENVIRONMENTAL DATA HANDLING
George B. Heaslip

THE MEASUREMENT OF AIRBORNE PARTICLES
Richard D. Cadle

ANALYSIS OF AIR POLLUTANTS
Peter O. Warner

ENVIRONMENTAL INDICES
Herbert Inhaber

URBAN COSTS OF WEATHER MODIFICATION
Terry A. Ferrar, Editor

CHEMICAL CONTROL OF INSECT BEHAVIOR: THEORY AND APPLICATION
H. H. Shorey and John J. McKelvey, Jr., Editors

# CHEMICAL CONTROL
# OF INSECT BEHAVIOR

# CHEMICAL CONTROL OF INSECT BEHAVIOR
## THEORY AND APPLICATION

EDITED BY

## H. H. SHOREY
UNIVERSITY OF CALIFORNIA
RIVERSIDE

## JOHN J. McKELVEY, JR.
ROCKEFELLER FOUNDATION
NEW YORK

A WILEY–INTERSCIENCE PUBLICATION

**JOHN WILEY & SONS**

**New York • Chichester • Brisbane • Toronto**

**Library of Congress Cataloging in Publication Data:**

Main entry under title:
   Chemical control of insect behavior.

   (Environmental science and technology)
   "A Wiley-Interscience publication."
   Includes index.
   1. Insects—Behavior. 2. Animal communication.
   3. Pheromones. 4. Insect control. I. Shorey,
Harry H. II. McKelvey, John J.

QL496.C43     595.7'05     76-46573
ISBN 0-471-78840-6

Printed in the United States of America

10 9 8 7 6 5 4 3 2

# SERIES PREFACE

## Environmental Science and Technology

The Environmental Science and Technology Series of Monographs, Textbooks, and Advances is devoted to the study of the quality of the environment and to the technology of its conservation. Environmental science therefore relates to the chemical, physical, and biological changes in the environment through contamination or modification, to the physical nature and biological behavior of air, water, soil, food, and waste as they are affected by man's agricultural, industrial, and social activities, and to the application of science and technology to the control and improvement of environmental quality.

The deterioration of environmental quality, which began when man first collected into villages and utilized fire, has existed as a serious problem under the ever-increasing impacts of exponentially increasing population and of industrializing society. Environmental contamination of air, water, soil, and food has become a threat to the continued existence of many plant and animal communities of the ecosystem and may ultimately threaten the very survival of the human race.

It seems clear that if we are to preserve for future generations some semblance of the biological order of the world of the past and hope to improve on the deteriorating standards of urban public health environmental science and technology must quickly come to play a dominant role in designing our social and industrial structure for tomorrow. Scientifically rigorous criteria of environmental quality must be developed. Based in part on these criteria, realistic standards must be established and our technological progress must be tailored to meet them. It is obvious that civilization will continue to require increasing amounts of fuel, transportation, industrial chemicals, fertilizers, pesticides, and countless other products and that it will continue to produce waste products of all descriptions. What is urgently needed is a total systems approach to modern civilization through which the pooled talents of scientists and engineers, in cooperation with social scientists and the medical profession, can be focused on the development of order and equilibrium to the presently disparate segments of the human environment. Most of the skills and tools that are needed are already in existence. Surely a technology that has created such manifold environmental problems is also capable of solving them. It is our hope that this Series in Environmental Sciences and Technology will not only serve to make this challenge more explicit to the established professional but that it also will help to stimulate the student toward the career opportunities in this vital area.

*Robert L. Metcalf*
*James N. Pitts, Jr.*
*Werner Stumm*

# PREFACE

Man, in large part because of his own rampant increase in population, seems to be running a collision course with insects in competition for choice foods and fibers. If, within the years to come, he is to cope successfully with insects, he must gain as precise a knowledge as is possible of their behavioral systems and apply that knowledge in pest management programs. The promise of success in this area of investigation, combined with recent spectacular advances in chemical methodology, have encouraged entomologists, biochemists, physiologists, and scientists in related disciplines to accelerate their research on insect behavior. They have made remarkable progress not only in devising novel ways of combating insects but also in bringing about substantial improvements in the traditional ways of managing insect populations.

The behavior of insects in response to the chemicals in and around them ramifies the broad areas of research in entomology which The Rockefeller Foundation has chosen to support: developing biodegradable pesticides; examining juvenile hormones, antihormones, and the like for their potential usefulness in combating insects; studying the chemical and morphological resistance of plants to insect attack; and exploring the use of sex and other pheromones for insect control. For the research on pheromones, Dr. H.H. Shorey, Dr. R.M. Silverstein, Dr. D.L. Wood, and Dr. W.L. Roelofs, leaders of laboratories engaged in the multidisciplinary, multiuniversity programs under Rockefeller Foundation auspices, felt a need to summarize their findings, after 6 years of cooperative endeavor, and to compare notes, so to speak, with preeminent colleagues throughout the world. The Foundation Study and Conference Center in Bellagio, Italy, provided the venue for the 5-day conference in May 1975 that was organized to accomplish this objective and led to the publication of this book.

This book deals primarily with those chemicals that insects perceive by smell and which constitute their most important stimuli when they are at a distance from the source of chemical. Secondarily, it considers taste substances which are perceived by insects when in contact with the source. The book opens with an assessment of the physiological characteristics of the nervous system of insects that enable them to perceive such chemicals. It then considers how chemicals normally function to mediate insect behavior. Next, it treats of the diversity of chemicals that stimulate behavior. Finally, it evaluates how behavior-modifying chemicals can be used in pest management systems.

The authors of the chapters present the state of the art for the topics of their special competence. They have cautiously drawn conclusions about the significance and usefulness of the body of knowledge they have helped to amass on insect behavior, especially as this knowledge may apply to the control of insects in practical ways. But in each chapter readers will find a wealth of stimulating ideas reinforced by elegant research which makes them relevant to the development of insect control strategies. The book provides new concepts and guidelines that will assist researchers in insect sensory physiology, behavior, chemical ecology, and pest management.

We are grateful to the distinguished scientists who participated in the Bellagio Conference and who prepared the manuscripts for this work. We appreciate the support and the participation of the directors of the foundation's programs: Dr. Ralph W. Richardson in Quality of the Environment, and Dr. John A. Pino in Conquest of Hunger, and we owe a debt of gratitude to Dr. and Mrs. Olson, Director and Assistant Director of the Bellagio Conference Center for their role in facilitating the conference.

H.H. Shorey
John J. McKelvey, Jr.

*Riverside, California*
*New York, New York*
*January 1977*

# CONTENTS

# CHEMICAL CONTROL
# OF INSECT BEHAVIOR

## CHAPTER 1

# Interaction of Insects with their Chemical Environment

H.H. Shorey
*Department of Entomology*
*University of California*
*Riverside, California*

In the early evolution of animal communication, chemical messages were probably the first signals put to use (1). Today chemical communication appears to be the primary mode of information transfer in most groups of animals (2) (3). Even in so-called non social animals—such as protozoans, annelids, molluscs, nematodes, and many arthropods—chemical communication is used for such diverse functions as location of prey, avoidance of predators, and signaling to species mates prior to mating or during times of danger. The sophistication of this communication mode has reached its peak in those social insects and mammals that live as interacting groups of individuals in colonies or societies. Indeed, in many insects the great diversity of behavioral interactions and physiologic responses that can result when individuals receive chemical messages emitted by others of the same species may have been in large part the factor that allowed the evolution of high levels in sociality (4).

Man, unlike most other animals, is microosmatic, and minimizes his use of chemical communication with other organisms, both with individuals of his own species and with those of other species. Unfortunately, man is also anthropocentric in his research and, because he has little intuitive feel for chemical communication systems, until recently he has minimized his studies of this communication mode and maximized studies of visual and sonic communication among insects and other animals. This situation is being rectified rapidly: our knowledge of chemical communication among insects, for example, increased exponentially during the last decade. These advances are documented in the various chapters of this book.

The surge of interest in insect chemical communication systems is due to a number of interacting factors: an increased awareness of our vast ignorance of this vitally important aspect of insect biology; advances in microanalytical chemistry that have allowed identification of the minute quantities of chemicals involved; and an appreciation that the manipulation of the chemical communication systems of certain insects may be a valuable tool in pest management. Research scientists from many disciplines have now focused on and often specialize in studies of insect chemical communication. These disciplines include chemistry, morphology, physiology, ecology, behavior, and pest management. Without doubt, some of the advances in knowledge will provide guidelines and stimulation for research workers interested in exploring the chemical communication systems of noninsect groups.

The term *communication* has defied a definition that is satisfactory to most biological scientists. A broad definition is used here, patterned after Wilson (1): biological communication entails the release of one or more stimuli by one organism that alter the likelihood of reaction by another organism; the reaction is of benefit to the stimulus emitter, the stimulus receiver, or both.

Chemicals used in communication have not been satisfactorily defined either. Most investigators now use the terms *pheromone, allomone,* and *kairomone,* which are defined as follows:

1.  A *pheromone* is a chemical or a mixture of chemicals that is released from one organism and that induces a response by another individual of the same species. The term was first coined by Karlson and Lüscher (5) and Karlson and Butenandt (6), and modifications to the definition were proposed by Kalmus (7). The most important modification is the omission of the implication in the original definition that a pheromone be synthesized *de novo* by the emitting organism. The chemicals may be either synthesized *de novo* or acquired intact by the organisms from their food or other aspects of the environment. This modification seems necessary because the biosynthetic pathways or the manner of acquisition of most chemicals regarded as pheromones are totally unknown.
2.  An *allomone* is a chemical or a mixture of chemicals that is released from one organism and that induces a response by an individual of another species; the response is adaptively favorable to the emitter (8). Examples of allomones are the defensive secretions that are released by many insects and that are poisonous or repugnant to attacking predators. The "secondary substances" of plants (9-12) are also allomones. These substances, which usually have no other apparent role in the physiology of the plant, have presumably evolved as defense mechanisms against herbivorous insects and other animals.
3.  A *kairomone* is a chemical or a mixture of chemicals that is released from one organism and that induces a response by an individual of another species; the response is adaptively favorable to the recipient (13). Blum (see Chapter 10) regards kairomones as pheromones or allomones that have evolutionarily backfired. The original development in the releasing individual of systems for the production and release of the chemicals presumably occurred in response to selective pressures favoring the releasing individual. Secondarily, individuals of other species evolved advantageous responses to the same chemical signals, often to the disadvantage of the releasing individual. For example, the secondary substances of plants, discussed

above as allomones conferring protection against herbivores, have in some cases been seized upon by certain herbivores that have evolved an ability to tolerate or even detoxify the chemicals and to use the chemicals as kairomonal stimulants for aggregation or feeding on the emitting plant species (see Chapter 2). Also, the various exudates from humans and other warm-blooded animals that attract blood-sucking insects serve as kairomones to the insects (see Chapters 7 and 18).

The responses induced by pheromones, allomones, or kairomones may be immediate behavioral reactions or long-lasting physiologic changes. The three classes encompass all chemicals used in communication between organisms. A fourth class of chemicals, discovered to be attractants or repellents through chemical screening, is not known to occur naturally. However, probably most chemicals in this latter class will eventually be found either to be naturally occurring chemicals that are reacted to by animals in their normal lives or to be sufficiently closely related to naturally occurring chemicals that they cause the same types of behavioral reactions as the natural chemicals.

Another way to categorize chemicals that modify animal behavior is in terms of the types of behavior they induce. Dethier et al. (14) erected the following six categories that encompass most behaviorally active chemicals:

1. A *locomotory stimulant* is a chemical that causes kinesis reactions (see Chapter 5) that, in the absence of orientation cues, often cause the animals to disperse from an area by increasing the speed of locomotion or appropriately affecting the rate of turning.
2. An *arrestant* is a chemical that causes kinesis reactions that, in the absence of orientation cues, often cause the animals to aggregate near the chemical source by decreasing the speed of locomotion or appropriately affecting the rate of turning.
3. An *attractant* is a chemical that causes animals to make oriented movements towards its source.
4. A *repellent* is a chemical that causes animals to make oriented movements away from its source.
5. A *feeding, mating, or ovipositional stimulant* is a chemical that elicits one of these behavioral reactions.
6. A *feeding, mating, or ovipositional deterrent* is a chemical that inhibits one of these behavioral reactions.

The same chemical may produce more than one of the foregoing reactions, and the classification of the reactions is in some respects oversimplified. For example, an attractant may induce much more elaborate behavioral reactions in animals than mere orientation to the source of a chemical gradient. A sensing of chemical cues, wind-direction cues, and visual cues from the environment may all be involved in the "attraction" of an animal to a distant odor source. Still, this classification is bound to produce some order in the field if it is adopted by most scientists. The term *attractant*, especially, has been very overworked and distorted by some writers: *sex attractant* is often used when *sex pheromone* will be more correct. A sex pheromone may be a locomotory stimulant, an attractant, an arrestant, and/or a sexual stimulant. Certainly *sex pheromone* is not synonymous with sex attractant (14) (15).

An insect does not think; it reacts. The reactions are usually triggered by external stimuli and modified by a host of environmental variables, internal physiologic variables, and some rudimentary learning. The reactions are often highly stereotyped and cause the insects to perform appropriate behaviors that enhance species survival when appropriate stimuli are encountered. Much of the sensory world of the insect involved in stimulation or inhibition of such behaviors as mating, feeding, and egg-laying is chemical. The reactions of the insects to these chemicals are so predictable that if man could learn enough about the attendant behaviors he could literally make the insects jump through a hoop.

It cannot be stressed too strongly that the key to devising efficient systems for the management of insect pests by chemically modifying their behavior is the acquisition of an intimate knowledge of the insects' own normal use of chemicals. This important factor is too often overlooked. Once a pheromone or other behaviorally active chemical is identified, there is a tendency to feel that the research is all over, and that the chemical can be used as a bait in traps or perhaps distributed through fields, causing insect control. Rather, the identification of the chemical should open the door to more necessary research to determine whether the normal behavior of insects can be interfered with and manipulated to our advantage.

## REFERENCES

(1)   E.O. Wilson. 1970. Chemical communication within animal species. In *Chemical Ecology*. E. Sondheimer and J.B. Simeone, Eds. Academic, New York, pp. 133-155.

(2)   M.C. Birch. 1974. *Pheromones*. American Elsevier, New York, pp. 1-495.

(3)   H.H. Shorey. 1977. Pheromones, In *How Animals Communicate*. T.A. Sebeok, Ed. Indiana University Press, Bloomington, In press.

(4)   M.S. Blum. 1974. *Pheromonal* bases of social manifestations in insects. In *Pheromones*. M.C. Birch, Ed. American Elsevier, New York, pp. 190-199.

(5)   P. Karlson and M. Luscher. 1959. "Pheromones": A new term for a class of biologically active substances. *Nature* **183**:55-56.

(6)   P. Karlson and A. Butenandt, 1959. Pheromones (ectcohormones) in insects. *Ann. Rev. Entomol.* **4**:39-58.

(7)   H. Kalmus, 1965. Possibilities and constraints of chemical telecommunication. *Proc. 2nd Int. Congr. Endocrinol., Lond.*, pp. 188-192.

(8)   W.L. Brown, 1968. An hypothesis concerning the function of the metapleural glands in ants. *Am. Natur.* **102**:188-191.

(9)   V.G. Dethier. 1970. Chemical interactions between plants and insects. In *Chemical Ecology*. E. Sandheimer and J.B. Simeone, Eds. Academic, New York, pp. 83-102.

(10)  T. Eisner, and J. Meinwald. 1966. Defensive secretions of arthropods. *Science* **153**:1341-1350.

(11)  R.H. Whittaker. 1970. The biochemical ecology of higher plants. In *Chemical Ecology*. E. Sondheimer and J.B. Simeone, Eds. Academic, New York, pp. 43-70.

(12)  R.H. Whittaker, and P.P. Feeny. 1971. Allelochemics: Chemical interactions between species. *Science* **171**:757-770.

(13)  W.L. Brown, Jr., T. Eisner, and R.H. Whittaker. 1970. Allomones and kairomones: Transspecific chemical messengers. *Bioscience* **20**:21-22.

(14)  V.G. Dethier, L. Barton Brown, and C.N. Smith. 1960. The designation of chemicals in terms of the responses they elicit from insects. *J. Econ. Entomol.* **53**:134-136.

(15)  J.S. Kennedy. 1972. The emergence of behavior. *J. Aust. Entomol. Soc.* **11**:19-27.

CHAPTER 2

# Insect Chemosensory Responses to Plant and Animal Hosts

L.M. Schoonhoven

*Department of Animal Physiology, Agricultural University*
*Wageningen, The Netherlands*

Animal behavior is intimately related with the surrounding world. When we try to explain behavior, we must know which factors the animal selects through its sense organs in the complexity of the physical and biologic environment. In addition, we have to know according to which rules the sensory system codes the selected information it transmits as nerve potentials to the brain. The entomologist seeking to explain feeding and oviposition behavior of insects has to pay special attention to the chemical senses since they play a decisive role in many phases of behavior.

Ethology has provided important insight into the nature of environmental cues that elicit certain behavioral responses in the animal. The "sign stimulus," which releases a specific behavioral response, is often of a remarkably simple structure (1). This view seems to be true not only with visual stimuli but also with chemical signals. The pheromones of many insects are striking examples of sign stimuli, although recently cases have been found in which delicate mixtures of some compounds are involved, indicating that pheromones have a greater degree of complexity than hitherto had been presumed. *Pieris brassicae* caterpillars feeding on nonhost plants treated with sinigrin (2), and *Papilio ajax* larvae eating from filter paper treated with essential oils from their host plants (3) may be considered classical examples supporting the concept of a simple sign stimulus. In spite of substantial efforts, food selection in several other insects (e.g., Colorado potato beetle and the silkworm) could not be explained by the presence of simple sign stimuli. These cases may be due either to the fact that we have not discovered as yet the essential component or that the optimal stimulus consists of a

I wish to thank Dr. T.H. Hsiao for his helpful comments on the manuscript.

complex aggregate of conditions. This chapter furnishes some evidence supporting the latter assumption.

## I. RECEPTORS

Chemoreceptors of insects are located on several strategic parts of the body. The antennae bear olfactory receptors, but often also contact chemoreceptors. Gustatory receptors (and some olfactory organs as well) are located on various mouthparts in many insects species, as has been established using morphologic, electrophysiologic, and behavioral methods. The tips of the maxillary and labial palps are covered by chemoreceptors, which may number in the hundreds (as in the case of the migratory locust) (4). Recently, chemoreceptors have been detected also in the buccal cavity of several insects. Some lepidopterous larvae were shown to have papillalike organs on the epipharynx, which by electrophysiologic experiments were found to be chemoreceptive (5). Locusts likewise possess fields of gustatory receptors at the inner face of their clypeo-labrum (6), and the epipharyngeal organ in aphids (7) probably also has a chemoreceptive function. Tarsal receptors in beetles (8, 9), butterflies [e.g., (10, 11)], and aphids play a role in food and oviposition site selection (12). Oviposition is in some cases also guided by chemoreceptors located on the ovipositor. Locusts perceive salt content and soil moisture with these organs (13), and the entomophagous wasp *Venturia* is able to discriminate between parasitized and nonparasitized hosts with its ovipositor (14).

The distinction between olfactory and gustatory receptors, which is usually made on morphologic bases, is probably not absolute. Blowfly taste receptors (15) and contact chemoreceptors in the larvae of *Manduca sexta* (16) have been found to react to some volatile substances.

## II. RECEPTOR SPECIFITY

Sensory physiologists, taking advantage of the specific reaction insects show to some chemicals at low concentration, have in several cases successfully located the receptors involved. Schneider and his co-workers made a very detailed study of pheromone receptors on the antennae of the male silkworm moth and have concluded that a few molecules may be sufficient to elicit an action potential (17). These "specialist" receptors show a high degree of specifity in that they respond to one or only a few closely related chemicals in combination with a low sensitivity threshold. Larvae of *Pieris brassicae* butterflies appear to have specialized receptors for mustard oil glucosides (18), and the beetle *Chrysolina brunsvicensis* has on its tarsi specialized receptors that monitor the presence of hypericin, a polynuclear quinone which typically occurs in its foodplant *Hypericum hirsutum* (9). These findings clearly corroborate the concept of simple sign stimuli. In all instances, however, many more receptors were found. Some of them showed reactions to many compounds, which often were chemically not related. Such "generalists," mediating olfactory information, have been analyzed in several insects [e.g., (19)(20)], and although the concept of "generalists" versus "specialists" has been disputed (21)(22), receptors with wider specificity ranges seem to be of general occurrence.

A great variety of gustatory receptor cell types are known, and the function of many of them, for example, those sensitive to sugars, salts, amino acids, or "deterrents," can easily be appreciated. The function of some fairly specific inositol receptors in various caterpillars (23), or the significance of a chlorogenic acid detector in the Colorado potato beetle larvae (24) is somewhat less clear. In both cases these cells may represent feeding-stimulant receptors, since inositol enhances food uptake in several insects and chlorogenic acid stimulates feeding in the Colorado beetle (25), like in another species feeding on potato (26).

In the case of insects specialized on animals hosts, less information is known. However, from the data available it seems that these insects operate along the same principles. The adults of *Aedes aegypti*, for instance, have on their probosci and tarsi receptor cells reacting to salts and cells responding to a number of sugars and amino acids (27).

## III. COMPLEXITY OF NATURAL STIMULI

We have mentioned that many if not all insects are equipped with a sensory system which includes cells sensitive to chemicals more or less commonly present in their plant or animal hosts. These chemicals serve as adequate stimuli for their "own" cells, but in addition may modify the reactions of other cells. Other chemicals, which do not stimulate a particular cell when tested alone, may also modulate the activity of certain cells. Thus some amino acids may enhance or decrease the reaction to sugar or to inositol (28) and salt. The mosquito *Aedes aegypti* is attracted by (+)-lactic acid, which occurs in human sweat (29). Several fatty acids, and also some essential oils, affect olfactory cells in antennal receptors in this insect, either by stimulating or by inhibiting the resting activity of these cells (30). In a detailed study, Bernard (31) found that lactic acid attracts the bug *Triatoma infestans,* which responds to low concentrations of the chemical by means of antennal receptors. Here, too, there appears to exist a limited specificity: the same receptor is also sensitive to pyruvic acid, whereas probably other cells are activiated by butyric acid and ammonia. Adenosine triphosphate (ATP) acts as a phagostimulant in several hematophagous insects, such as *Rhodnius prolixus* (32), and also in the phytophagous aphid *Myzus persicae* (33). In many cases sodium salts decrease the responsiveness of sugar receptors. Calcium, likewise, depresses in *Pieris brassicae* caterpillars the reactivity of the sugar receptors, which consequently leads to a diminished food uptake (Fig. 1) (5). Such inhibitory and synergistic effects greatly complicate the sensory reaction patterns that arise from complex natural stimuli and, as a result, decrease the predictability of the resulting behavior. Also, in the case of hematophagous insects, mixtures of known attractants never completely duplicate the host odors. Apparently, the insects react optimally to the very complex natural stimulus (33). Again, synergistic effects are probably involved. For instance, the response of *Aedes aegypti* to lactic acid is considerably stronger in the presence of carbon dioxide than to lactic acid alone (29).

The complexity of the chemical composition of the food becomes still more difficult to grasp and to clarify when variations among individuals and fluctuations due to daily or seasonal rhythms are taken into consideration. The insect is able to notice, probably by way of its chemical senses, variations of such subtle fluctuations and may react, not only with behavioral responses, but also through morphogenetic changes. The peach aphid starts to produce alate offspring when the amount of methionine decreases in the diet (34); the rabbit flea enters the reproductive phase when the hormonal balance in the blood of its host changes (35).

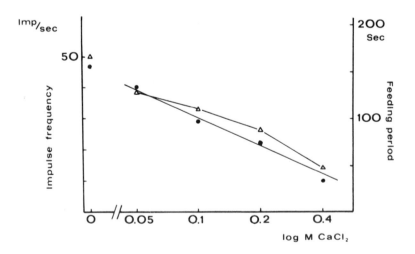

Fig. 1.    Inhibiting effect of CaCl₂ on the uptake by *Pieris brassicae* larvae of 4% agar substrate, containing 0.1 *M* sucrose. The feeding periods (*triangles*) represent the duration of single meals. The impulse frequency of the sucrose receptors (when stimulated with 0.1 *M* sucrose and 0.1 *M* NaCl) is reduced when CaCl₂ is added to the stimulus solution (*dots*). After data from Ma (5).

## IV.  VARIATIONS IN THE CHEMORECEPTIVE SYSTEM

The chemosensory system of an insect may vary in its sensitivity because of a number of factors. It seems likely, and in some cases it has been proven, that such variations are reflected in behavior.

1. Age. The sensitivity of some chemoreceptors alters significantly within one larval instar in the caterpillar of *Manduca sexta,* although the differences are not spectacular (36).
2. Feeding history. The composition of the food on which *Manduca sexta* larvae are grown influences considerably the sensitivity levels of various taste receptors. That food preferences are also influenced by the insect's feeding history is probably partly based on changes in the peripheral sensory system (36).
3. Effect of food deprivation. Immediately after a feeding period the sensilla on the domes of the palps of *Locusta migratoria* close, but they open gradually within the next 2 hrs. (37).
4. Adaptation rate. Feeding intensity on a certain food is correlated with the taste of it. For example, in the case of the larvae of *Pieris brassicae* when the food composition differs from the optimal conditions, the duration of the meals becomes correspondingly shorter. This could be explained by. differences in the adaption rates of the various taste cells involved.

   The adaptation rate of the sugar-sensitive cell resembles those found in the sinigrin cell and the amino acid cell. The "deterrent" cell, however, adapts more slowly. In a typical case 0.01 M sucrose was found to elicit 75 impulses during the third second of stimulation and decreased to a level of 25 pulses/sec after 30 sec, whereas the corresponding frequencies of the deterrent cell, stimulated by $3 \times 10^{-6}$ M strychnine, amounted to 60 and 45 pulses/sec, respectively. The "taste" of a mixture of the two chemicals thus changes gradually, becoming more repulsive as time passes. Low levels of deterrents, though not preventive of feeding, may reduce the lengths of the feeding periods (Fig. 2).
5. Individual variability. It is generally known that striking differences exist between individuals of the same species with regard to food acceptance [(e.g., (38)]. Conceivably such differences originate in different sensory information reaching the central nervous system. In one example, in which behavioral differences could be correlated with sensitivity levels in some specific taste receptors, this explanation could be verified (36). Moreover, the sensitivities of different receptor cell types vary independently of each other. Consequently, when different individuals are stimulated by complex stimuli, different action-potential patterns may arise. Especially in cases where the number of receptor cells is very small, as in caterpillars, such deviations from average sensitivity levels could be significant and explain individual differences in feeding behavior.

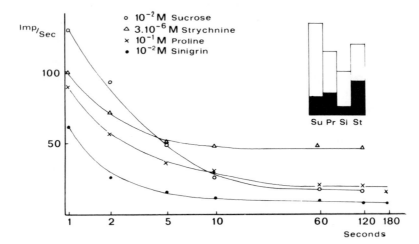

Fig. 2. Adaptation rates of four different receptor cells in the maxillary sensilla styloconica of a *Pieris brassicae* larva when stimulated with their typical stimuli. Abscissa: time in log scale. Ordinate: impulse frequencies. The white columns at the right indicate frequency ratios of the various cells during the first second of stimulation, the black columns represent the frequency ratios after 30 sec.

## V. CONCLUSION

Chemoreceptors of insects form a complicated and subtle sensory system that enables them to differentiate between many natural stimuli of fairly great diversity. When we think of key-lock systems releasing certain behavior, it should be stressed that the key (and consequently the lock, too) in several cases may be very intricate. Even the sign stimulus example par excellence, namely, sinigrin for *Pieris brassicae,* plays only a relatively minor role in food-acceptance behavior. Several other chemicals and physical characters have to be present at the same time to ensure a normal feeding reaction. In many other cases the relative importance of simple sign stimuli (such as sinigrin) may be small and therefore more difficult to find. Each insect species has evolved a sensory machinery that (by way of subtle inhibitory and synergistic interactions) is optimal in discriminating between acceptable and unacceptable food. Such machinery compares the incoming information with an inate (and in some cases modified by learning) ideal pattern. When the incoming information differs too much (and the permissible difference is greater in polyphagous than in monophagous species) from the desired pattern, the food is rejected. Thus it is usually difficult to design an artificial diet that in a dual-choice situation has a stimulus value equivalent to natural food. However, it is difficult to reconcile the finding that copper ions in some species of caterpillars severely alter sensory input, without preventing food uptake (18).

In evolutionary terms, it can be inferred that the chemical senses are highly plastic. For example, larvae of *Hyponomeuta malinellus,* which feed on apple, have in their maxillary sensilla a receptor that is sensitive to sorbitol, whereas the very closely related *H. rorellus,* occurring on willow trees, lacks this receptor (18). *Chrysolina brunsvicensis* beetles possess, in contrast to some other *Chrysolina* species, a specific hypericin detector in their tarsal receptors, enabling them to recognize their food plants (9).

## REFERENCES

(1)   N. Tinbergen. 1951. *The Study of Instinct.* Oxford University press, Oxford.

(2)   E. Verschaffelt. 1910. The cause determining the relation of food in some herbivorous insects. *Proc. Acad. Sci. Amster.* 13:536-542.

(3)   V.G. Dethier. 1941. Chemical factors determining the choice of food plants by *Papilio* larvae. *Am. Natur.* 75:61-73.

(4)   W.M. Blaney 1974. Electrophysiological responses of the terminal sensilla on the maxillarly palps of *Locusta migratoris* (L.) to some electrolytes and non-electrolytes. *J. Exp. Biol.* 60:275-293.

(5)   W-C. Ma. 1972. Dynamics of feeding responses in *Pieris brassicae* Linn. as a function of chemosensory input: A behavioural, ultrastructural and electro-physiological study. *Meded. Landbouwhogeschool, Wageningen* 72(11):1-162.

(6)   P.T. Haskell, and L.M. Schoonhoven. 1969. The function of certain mouth part receptors in relation to feeding in *Schistocerca gregaria* and *Locusta migratoria migratorioides. Entomol. Exp. Appl.* 12:423-440.

(7)   R.J. Wensler and B.K. Filshie. 1969. Gustatory sense organs in the food canal of aphids. *J. Morphol.* 129:473-491.

(8)   B. Stürckow. 1959. Über den Geschmacksinn und den Tastsinn von *Leptinotarsa decemlineata* Say. *Z. Vergl. Physiol.* 42:255-307.

(9)   C.J.C. Rees. 1969. Chemoreceptor specificity associated with choice of feeding site by the beetle *Chrysolina brunsvicensis* on its hostplant *Hypericum hirsutum. Entomol. Exp. Appl.* 12:565-583.

(10)  W.H. Calvert. 1974. The external morphology of foretarsal receptors involved with host discrimination by the nymphalid butterfly, *Chlosyne lacinia. Ann. Entomol Soc. Am.* 67:853-856.

(11)  W-C. Ma. and L.M. Schoonhoven. 1973. Tarsal contact chemosensory hairs of the large white butterfly, *Pieris brassicae,* and their possible rôle in oviposition behaviour. *Entomol. Exp. Appl.* 16:343-357.

(12)  F. Klingauf. 1974. Die Wirkung des Glucosids Phlorizin auf das Wirtswahlver-halten von *Rhopalosiphum insertum* (Walk.) und *Aphis pomi,* De Geer (Homoptera: Aphididae). *Z. Angew. Entomol.* 68:41-55.

(13)  M.J. Norris. 1968. Laboratory experiments on oviposition responses of the Desert Locust, *Schistocerca gregaria* (Forsk.). *Anti-Locust Bull.* 43:1-47.

(14)  W.K. Ganesalingam. 1974. Mechanism of discrimination between parasitized and unparasitized hosts by *Venturia canescens* (Hymenoptera: Ichneumonidae). *Entomol. Exp. Appl.* 17:36-44.

(15)  V.G. Dethier. 1972. Sensitivity of the contact chemoreceptors of the blowfly to vapors. *Proc. Natl. Acad. Sci. USA* 69:2189-2192.

(16)  E. Städler. 1974. Personal communication.

(17)  D. Schneider. 1970. Olfactory receptors for the sexual attractant (bombykol) of the silk moth. In *The Neurosciences: Second Study Program.* F.O. Schmitt, Ed.

Rockefeller University Press, New York, pp. 511-518.

(18) L.M. Schoonhoven. Unpublished results.

(19) V. Lacher. 1964. Elektrophysiologische Untersuchungen an einzelnen Rezeptoren für Geruch, Kohlendioxyde, Luftfeuchtigkeit und Temperatur auf den Antennen der Arbeitsbiene und der Drohne *(Apis mellifica L.)*. *Z. Vergl. Physiol.* 48:587-623.

(20) L.M. Schoonhoven, and V.G. Dethier. 1966. Sensory aspects of host-plant discrimination by lepidopterous larvae. *Arch. Néerl. Zool.* 16:497-530.

(21) E. Vareschi. 1971. Duftunterscheidung bei Honingbiene Einzelzell-Ableitungen und Verhaltensreaktionen. *Z. Vergl. Physiol.* 75:143-173.

(22) H. Sass. 1973. Das Zusammenspiel mehrerer Rezeptortypen bei der nervösen Codierung von Geruchsqualitäten. *Verh. Dtsch. Zool. Ges.* 66:198-201.

(23) L.M. Schoonhoven. 1972. Plant recognition by lepidopterous larvae. *Proc. Symp. Roy. Entomol. Soc. Lond.* 6:87-99.

(24) B.K. Mitchell and L.M. Schoonhoven. 1974. Taste receptors in Colorado beetle larvae. *J. Insect Physiol.* 20:1787-1793.

(25) T.H. Hsiao and G. Fraenkel. 1968. Isolation of phagostimulative substances from the host plant of the Colorado potato beetle. *Ann. Entomol. Soc. Am.* 61:476-484.

(26) J. Meisner, K.R.S. Ascher, and D. Lavie. 1974. Phagostimulants for the larva of the potato tuber moth *Gnorimoschema operculella* Zell. *Z. Angew. Entomol.* 77:77-106.

(27) Yu.A. Yelizarov and Ye.Ye. Sinitsyna. 1974. The contact chemoreceptors of *Aedes aegypti* (Dipt. Culicidae). (In Russian.) *Zool. Zh.* 53:577-584.

(28) V.G. Dethier. 1971. A surfeit of stimuli: A paucity of receptors. *Am. Sci.* 59:706-715.

(29) F. Acree, R.B. Turner, H.K. Gouck, M. Beroza, and N. Smith. 1968. L-lactic acid: A mosquito attractant isolated from humans. *Science* 161:1346-1347.

(30) V. Lacher, 1967. Elektrophysiologische Untersuchungen an einzelnen Geruchsrezeptoren auf den Antennen weiblicher Moskitos *(Aedes aegypti L.)*. J. Insect Physiol. 13:1461-1470.

(31) J. Bernard. 1974. Étude electrophysiologique de récepteurs impliqués dans l'orientation vers l'hôte et dans l'acte hématophage chez un Hémiptère: *Triatoma infestans. Unpublished thesis, University of Rennes.*

(32) W.G. Friend and J.J.B. Smith. 1971. Feeding in *Rhodnius prolixus:* Potencies of nucleoside phosphates in initiating gorging. *J. Insect Physiol.* 17:1315-1320.

(33) R. Hertel. 1974. Einfluss von ATP in einer holidischen Diät auf *Myzus persicae* (Sulz.) (Aphidina). *Experientia* 30:775-776.

(34) T.E. Mittler, and J.E. Kleinjan. 1970. Effect of artificial diet composition on wing production by the aphid *Myzus persicae. J. Insect Physiol.* 16:833-850.

(35) M. Rothschild, B. Ford, and M. Hughes. 1970. Maturation of the male rabbit flea *(Spilopsyllus cuniculi)* and the oriental rat flea *(Xenopsylla cheopsis)*: Some effects of mammalian hormones on development and impregnation. *Trans. Zool. Soc. Lond.* 32:105-188.

(36) L.M. Schoonhoven. On the variability of chemosensory information. *Proc. Symp. Hostplant in Relation to Insect Behaviour and Reproduction,* Budapest (in press).

(37) E.A. Bernays, W.M. Blaney, R.F. Chapman. 1972. Changes in chemoreceptor sensilla on the maxillary palps of *Locusta migratoria* in relation to feeding. *J. exp. Biol.,* 57:745-753.

(38) E. Merz. 1959. Pflanzen und Raupen. Über einige Prinzipien der Futterwahl bei Grosschmetterlingsraupen. *Biol. Zentr.* 78:152-188.

CHAPTER 3

# Insect Chemosensory Responses to Other Insects

W. D. Seabrook
*Department of Biology*
*University of New Brunswick*
*Fredericton, Canada*

Concern over the use of insecticides has given great impetus to the investigation of noninsecticidal insect control techniques. Among these new techniques, the use of pheromones appears to have great potential in that they are environmentally safe and have a considerable degree of species specificity. This potential use of pheromones in control, however, has yet to be realized. For although the chemical isolation and identification of pheromones from many species has taken place in recent years (1-5), sound strategies for the use of these compounds have yet to be developed. This is, in part, because investigations of the physiologic mechanisms for the perception of pheromones have not kept abreast of the chemical and behavioral studies.

The interface between the insect and the pheromone is the receptor organ; it represents the first opportunity for behavioral manipulation by man. Receptor organs present certain advantages as a basis upon which to develop control strategies. The neurons of these structures pass without synaptic interruption directly to the olfactory lobe (deutocerebrum) of the brain (6), and the antennal response to sex pheromones is independent of several environmental factors, as well as of age (7). Thus any control strategies that depend directly upon the responses of these sensory neurons will be independent of many environmental factors.

It is not the intent of this paper to consider the entire topic of insect olfaction, particularly in view of the excellent review by Kaissling (8). I shall

The author is indebted to Mrs. B. Ponder and Mr. P. Nitishin for technical assistance, to Mrs. L. Dyer for reviewing the manuscript, and to Dr. K.-E. Kaissling for his helpful criticisms and suggestions.

concentrate instead upon certain problems of insect olfaction on which light has been shed since Kaissling's review. In this regard, I shall draw heavily upon results that have come from investigations carried out in our laboratories. The following papers contain extensive bibliographies on olfaction: Schneider (9), Boeckh et al. (10), and a review of odor theory by Dravnieks (11). In addition there are a number of comprehensive papers on the particular subject of pheromone perception: Boechk et al. (10), Priesner (12), Schneider (13) (14), Kaissling (8), Karlson and Schneider (15), and Payne (16).

## I. STRUCTURE AND PLACEMENT OF RECEPTORS

The main site of olfactory receptors in insects is on the antenna, although such receptors have also been located on the maxillary palps (17) (18) and on the labial palps (19). In the Lepidoptera, sensilla responsive to pheromones are the sensilla trichodea (olfactory hairs) located on the antenna (20). The sensilla trichodea are found on both sexes of *Bombyx mori* (21), but in many Lepidoptera are found only on the male antennae (22) (23). More than one type of sensilla trichodeum are found on the antennae of *Trichoplusia ni* (24). Four distinct types of sensilla trichodea have been described from the antenna of the male eastern spruce budworm *Choristoneura fumiferana* (25), of which only one type acts as a pheromone receptor (26).

The pheromone-sensitive sensilla trichodea are innervated by a variable number of primary sensory neurons. Two neurons innervate the pheromone receptors of *Bombyx* (27). In the red-banded leaf roller, *Argyrotenia velutinana,* three neurons respond in the sex-pheromone-sensitive sensillum, however only two of these are thought to be chemosensory (28). Five neurons innervate the sex pheromone receptor of the eastern spruce budworm (26); one is a mechanoreceptor and four respond to the sex pheromone (29).

There is a considerable diversity in the number of pheromone-sensitive sensilla found in different species of Lepidoptera. The plumose antenna of *Bombyx* possesses 16,000 trichodea sensitive to the sex pheromone (14). The much simpler filamentous antenna of the male budworm *Choristoneura fumiferana* possesses 2300 sensilla trichodea (25), of which only 300 possess cells that respond to the sex pheromone (26).

In the Coleoptera, studies of sex-pheromone perception have concentrated on bark beetles of the family Scolytidae and have used the electroantennogram (EAG) technique (30) (31). This technique offers little opportunity for attributing a particular function to a specific type of sensillum. However, based upon extirpation experiments, it was suggested that the sensilla trichodea type-II are responsive to beetle-produced odors and, through gross electrophysiologic studies, that the sensilla basiconica are responsive to the aggregation pheromones frontalin and brevicomin (32). Payne (30) suggests that the short sensilla basiconica may serve an olfactory function. Sensilla possessing cells responding to the aggregation pheromone are located on the antenna of both sexes in Scolytidae.

In the Hymenoptera the sensilla placodea and *coeloconica* (pit pegs) have an olfactory function (33). The poreplate receptors (sensilla placodea) are sensitive to queen substance in *Apis mellifera* (10) (34). Dumpert (35) has suggested an olfactory role for two types of sensilla trichodea and/or sensilla basiconica on the antenna of the ants *Lasius fulginosus.* The sensillum trichodeum curvatum of *Lasius* has been shown to respond to the alarm pheromone (36).

The Diptera have a considerable range of olfactory sensitivity, and possess several types of sensilla trichodea and sensilla basiconica, as well as sensilla styloconica, clavate receptors, and olfactory pits (8) (37) (38). However, no specific receptor has been localized as a pheromone receptor in this order.

Levinson et al. (39) described the sensory structures on the antenna of the bed bug *Cimex* (Hemiptera) and isolated two types of sensilla (E1 and E2) that respond to the two major components of the bed bug's alarm pheromone.

The structure of insect olfactory sensilla has been reviewed by Slifer (40) (41), Schneider and Steinbrecht (20), Steinbrecht (42), and Kaissling (8). The structure of a generalized olfactory sensillum, shown in Fig. 1 is typical of a lepidopteran sex pheromone receptor sensillum. These sensilla range in length from 150 to 370 $\mu$ in the males of *Antheraea pernyi,* from 45 to 110 $\mu$ in the males of *Bombyx mori* (20), and from 50 to 70 $\mu$ in the males of *Choristoneura fumiferana* (25). They are thin-walled hairs, and their surfaces are perforated by large numbers of pores. In *Antheraea* there are 20,000 to 50,000 pores per sensillum, each with a diameter of approximately 100 Å. In *Bombyx* there are 1000 to 3000 pores with a diameter of 100 to 150Å (20) and with a density of 2-7 pores/$\mu^2$ in the male and 2-5 pores/$\mu^2$ in the female (27). These pores may open into a pore kettle from which a number of pore tubules extend toward the surface of the dendrite, a distance of about 4000 Å, making direct contact with the dendritic surface (43). Slifer (40) has suggested that these "pore filaments" are of dendritic origin and extend out through the cuticular pores to the surface, bringing the dendritic membrane directly in contact with the surface of the hair. Ernst (44) reported that the pore tubules are secreted by the trichogen cell. These tubules conduct the stimulatory molecules to the sensillum liquor, where they eventually come in contact with the dentrite membrane (27). Scott and Zacharuk (45), in a study of the sensilla basiconica of an elaterid beetle, found that the pore kettles do not make contact with the dendritic surface membrane. Zacharuk (46) reported that the pore tubules are secreted by the dendrite and are lipoidal in nature, with a core of nonlipoidal material. He speculates that stimulatory molecules diffuse through this plasma in the receptor cavity and that the selectivity of the sensillum may be related to the selectivity of the tubule contents. The portion of the tubule that traverses the cuticle of the sensillum possesses a solid core of material, which may be protein, polysaccharide, or glycoprotein.

Steinbrecht and Muller (43) found these pore tubules have an electronlucid core, thus the stimulatory molecules can pass by surface diffusion alone to the acceptor sites, which Kaissling (47) postulates are located on the cell membrane. In freeze-etch replicas of sensillum surfaces showing pores (48), the pores appear to be extremely shallow. A similar condition is found in scanning electron micrographs of pores on an olfactory receptor of the eastern

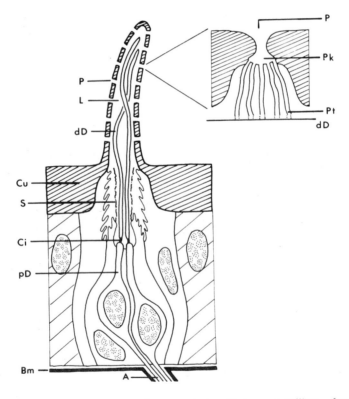

Fig. 1. Schematic diagram of a typical olfactory sensillum [modified after Steinbrecht (42)]. Inset: detail of pore kettle.

A: axon; Bm: basement membrane; Ci: ciliary body; Cu: cuticle; dD: distal dendrite; pD: proximal dendrite; L: lumen of sensillum; P: pore; Pk: pore kettle; Pt: pore tubule; S: sheath of dendrite.

spruce budworm (Fig. 2a,b). It is suggested that these tubules are not empty but contain a proteinaceous fluid. Steinbrecht and Kasang (48) demonstrate the presence of a thin surface coat on the sensillum with a total thickness of about 185 Å, which is probably a "stabilized lipid." Figures 3a and 3b are scanning electron photomicrographs of an olfactory trichodium of the eastern spruce budworm; Fig. 3a is the control sensillum, Fig. 3b is a sensillum treated with protease [using the technique of Seabrook and Jaeger (49)]. The

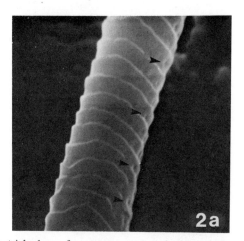

Fig. 2a.    Sensillum trichodeum from eastern spruce budworm antenna. Arrows point to cuticular pores opening into pore kettles. The pores are located only along that surface of the sensillum facing forward and toward the antennal flagellum. 20,480 X

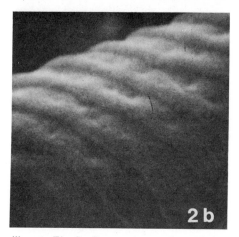

Fig. 2b.    Similar sensillum to Fig. 2a. Pores appear to be occluded by material contained within them. 51,200 X

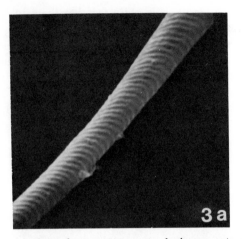

Fig. 3*a*.    Sensillum trichodeum from eastern spruce budworm antenna not treated with protease. Note soft, rounded nature of ridges. 10,240 X (All specimens for scanning electron microscope are coated with gold.)

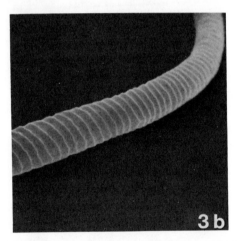

Fig. 3*b*.    Sensillum trichodeum from eastern spruce budworm antenna after digestion with protease. Note sharp detail of ridges. 10,880 X

ridges and surface features on the treated sensillum are considerably sharper than on the control hair, indicating that a surface layer has been digested from the hair. Since lipid solvents such as acetone and chloroform did not cause a similar sharpening of the image, it is proposed that the surface layer is not a lipid, but contains a protein, which may be extruded through the pores in the sensillum.

## II. FUNCTION OF ACCEPTORS

Kaissling (8) has demonstrated that the antennae of the silk moth *Bombyx* are extraordinarily efficient filters of odorant molecules, capable of absorbing approximately 27% of the odorant molecules from an airstream with a cross-sectional area similar to that of the outline area of the *Bombyx* antenna. He points out that those insects with smaller, more filamentous types of antennae will be considerably less effective at screening odorant molecules from an airstream.

The molecules filtered out of the airstream are adsorbed onto the surface of antennal sensilla and possibly onto the antenna itself. Kaissling (47) maintains that the entire sensillum acts as a catchment area for stimulatory molecules, as opposed to the view that only those molecules that make a direct hit on the olfactory pore are stimulatory. The adsorption phenomenon is rather unspecific (50) and does not contribute to the high specificity of the sensory response. Kasang and Kaissling (50) maintain that the metabolic processes located in the antenna, which break down the bombykol molecules into fatty acids and fatty esters, are equally unspecific and that the specificity of the response must be associated with the receptor molecule. It is difficult to associate the slow penetration half-time of 2 min and even slower enzymatic degradation of the bombykol (50% in 4 min) with the rapid responses of the electroantennogram, which may reach its maximum voltage within 0.5 sec and return to the resting membrane potential within 3 sec after the termination of the stimulus (10).

Kaissling (47) concludes that the binding sites for the odorant molecules or acceptors are located on the cell membrane and that, at least for the honeybee drone and for *Bombyx*, the number of acceptors at least equals the number of pore tubules. Kasang and Kaissling (50) attribute the specificity of the neuron to these acceptor sites on the membrane. However, Zacharuk (46) suggests that the specificity of the sensillum may be a property of the nonlipoidal fluid of the pore tubule. Evidence (Figs. 3a,b) that the sensillum is covered with a protein coat opens the possibility that the binding sites or acceptors for the odorant molecules may be located on the surface of the sensillum. Riddiford (51) has extracted two proteins common to the antennae of three saturnid moth species by elution with Ringer's solution. Elution in Ringer's solution blocked the sensitivity of these moths to their pheromones, and this block remained until the reappearance of the protein. Riddiford believes that the source of these proteins is the pore tubules, but that they are "externalized" and subject to constant resynthesis and replacement. However, Kaissling (8) did not find a significant decrease of the EAG in *Bombyx* and *Antheraea* after dipping the antennae for up to 12 hr in Ringer's solution. In studies of the fate of the sex pheromone molecules of *Trichoplusia ni*, *cis*-7-dodecen-1-ol acetate, in reaction with antennal proteins, Ferkovitch et al.

(52,53) discovered that both enzymatic and nonenzymatic binding occurred. The pheromone was degraded to its alcohol, cis-7-dodecen-1-ol, which is a potent inhibitor of sex pheromone behavior in this species. They suggest that the inhibitor and pheromone bind to different parts of the protein or to different proteins. These proteins, which bind and degrade the sex pheromone, were located in the hemolymph and legs as well as on the antenna (50) (52) (53). Schneider (14) has suggested that enzymes capable of degrading pheromones may play a double role—first in olfaction on the antenna, and second on the bodies and legs of insects—to prevent a buildup of pheromone molecules in these regions. A buildup of these molecules on the body of the insect would interfere with the chemical communication system. Mayer (54) found that the male antenna degraded the pheromone to alcohol twice as fast as did the female antenna of *T. ni,* and that the pheromone was hydrolyzed more rapidly than six closely related isomers and analogues, thus indicating some specificity in the acceptor proteins for the sex pheromone. Kasang (55) calculated that 99% of the pheromone molecules of *Bombyx* are adsorbed noncovalently onto micromolecular structures and that large amounts of the pheromone are enzymatically converted into acids.

A number of workers have speculated upon the origin of the stimulus emitter-acceptor system (15). In those insects that use host-plant compounds in their pheromone systems, the receptor system is thought to have evolved first for orientation to host plants, and the insect-produced pheromone to have evolved secondarily. In the Lepidoptera, which use straight-chain molecules, the emitter system is thought to have evolved first. The waxy epicuticle of insects contains higher fatty alcohols, and areas of cuticle may have become specialized for the elimination of these materials. The receptor system in this case may have developed as a receptor for body odor and subsequently become specialized for particular waxy alcohols.

Little is known of the fate of the pheromone molecules once they have reacted with the antennal proteins and have been enzymatically degraded. Kasang (55), however, reports that after 30 min incubation with tritium-labeled bombykol, $H^3$ molecules had penetrated to the inner parts of the *Bombyx* antenna. This may indicate an uptake of metabolic byproducts by the dendritic branches of the neurons. Kaissling (56) has proposed a two-step process for the inactivation of the pheromone molecule. He refers to the first step as "early inactivation" of the molecule, a process that would physically or chemically reduce the concentration of stimulatory molecules nearer the acceptor. The second step, or "late inactivation," is a process that removes the odor molecules or their reaction products from the sensory hairs over a period of minutes. Early inactivation accounts for certain characteristics of the EAG waveform; late inactivation takes into account the migration of bombykol molecules within the antenna.

Six steps are considered critical in the transduction of the olfactory stimulus to the receptor potential (57):

1. Adsorption on the antennal surface
2. Diffusion to the receptor molecules
3. Binding
4. Inactivation of the receptor molecule
5. Change of membrane conductance
6. Early inactivation of the odor molecule

The evidence presented indicates that the initial reaction in the transduction of pheromone information to sensory neurons is nonenzymatic binding. This may be followed in some cases by an enzymatic alteration of the pheromone substrate. Such reactions would rely heavily on the stereochemical fit of the stimulatory molecule and its acceptor (58).

## III. SPECIFICITY OF RECEPTORS

There is some controversy (59) concerning the specificity of insect olfactory receptors. Much of the research in this area centers on the perception of sex pheromones. In recording electroantennograms from the antenna of *Bombyx*, Schneider (60) noted that there were no qualitative differences in the response to the odor of pheromone glands of various saturnid moths on this system. The only differences were in the relative amplitudes of the responses. Schneider et al. (61) subsequently reported, using extracellular recording techniques, that two to three neurons were constantly firing as background activity in the long trichodea of *Antheraea* and that these cells responded by an increased rate of discharge to stimulation by the sex pheromone. They further noted that in the sensilla basiconica, the longer the hair, the fewer the compounds that stimulated it. These findings led to the concepts of *odor-generalist* and *odor-specialist* neurons (10).

Odor generalists are neurons of the same sensillum type in one individual that have different but partly overlapping reaction patterns in all the reacting cells. Examples of such odor generalists are sensilla basiconica.

Odor specialists are olfactory neurons belonging to a recognizable type of sensillum giving stereotyped responses to a series of compounds. Examples of odor specialists are the receptors for sex attractants. Schneider *et al.* (61) demonstrated in *Antheraea* that, of three cells in a common sensillum basiconicum, one may be excited, one may be inhibited, and one may show no response to a common olfactory stimulus. A similar phenomenon has been found in the sex-pheromone receptor sensillum of *Trichoplusia ni* (54). Kaissling and Renner (34), using single-unit recordings, identified two cell types responding to pheromones of the honeybee—one cell responding to the queen substance (9-oxo-*trans*-2-decenoic acid) and the second cell responding to scent from the Nasanov gland. Vareschi (62) further separated seven cell types with no overlap in specificity; there was, however, some variation of specificity within each of these cell types.

Schneider et al. (61) noted that geometric isomers of bombykol were up to 1000 times less effective at the bombykol receptor than the true pheromone, bombykol (*cis*-10, *trans*-12-hexadecadienol). The same rule applied to *Trichoplusia ni:* Payne et al. (32) reported that the more a parapheromone (a nonpheromone chemical that elicits typical pheromone responses) varies in its structure from the structure of the true pheromone, the greater the amount needed to elicit a significant EAG.

The information that has been presented so far indicates a considerable degree of receptor-cell specificity to compounds of closely related chemical structure, and lends support to the concept of odor-specialist neurons. Recently, O'Connell (59), using computer analysis of single-unit neuronal responses in sex-pheromone receptor sensilla of the red-banded leaf roller, has indicated that the two cells responsive to the sex pheromone and a number of its analogues do not respond in an identical fashion to each of these various compounds. He found that the response frequency to the pheromone and to six other behaviorally active compounds varied between the two cells, and claims that these differences in response were due to intrinsic factors within the neurons. He argues that odor quality is not encoded in the simple presence or absence of activity in any one neuron but is encoded by a pattern that may vary across an ensemble of receptor neurons to produce a unique total response. O'Connell reasons that as some neurons can respond to eight out of nine behaviorally active compounds, there could be as many as eight "functionally" different receptor sites. He further states that because one cell responds preferentially to *cis*-11-tetradecenyl acetate (the pheromone) and the second cell responds preferentially to the *trans* isomer (an inhibitor), that there must be at least two "functionally" distinct receptor sites.

In the eastern spruce budworm the number of sex-pheromone-responsive sensillae is extremely limited (300 per antenna) and restricted to one morphologically distinct type of sensillum trichodeum (26). Based upon histologic studies, this sensillum contains five neurons. Four of these respond to *trans*-11-tetradecenal (29), which is the sex pheromone (63), and one acts as a mechanoreceptor (29). These same four neurons respond to *trans*-11-tetradecynl acetate (29), a known inhibitor of the sex pheromone (64).

If the four pheromone-sensitive neurons are adapted to the sex pheromone and are then immediately exposed to the acetate inhibitor, they do not respond to the acetate. When the converse test is applied, adapting them to the acetate, they remain adapted when exposed to the pheromone (aldehyde) (Fig. 4). These cross-adaptation experiments indicate that both the aldehyde and the acetate are reacting with the same acceptor site. Boeckh et al. (10)

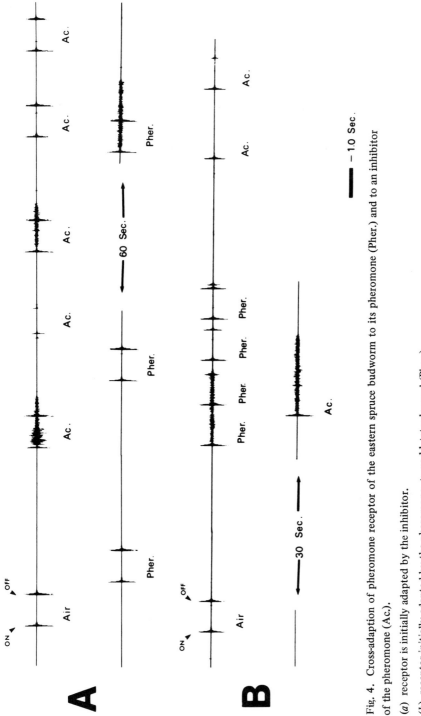

Fig. 4. Cross-adaption of pheromone receptor of the eastern spruce budworm to its pheromone (Pher.) and to an inhibitor of the pheromone (Ac.).

(*a*) receptor is initially adapted by the inhibitor.

(*b*) receptor initially adapted by the pheromone, *trans*-11-tetradecenal (Pher.).

Stimulus duration: 1 second.

Recording with tungsten electrodes using the technique of Albert et al. (26).

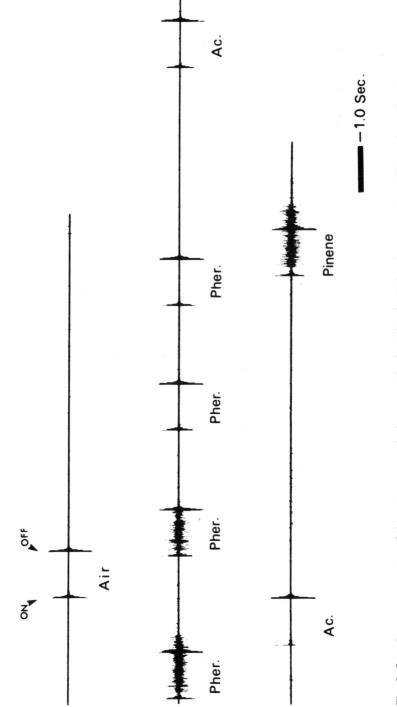

Fig. 5. Sex pheromone receptor of the eastern spruce budworm moth adapted to both the pheromone (*trans*-11-tetradecenal) and an inhibitor (*trans*-11-tetradecenyl acetate) responds to the structurally unrelated compound α-pinene. All trichodes on the antenna of the spruce budworm moth responded to ε-pinene (26).

have demonstrated a similar phenomenon with reduced responsiveness of the honeybee pheromone receptor to caproic acid after exposure to queen substance. In this instance they cited three possible locations for this cross-adaptation: the adaptation could arise in the receptors, in the dendrites, or in the impulse-generating mechanism.

However, if the budworm receptor is adapted to the aldehyde and to the acetate, and subsequently exposed to α -pinene, a response is generated in the neuron (Fig. 5). In this instance the adaptation is in neither the dendrite nor the impulse-generating membrane—since the neuron responds readily to pinene—but must be in the acceptor. The aldehyde and the acetate are reacting with the same acceptor protein; the pinene is probably reacting with a different acceptor. The above results would indicate that in this instance a single type of acceptor is able to react with two closely related molecules.

## IV. RESPONSE TO PHEROMONE INHIBITORS AND SYNERGISTS

Compounds that inhibit the behavioral response of the recipient to its sex pheromone have been described for a number of insect species (65). In some cases the inhibitor is produced by the same individual as the one that produced the pheromone, and acts to mask the effect of the pheromone (64) (66)-(74). In some instances the inhibitor may function reproductively to isolate two species having a common pheromone, and with geographic overlapping ranges (75).

Three mechanisms have been proposed to explain the perception of the sex-pheromone inhibitors: reception by separate neurons, separate acceptors on a common neuron, and common acceptors. Males of *Plodia interpunctella* or *Cadra cautella* can be habituated to behavioral inhibitors of the sex-pheromone response, while subsequently showing responsiveness to a mixture of the sex pheromone and the inhibitor. A 3-min preexposure to the inhibitor (*trans, trans*-9, 12-tetradecadien-1-ol acetate) produced increased responses in both species on subsequent exposure to the pheromone (*trans, cis* -9, 12-tetradecadien-1-ol) plus inhibitor than the response of individuals that had not been pretreated. Sower et al. (76) explained these results by suggesting that the pheromone and the inhibitor are perceived by separate sensors.

Mayer (77) demonstrated that in the males of *Trichoplusia ni*, EAGs elicited by exposure to the sex-pheromone inhibitor (*cis*-7-dodecen-1-ol) were similar to EAGs resulting from exposure of the antenna to the pheromone (*cis*-7-dodecen-1-ol acetate). In *T. ni* and *Pseudoplusia includens*, both of which use the same sex pheromone, McLaughlin et al. (78) found that the inhibitor did not interfere with sex-pheromone behavior in that the males exhibited an upwind flight behavior and clasper extension when exposed to the sex pheromone. It did, however, prevent the moths from orienting to a locus of attraction provided that the inhibitor source was within 30 cm of the pheromone source. They conclude from these experiments that the inhibitor is perceived by a sensory neuron separate from the pheromone receptor neuron, and that modification of pheromone-elicited behavior is not a peripheral but a central phenomenon. Such a possibility has been suggested also by Kaissling (8).

Roelofs and Comeau (1)(58) reported that compounds structurally related to the sex attractant for the red-banded leaf roller (RiBLuRE, *cis*-11-tetradecenyl acetate) can have either a synergistic or inhibitory effect on attractancy in field studies. They propose that these structurally related molecules are interacting with the binding sites on similar receptors, with varying degrees of affinity.

In single receptor recordings from the sex-pheromone receptor of the red-banded leaf roller, O'Connell (28) found that out of three neurons present the same two responded, although differently, to the sex pheromone; to its inhibitor (*trans*-11-tetradecenyl acetate); and to a synergist (dodecyl acetate). The result of stimulation with a mixture of the pheromone plus its inhibitor is a decrease in the activity of a type-I cell when compared to the response to an equivalent amount of the pheromone. The response to a mixture of the pheromone plus the synergist is increased considerably over the observed response to the pheromone alone. He concludes that the mechanisms are operating at the receptor molecule level, and support the concept proposed by Roelofs and Comeau (1)(58). However, in a subsequent study, differential responses in the two chemosensory cells of the pheromone receptor to seven behaviorly active compounds, including the pheromone and its inhibitor, lead O'Connell (59) to conclude that there must be separate acceptors for the pheromone and for the inhibitor.

The tortricid moth *Adoxophyes orana* uses *cis*-9- and *cis*-11-tetradecenyl acetate as its sex pheromone and the geometric isomers *trans*-9- and *trans*-11-tetradecenyl acetate as its inhibitors. EAG responses to mixtures of the pheromone and its inhibitor showed no additive effects when compared with EAG responses to the pheromone alone. Minks et al. (79) suggested that the pheromone and its inhibitor may be acting on the same receptors.

In Section III evidence was presented that in the eastern spruce budworm, the pheromone (*trans*-11-tetradecenal) and its inhibitor (*trans*-11-tetradecenyl acetate) excite the same neurons and react with the same acceptor (3)(4). How does the insect code the excitatory stimulus and the inhibitory stimulus in the same neuron? Preliminary data presented in Table 1 suggest that the neuron adapts more rapidly to the inhibitor than to the sex pheromone.

Based on these preliminary data, and upon the fact that these neurons will occasionally adapt to the inhibitor prior to the conclusion of a 1-sec exposure,

TABLE 1.

Duration of Adaptation to a 1-sec Stimulus of
Trans-11-Tetradecenyl Acetate by
Pheromone Receptor Neurons of the
Eastern Spruce Budworm

| Concentration (mg) | Duration of Adaptation (sec) |
|---|---|
| $1 \times 10^{-3}$ | 60-70 |
| $1 \times 10^{-4}$ | 40 |
| $1 \times 10^{-5}$ | 20-30 |
| $1 \times 10^{-6}$ | 20-30 |
| $1 \times 10^{-7}$ | ??? |

A 1-sec stimulus of trans-11-tetradecenal does not
result in adaptation of the neurons in the concentration range $10^{-8}$ to $10^{-5}$ mg.

we suggest that inhibition in the eastern spruce budworm is the result of the differential adaptation of the receptor neurons to the pheromone and its inhibitor.

The differential adaptation of a nueron, by these two compounds, acting on a common acceptor probably indicates a different affinity of each of these molecules for the common acceptor. This would support the mechanism proposed by Roelofs and Comeau (1)(58) for the responses of a neuron to structurally similar compounds.

Current evidence indicates that three different mechanisms may be involved in the perception of the inhibitors of sex pheromone behavior in the five species in which this phenomenon has been investigated.

Recently, considerable attention has been devoted to the study of minor components that may occur as a natural part of the insect's pheromone system (79)-(81). In addition, there may be synthetic structural analogues of the pheromone molecule that have not been reported as a component of the natural pheromone (1)(58)(82), but which, when mixed with the pheromone, increase the behavioral response of the insect to a mixture of the pheromone and the synergist over that observed to the pheromone alone. In bark beetles certain host-tree terpenes have been shown to synergize aggregation pheromone behavior (83)(84). The specific ratio of the geometric isomers has been shown to be important to attraction in several insect species whose pheromone systems use the same compounds but in different ratios (85).

Roelofs and Comeau (58) reported that dodecyl acetate was strongly synergistic to the pheromone RiBluRe in trapping tests with the red-banded leaf roller, and they subsequently listed other 11- and 12-carbon acetates as synergistic to RiBLuRe (1). Because of similarities in molecular structure they suggested that these compounds may be reacting with the same receptor site as does the pheromone. Little electrophysiologic evidence is available concerning the perception of the minor pheromone components and synergists. O'Connell (28) reported that dodecyl acetate had little effect on pheromone-receptor activity in the red-banded leaf roller when applied alone, but when applied (50:50) with the attractant, *cis*-11-tetradecenyl acetate, the action-potential frequency as considerably greater.

## V. RESPONSE TO CONCENTRATION

An increase in the concentration of the stimulatory molecules results in an increase in the amplitude of the EAG signal derived from the stimulated antenna (9)(23)(31)(86)(87). Furthermore, an increase in the frequency of action potentials in pheromone receptor neurons increases with an increase in the stimulus concentration (34)(88). The increase in receptor response for increasing stimulus concentrations holds true for low-stimulus concentrations. In *Bombyx* the EAG response to bombykol increases in the concentration range $10^{-3}$ through $10^{3}$ $\mu$g (8). In the eastern spruce budworm *Choristoneura fumiferana,* single-receptor responses have been recorded to the sex pheromone (*trans*-11-tetradecenal) in the concentration range $10^{-8}$ to $10^{-5}$ mg, and a maximum frequency of discharges was observed at $10^{-6}$ mg, suggesting that sensory adaptation was occurring at the higher concentration (Fig. 6). Gesteland and Muller (89) have noted a similar phenomenon.

The response of the pheromone receptors in the eastern spruce budworm can be correlated with the concentration of the pheromone only during the first 400 msec of the response (26). In Fig. 6 the stippled area represents that portion of the response that shows a significant correlation with concentration. A similar phenomenon is apparent in photographs showing single-receptor responses to the queen substance of *Apis* (34). A definite increase in response takes place with an increase in the concentration of queen substance, but this correlation is most obvious during the initial phase of the response.

The concentration of pheromone at source is of little significance, however, for determining the lower threshold of concentration for receptor activity. Schneider and his colleagues have calculated molecular densities in the region of the receptor trichodeum required to elicit both electrophysiologic and behavioral responses (10). It has been determined that the olfactory receptors for bombykol on *Bombyx* will respond with a single-action potential to a single pheromone molecule (88).

The data presented in Fig. 6 and the responses of the pore plate receptor to queen substance in the honeybee indicate that pheromone receptors begin

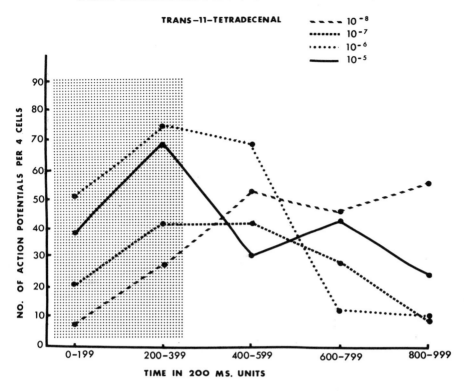

Fig. 6. Responses of a sex pheromone receptor of the eastern spruce budworm to a range of concentrations of the sex pheromone (*trans*-11-tetradecenal). The stippled area represents that period of the response during which there is a correlation between the concentration and the response. After 400 msec there is no correlation between stimulus and response. Data from Albert et al. (26).

to adapt within a few hundred milliseconds of the onset of the stimulus. In the case of the eastern spruce budworm, the response is independent of the stimulus concentration after this period. This indicates that pheromone receptors adapt readily to the pheromone. A 7-sec pheromone stimulus at a concentration of $10^{-6}$ mg applied to the antenna of the eastern spruce budworm adapts individual receptors for a period of greater than 2 min (29). In *Trichoplusia ni* EAG responses in the antenna recovered from sensory adaptation within 30 to 60 sec (7). Kaissling (56) reported a time of 256 min for complete recovery of EAG responses in *Bombyx*. According to Gesteland and Müller (89), olfactory cells may require up to 30 min to return to normal excitability.

Increased pheromone concentrations have been shown to bring about successive stages of behavioral response, from antennal elevation to copulatory movements. Bartell and Shorey (90) report a decrease in these responses at high pheromone concentrations, which may be attributable to receptor adaptation. Pheromone concentration has been implicated as a species-isolating mechanism between *Trichoplusia ni* and *Autographa californica* (91). Both species use the same pheromone, *cis*-7-dodecenyl acetate, but *A. californica* is trapped at a low pheromone-release rate while *T. ni* is trapped at a high release rate.

Several authors have described upwind flight by male moths, apparently in response to the female scent. On losing the scent the moths made crosswind casts (92) (93). Farkas and Shorey (94) claim that pink bollworm moths can steer toward an odor source without sensing wind direction and that, with increasing sex-pheromone concentration, the moth's airspeed decreases due to a decrease in wing-beat frequency (95). Apparently, the loss of the odor is identical to adaptation of the odor receptors from the viewpoint of sensory input and the casting flight is a response to loss of perception of the odor stimulus. Upon disadaptation of the receptors the upwind angle would increase. The data presented concerning the lack of correlation between receptor response and pheromone concentration casts doubt on the ability of a moth to sense a concentration gradient, and thus to orientate in a pheromone concentration gradient by straight flight up the gradient.

Species of male Lepidoptera from a number of families have been shown to possess pheromones that, when released, have an aphrodisiac effect on the calling female. These compounds work at short range, stimulating the female to mate (96). EAG responses indicate that the pheromone is perceived by receptors located on the antennae of both male and female moths (97)-(99). In the butterfly *Danaus chrysippus,* the receptors for the male pheromone are thin-walled sensilla basiconica (100).

Little species specificity is attributed to the EAG response for the male pheromone: pheromones from different species elicit EAGs of similar amplitude in a common recipient (98). Grant et al. (99) obtained EAGs of similar amplitude when they stimulated the antenna of the armyworm *Pseudaletia unipuncta* with benzaldehyde, benzyl alcohol, and benzoic acid—all components of the armyworm's male pheromone—or with the related aromatic compounds 2-phenyl ethanol and benzyl acetate. However, Seibt et al. (100) found that the antenna of *Danaus chrysippus* is able to discriminate between two components of its male pheromone: the major component, pyrrolizi-dinone, gives rise to large-amplitude EAGs; and a second component, a diol, gives rise to small EAGs.

Little is known of the response of the female receptors to male pheromones. The available data suggest that the species specificity attributable to the female sex pheromone is not present in this system. The reception of male pheromones by sensilla basiconica is in accord with the role of generalist receptors attributed to such sensilla (10).

## VI. PERCEPTION OF KAIROMONES AND ALLOMONES

Although a considerable literature is developing concerning behavioral responses to kairomones and allomones [see Brown (101)], there has been virtually no work involving the chemosensory basis of the response. Certain conclusions can, however, be drawn from the literature on behavior.

Pheromones of the host insect may on occasion be used to attract the parasite or predator to its host, as in the bark beetles *Ips* (102) and *Dendroctonus* (83) (103), in the lepidopteran *Gnorimoschema* (104) and in the homopteran *Aonidiella* (105). Egg parasites are attracted to secretions from host scales, as in the attraction of *Trichogramma evanescens* to the eggs of the corn earworm *Heliothis zea* (106) and *Cadra cautella* (107), or of *Cardiochiles nigriceps* to its host *Heliothis virescens* (108).

Parasites may be attracted to the frass of host larvae. *Microplitis croceipes* is attracted to 13-methylhentriacontane, a constituent of the frass of *Heliothis zea* (108) (109). The parasite wasp, *Orgilus lepidus,* is attracted to heptanoic acid, a volatile in the frass of the potato tuberworm, *Gnorimoschema* (110).

All of the above-mentioned kairomones are volatile substances and elicit host-seeking behavior. As they can operate at a distance from the host, it may be assumed that they act as olfactory stimulants. Kairomones, however, appear to act at several levels in the behavioral response of the parasite to its host. They may act as contact stimulants, secreted upon the larval cuticle and perceived through the parasite's tarsi, which enable the parasite to orientate to its host (111), or they may be frass constituents that release a probing behavior when perceived through contact with the antennae.

The final behavioral level is oviposition. Constituents of the hemolymph have been implicated in the release of ovipositional behavior. These constituents are perceived by contact chemoreceptors on the parasite's ovipositor. Lewis and Jones (109) have demonstrated the presence of a kairomone in the hemolymph of *Heliothis zea* to which the parasite *Microplitis* responds. Final acceptance of *Heliothis virescens* as a host by the braconid *Nigriceps viereck* is by means of sampling the internal constituents of the host by the parasite's ovipositor (111). A number of hemolymph amino acids have been implicated in acceptance of *Galleria melonella* and *Celerio euphorbiae* as hosts by the hymenopteran *Itoplectis*. In this instance a hexose may act as a synergist to the amino acids (112).

Salivary secretions operate as kairomones (108). Corbet (113) has established that the mandibular-gland secretions of the flour moth, *Anagasta kuehniella,* elicit jabbing reactions by the ichneumonid parasite, *Venturia conescens,* the receptors for this response being located on the antenna (114).

To date no specific receptors have been isolated for the perception of kairomones. However, by means of appendage-ablation studies, Hays and Vinson (111) determined that in the parasite *Cardiochiles nigriceps* receptors on the antenna are involved in host-finding, receptors on the tarsi are responsible for host orientation, and receptors on the ovipositor are involved in the final host acceptance.

The duration of exposure and concentration of the kairomone can effect the behavioral response of the parasite (109). Corbet (114) has determined that preexposure of the parasite *Nemeritis canescens* to the mandibular secretion of its host *Ephestia kuehniella* influences its subsequent response to host larvae or to a crude extract of the glands. The effect of the preexposure depends on the time of exposure and on the concentration of the stimulus. She suggests that this phenomenon may be the result of events occurring either at the sense organs or within the central nervous system.

Based on behavioral assays, Jones et al. (108) have determined a considerable degree of specificity in those compounds that will direct the parasite *Microplitis* to its host. The kairomone was found to be 13-methylhentriacontane. Any alteration in chain length caused a reduction in response, as did moving the methyl group to an adjacent carbon atom. Precise chemical constituents of the hemolymph are required to induce oviposition in *Itoplectis* (112).

Mandibular-gland secretions, while working as kairomones at relatively low concentrations, can have a disruptive effect on parasite behavior at high concentrations (115). Larvae under attack by a parasite may on occasion hit the parasite with mandibular-gland secretions. Under these conditions the parasite ceases its attack and undergoes a period of grooming behavior (111) (117). This behavior may result from adaptation of the receptors to a high concentration of the kairomone; or it may be that at such concentrations the kairomone is acting as a repellent—that is, as an allomone.

Many alarm pheromones of ants have been shown to have a secondary role as defensive secretions (116) (117). As defensive secretions, these compounds act as allomones. A defensive role has been attributed to 2-heptanone. The ants *Cononmyrma pyramica* and *Iridomyrmex pruinosis* have been shown to use this compound to deter fire ants (118) (119), and it may be used by ants to defend against the genus *Atta* (120). Secretions of the mandibular glands of *Lasius umbratus* contain both the aldehyde citronellal, which has been implicated in alarm behavior, and the alcohol citronellol, which may be employed as a defensive secretion (121). Formic acid, a major component of ant venom, and undecane act as defensive secretions for the ant *Acanthomyops claviger* (122); the undecane helps to spread the formic acid over the agressor's cuticle. In the above examples there is no indication of the sensory basis for the response in the recipient to a defensive secretion. In most cases the secretion appears to act by contact with, and possible penetration through, the cuticle.

Ants of the *Formica sanguinea* group carry out slave raids on colonies of other species of *Formica.* The attacking ants discharge copious amounts of decyl, dodecyl, and tetradecyl acetates into the colony under attack. These compounds disperse the defending workers, and have been called propaganda pheromones (123) (124). They appear to act as allomones and to be perceived olfactorily.

Coccinellin $(C_{13}H_{23}NO)$, a defensive secretion of the ladybug beetle *Coccinella septempunctata,* has a bitter taste and repels the ant *Myrmica rubra* both by taste and smell (125) (126).

Reflex bleeding in the Coleoptera results in the release of a viscous blood which in the larvae of both the spotted cucumber beetle (*Diabrotica unidecimpunctata*) and the banded cucumber beetle (*D. balteata*) (127), and in the adult lampyrid *Photinus* (128) is a potent deterrent to predating ants. The viscous blood has a bitter taste and a pungent odor, which may induce both an olfactory and gustatory response in the predator.

Cornicle secretions of aphids contain an alarm pheromone, *trans*-β-farnasene. This compound also serves as a defensive secretion. When predators contact the secretion they break off the attack and undertake extensive cleaning of the contaminated area (129).

## VII.   CONTROL STRATEGIES

Intensive research on the sensory physiology of a limited number of insect species, belonging to an even fewer number of families, has taken place. There is sometimes agreement among the various species: at high pheromone concentration levels the response of the receptor neurons to the stimulus decreases or even disappears. On the other hand, consider the results of research on sensory responses to pheromone inhibitors: there are three potential mechanisms, each one of which would require the development of a different strategy for its use in control. It is apparent that before any generalizations can be made, considerably more comparative research is required.

Present knowledge can, to some extent, be applied to the strategic use of pheromones. Trapping—both as a control technique and for pest survey—is facilitated by pheromone lures. Certainly information concerning the effect of pheromone concentration on the sensory response is invaluable. At high concentrations the response of the sensory neurons to the stimulus decreases or completely disappears. There are further indications that receptor neurons adapt readily to the pheromone. These facts should have a bearing upon trap design, for if pheromone concentrations are too high in the entry port of the trap a significant number of potential victims will never enter the trap. Trapping efficiency may in some cases be suboptimal due to this factor.

Compounds that inhibit the behavioral response to the sex pheromone may have a great application in insect control. This is particularly true in the case of the spruce budworm, in which both the pheromone and the inhibitor appear to compete for the same acceptor site. In this instance blocking of the acceptor site by the inhibitory molecule completely eliminates the response to the pheromone in the sensory neurons. However, in the case of *Plodia* and *Cadra* a different mechanism may operate. If separate neurons perceive the two compounds, then the response to the inhibitor can be adapted or habituated, leaving the response to the pheromone untouched. The same problem could apply in the case of the red-banded leaf roller; here the pheromone and inhibitor find at different acceptor sites on the same neuron. Further research is needed on the sensory basis of pheromone inhibition.

In some species the sensory neurons adapt readily to the sex pheromone. This feature could be exploited in developing control strategies, particularly in forest insect control where very large land areas are treated and moth trapping is quite impractical. Sufficient microencapsulated pheromone could be laid down, raising the concentration of pheromone molecules in the air to a level adequate to adapt the receptors. In the case of the spruce budworm, relatively brief exposure to sex-pheromone concentrations as low as $10^{-7}$ mg can result in complete adaptation of the receptors.

The use of parasites for biologic control of insect pests ia a well-established technique. Considerable effort is expended in the worldwide collection of potential parasite species for any given host insect. The use of pheromones and kairomones to select from the field those parasites that have an interest in the host under study could prove to be a most efficient way of identifying and collecting potential parasites.

## REFERENCES

(1)   W.L. Roelofs, and A. Comeau. 1971. Sex pheromone perception: Synergists and inhibitors for the red-banded leaf roller attractant. *J. Insect Physiol.* 17:435-448.
(2)   D.A. Evans, and C.L. Green. 1973. Insect attractants of natural origin. *Chem. Soc. Rev.* 2:75-97.
(3)   R.M. Silverstein, 1971. Recent and current collaborative studies of insect pheromones. *Proc. 2nd Int. IUPAC Congr.* (Israel) 3:69-89.
(4)   J.G. MacConnell, and R.M. Silverstein. 1973. Recent results in insect pheromone chemistry. *Angew. Chem. Int. Ed. Engl.* 12:644-654.
(5)   M. Jacobson. 1974. Insect pheromones. In *The Physiology of Insecta,* Vol. 3, 2nd ed. Morris Rockstein, Ed. Academic, New York, pp. 229-276.
(6)   R.A. Steinbrecht. 1969. On the question of nervous syncitia: Lack of axon fusion in two sensory nerves. *J. Cell Sci.* 4:39-53.

(7) T.L. Payne, H.H. Shorey, and L.K. Gaston. 1970. Sex pheromones of noctiud moths, factors influencing antennal responsiveness in males of *Trichoplusia ni*. *J. Insect Physiol.* **16**:1043-1055.

(8) K.-E. Kaissling. 1971. Insect olfaction. In *Handbook of Sensory Physiology*. Vol. 4, Part 1. L.M. Beidler, Ed. Springer-Verlag, New York, pp. 351-431.

(9) D. Schneider. 1957. Elektrophysiologische Untersuchungen von Chemo und mechanorezeptoren der Antenne des Seidenspinners *Bombyx mori* (L.). *Z. Vgl. Physiol.* **40**:8-41.

(10) J. Boeckh, K.-E. Kaissling, and D. Schneider. 1965. Insect olfactory receptors. In *Cold Spring Harbour Symposia on Quantitative Biology: Sensory Receptors*, Vol. 30, pp. 263-280.

(11) A. Dravnieks. 1966. Current status of odor theories. *Adv. Chem. Ser.* **56**:29-52.

(12) E. Priesner. 1968. Die Interspezifischen Wirkungen der Sexuallokstoffe der Saturniidae (Lepidoptera). *Z. Vgl. Physiol.* **61**:263-297.

(13) D. Schneider. 1969. Insect olfaction: Deciphering system for chemical messages. *Science* (Wash.) **163**:1031-1037.

(14) D. Schneider. 1970. Olfactory receptors for the sexual attractant (Bombykol) of the silk moth. In *The Neurosciences: Second Study Program*. Francis O. Schmitt, Ed. Rockefeller University Press, New York, pp. 511-518.

(15) P. Karlson, and D. Schneider. 1973. Sexualhormone der Schmetter-linge als Modelle Chemischer Kommunikation. *Naturwissenschaften* **60**:113-121.

(16) T.L. Payne. 1974. Pheromone perception. In *Pheromones*. M. Birch, Ed. North-Holland, Amsterdam, pp. 35-61.

(17) U. Bässler. 1958. Versuche zur Orientierung der Stechmüken die Schwarmbildung und die Bedeutung der Johnstonschen Organs. *Z. Vgl. Physiol.* **41**:300-330.

(18) F.E. Kellogg. 1970. Water vapour and carbon dioxide receptors in *Aedes aegypti*. *J. Insect Physiol.* **16**:99-108.

(19) R.A. White, U. Paim, and W.D. Seabrook. 1974. Maxillary and labial sites of carbon dioxide-sensitive receptors of larval *Orthosoma brunneum* (Forster) (Coleoptera:Cerambycidae). *J. Comp. Physiol.* **88**:235-246.

(20) D. Schneider, and R.A. Steinbrecht. 1968. Checklist of insect olfactory sensilla. *Symp. Zool. Soc. Lond.* **23**:279-297.

(21) D. Schneider, and K.-E. Kaissling. 1957. Der Bau der Antenne des Seidenspinners *Bombyx mori* L. II. Sensillen, cuticulare Bildungen und innerer Bau. *Zool. Jahrb.* **76**:223-250.

(22) J. Boeckh, K.-E. Kaissling, and D. Schneider. 1960. Sensillen und Bau der Antennengeissel von *Telea polyphemus* (Vergleiche mit Weiteren Saturniden: Antheraea, Platysamia, und Philosamica). *Zool. Jahrb.* **78**:559-584.

(23) J. Boeckh, H. Sass, and D.R.A. Wharton. 1970. Antennal receptors: Reactions to female sex attractant in *Periplaneta americana*. *Science* (Wash.) **168**:589.

(24) R.N. Jefferson, R.E. Rubin, S.U. McFarland, and H.H. Shorey. 1970. Sex pheromones of noctuid moths. XXII. The external morphology of the antennae of *Trichoplusia ni, Heliothis zea, Prodenia ornothogalli*, and *Spodoptera exigua*. *Ann. Entomol. Soc. Am.* **63**:1227-1238.

(25) P.J. Albert, and W.D. Seabrook. 1973. Morphology and histology of the antenna of the male eastern spruce budworm, *Choristoneura fumiferana* (Clem) (Lepidoptera: Tortricidae). *Can. J. Zool.* **4**:433-448.

(26) P.J. Albert, W.D. Seabrook, and U. Paim. 1974. Isolation of a sex pheromone receptor in males of the eastern spruce budworm, *Choristoneura fumiferana* (Clem) (Lepidoptera: Tortricidae). *J. Comp. Physiol.* **91**:79-89.

(27)  R.A. Steinbrecht, 1973. Der Feinbau olfacotischer Sensillen des Seidenspinners (Insecta, Lepidoptera). *Z. Zellforsch.* **139**:533-565.

(28)  R.J. O'Connell. 1972. Responses of olfactory receptors to the sex attractant, its synergist and inhibitor in the red-banded leaf roller, *Argyrotaenia velutinana.* In *International Symposium on Olfaction and Taste.* IV. D. Schneider, Ed. Wissenschaftliche Verlagsgesellschaft, Stuttgart, pp. 180-186.

(29)  W.D. Seabrook. Unpublished.

(30)  T.L. Payne. 1970. Electrophysiological investigations on response to pheromones in bark beetles. *Contrib. Boyce Thompson Inst.* **24**:275-282.

(31)  T.L. Payne. 1971. Bark beetle olfaction. I. Electroantennogram responses of the southern pine beetle (Coleoptera: Scolylidae) to its aggregation pheromone, frontalin. *Ann. Entomol. Soc. Am.* **64**:266-268.

(32)  T.L. Payne, H.A. Moeck, C.D. Wilson, R.N. Coulson, and W.J. Humphreys. 1973. Bark beetle olfaction. II. Antennal morphology of sixteen species of *Scolytidae* (Coleoptera). *Int. J. Insect Morphol. Embryol.* **2**:177-192.

(33)  V. Lacher. 1964. Elektrophysiologische Untersuchungen an einzelin Rezeptoren für Geruch, Kohlendioxyd, Luftfeuchtigkeit und Temperatur auf der Antennen der Arbeitsbiene und der Drohne (*Apis mellifica*). *Z. Vgl. Physiol.* **48**:587-623.

(34)  K.-E. Kaissling, and M. Renner. 1968. Antennale Rezeptoren für Queen Substance und Sterzelauft bei der Honigbiene. *Z. Vgl. Physiol.* **59**:357-361.

(35)  K. Dumpert. 1972. Bau und Verteilung der Sensillen auf der Antennengeißel von *Lasius fulginosus* (Latr.) (Hymenoptera: Formicidae). *Z. Morphol. Tiere* **73**:95-116.

(36)  K. Dumpert. 1972. Alarmstoffrezeptoren auf der Antennae von *Lausis fulginosis* (Latr.) (Hymenoptera: Formicidae). *Z. Vgl. Physiol.* **76**:403-425.

(37)  C.T. Lewis. 1971. Superficial sense organs of the antenna of the fly, *Stomoxys calcitrans. J. Insect Physiol.* **17**:449-461.

(38)  C.T. Lewis. 1972. Chemoreceptors in haematophagus insects. In *Behavioural Aspects of Parasite Transmission.* E.U. Canning and C.A. Wright, Eds. Suppl. 1 to the *Zool. J. Linn. Soc.* **51**:201-213.

(39)  H.Z. Levinson, A.R. Levinson, B. Muller, and R.A. Steinbrecht. 1974. Structure of sensilla, olfactory perception, and behaviour of the bedbug, *Cimex lectularius,* in response to its alarm pheromone. *J. Insect Physiol.* **20**:1231-1248.

(40)  E.H. Slifer. 1967. Thin walled olfactory sense organs on the insect antenna. In *Insects and Physiology.* J.W.L. Beaument and J.E. Treherne, Eds. Oliver and Boyd, Edinburgh, pp. 232-245.

(41)  E.H. Slifer. 1970. The structure of arthropod chemoreceptors. *Ann. Rev. Entomol.* **15**:121-142.

(42)  R.A. Steinbrecht. 1969. Comparative morphology of olfactory receptors. In *International Symposium on Olfaction and Taste.* III. Carl Pfaffman, Ed. Rockefeller University Press, New York, pp. 3-21.

(43)  R.A. Steinbrecht and B. Müller. 1971. On the stimulus conducting structures in insect olfactory receptors. *Z. Zellforsch.* **117**:570-575.

(44)  K.D. Ernst. 1969. Die Feinstruktur von Richsensillen auf der Antenne des Aaskäfers, *Necrophorus. Z. Zellforsch Mikroskop. Anat.* **94**:72-102.

(45)  D.A. Scott, and R.Y. Zacharuk. 1971. Fine structure of the antennal sensory appendix in the larva of *Ctenicerca destructor* (Brown) (Elateridae: Coleoptera). *Can. J. Zool.* **49**:199-210.

(46)  R.Y. Zacharuk. 1971. Fine structure of peripheral terminations in the porous sensillar cone of larvae of *Ctenicerca destructor* (Brown) (Coleoptera: Elateridae)

and probable fixation artifacts. *Can. J. Zool.* **49**:789-799.

(47)  K.-E. Kaissling. 1969. Kinetics of olfactory receptor potentials. In *International Symposium on Olfaction and Taste.* III. Carl Pfaffman, Ed. Rockefeller University Press. New York, pp. 52-70.

(48)  R.A. Steinbrecht, and G. Kasang. 1971. Capture and conveyance of odour molecules in an insect olfactory receptor. In *International Symposium on Olfaction and Taste.* IV. D. Schneider, Ed. Wissenschaftliche Verlagsgesellschaft, Stuttgart, pp. 193-199.

(49)  W.D. Seabrook and V.A. Jaeger. Unpublished.

(50)  G. Kasang, and K.-E. Kaissling. 1972. Specificity of primary and secondary olfactory processes in *Bombyx* antennae. In *International Symposium on Olfaction and Taste.* IV. D. Schneider, Ed. Wissenschaftliche Verlagsgesellschaft, Stuttgart, pp. 200-206.

(51)  L.M. Riddiford. 1970. Antennal proteins of saturnid moths, their possible role in olfaction. *J. Insect Physiol.* **L6**:653-660.

(52)  S.M. Ferkovich, M.S. Mayer, and R.R. Rutter. 1973. Conversion of the sex pheromone of the cabbage looper. *Nature* (Lond.) **242**:53-55.

(53)  S.M. Ferkovich, M.S. Mayer, and R.R. Rutter. 1973. Sex pheromone of the cabbage looper: Reactions with antennal proteins in vitro. *J. Insect Physiol.* **19**:2231-2243.

(54)  M.S. Mayer. Personal communication.

(55)  G. Kasang. 1974. Uptake of the sex pheromone, 3H-bombykol and related compounds by male and female *Bombyx* antennae. *J. Insect Physiol.* **20**:2407-2422.

(56)  K.-E. Kaissling. 1972. Kinetic studies on olfactory receptors of *Bombyx mori.* In *International Symposium on Olfaction and Taste.* IV. D. Schneider, Ed. Wissenschaftliche Verlagsgesellschaft, Stuttgart, pp. 207-213.

(57)  K.-E. Kaissling. 1974. Sensory transduction in insect olfactory receptors. In *Biochemistry of Sensory Function.* L. Jaenick, Ed. Springer-Verlag, New York, pp. 243-270.

(58)  W.L. Roelofs, and A. Comeau. 1968. Sex pheromone perception. *Nature* (Lond.) **220**:600-601.

(59)  R.J. O'Connell. 1975. Olfactory receptor responses to sex pheromone components in the red-banded leaf roller moth. *J. Gen. Physiol.* **65**:179-205.

(60)  D. Schneider. 1962. Electrophysiological investigation on the olfactory specificity of sexual attracting substances in different species of moths. *J. Insect Physiol.* **8**:15-30.

(61)  D. Schneider, V. Lacher, and K.-E. Kaissling. 1964. Die Reaktionweise und das Reaktionsspektrum von Riechzellen bei *Antheraea pernyi* (Lepidoptera, Saturniidae). *Z. Vgl. Physiol.* **48**:632-662.

(62)  E. Vareschi. 1971. Duftienterscheidung bei der Honigbiene-Einzelzell-Ableitungen und Verhaltensreaktionen. *Z. Vgl. Physiol.* **75**:143-173.

(63)  J. Weatherson, W. Roelofs, A. Comeau, and C.J. Sanders. 1971. Studies of physiologically active arthropod secretions. X. Sex pheromone of the eastern spruce budworm *Choristoneura fumiferana* (Lepidoptera: Tortricidae). *Can. Entomol.* **103**:1741-1747.

(64)  C.J. Sanders, and G.S. Lucuik. 1972. Masking of female sex pheromone of the eastern spruce budworm by excised female abdominal tips. *Can. Dept. Environ. Bi-mon. Res. Notes* **28**:10-11.

(65)  D.O. Hathaway, T.P. McGovern, M. Beroza, H.R. Moffitt, L.M. McDonough, and

B.A. Butt. 1974. An inhibitor of sexual attraction of male codling moths to a synthetic sex pheromone and virgin females in traps. *Environ. Entomol.* 3:522-524.

(66) J.E. Casida, H.C. Coopel, and T. Watanabe. 1963. Purification and potency of the sex attractant from the introduced pine sawfly, *Diprion similis. J. Econ. Entomol.* 56:18-24.

(67) R.S. Berger, J.M. McGough, and D.F. Martin. 1965. Sex attractants of *Heliothis zea* and *H. virescens. J. Econ. Entomol.* 8:1023-1024.

(68) M. Jacobson, and L.A. Smalls. 1966. Masking of the American cockroach sex attractant. *J. Econ. Entomol.* 69:414-416.

(69) M. Jacobson, and L.A. Smalls. 1967. Sex attraction masking in the cynthia moth. *J. Econ. Entomol.* 60:296.

(70) H.H. Shorey, and L.K. Gaston. 1967. Sex pheromones of noctuid moths. XII. Lack of evidence for masking of sex pheromone activity in extracts from females of *Heliothis zea* and *H. virescens. Ann. Entomol. Soc. Am.* 60:847-849.

(71) M. Beroza. 1967. Nonpersistent inhibitor of the gypsy moth sex attractant in extracts of the insect. *J. Econ. Entomol.* 60:875-876.

(72) C.J. Sanders, R. Bartell, and W. Roelofs. 1972. Field trials for synergism and inhibition of trans-11-tetradecenal, the sex pheromone of the eastern spruce budworm. *Can. Dept. Environ. Bi-mon. Res. Notes.* 28:10.

(73) !. Weatherston and W. MacLean. 1974. The occurrence of (E)-11-tetradecen-1-ol, a known sex attractant inhibitor, in the abdominal tips of virgin female eastern spruce budworm, *Choristoneura fumiferana* (Lepidoptera: Tortricidae). *Can. Entomol.* 106:281-284.

(74) J.H. Borden. 1974. Pheromone mask produced by male *Trypodendron lineatum* (Coleoptera: Scolytidae). *Can. J. Zool.* 52:533-536.

(75) L.L. Sower, K.W. Vick, and J.H. Tumlinson. 1974. (Z,E)-9,12-tetradecadien-1-ol: A chemical released by female *Plodia interpunctella* that inhibits the sex pheromone response of male *Cadra cautella. Environ. Entomol.* 3:120-122.

(76) L.L. Sower, K.W. Vick, and K.A. Ball. 1974. Perception of olfactory stimuli that inhibit the responses of male phycitid moths to sex pheromones. *Environ. Entomol.* 3:277-279.

(77) M.S. Mayer. 1973. Electrophysiological correlates of attraction in *Trichoplusia ni. J. Insect Physiol.* 19:1191-1198.

(78) J.R. McLaughlin, E.R. Mitchell, D.L. Chambers, and J.H. Tumlinson. 1974. Perception of Z-7-dodecen-1-ol and modification of the sex pheromone response of male loopers. *Environ. Entomol.* 3:677-680.

(79) A.K. Minks, W.L. Roelofs, E. Schuurmans-Van Dijk, C.J. Persoons, and F.J. Ritter. 1974. Electroantennogram responses of two tortricid moths using two-component sex pheromones. *J. Insect Physiol.* 20:1659-1665.

(80) J.C. Young, R.M. Silverstein, and M.C. Birch. 1973. Aggregation pheromone of the beetle *Ips confusus*: Isolation and identification. *J. Insect Physiol.* 19:2273-2277.

(81) W.L. Roelofs. Unpublished [cited in R.J. O'Connell (59)].

(82) W. L. Roelofs, and R.T. Cardé. 1974. Oriental fruit moth and lesser appleworm attractant mixtures refined. *Environ. Entomol.* 3:586-588.

(83) W.D. Bedard, P.E. Tilden, D.L. Wood, R.M. Silverstein, R.G. Brownlee, and J.O. Rodin. 1969. Western pine beetle: Field response to its sex pheromone and a synergistic host terpene, myrcene. *Science* (Wash.) 164:1284-1285.

(84) G.B. Pitman, 1969. Pheromone response in pine bark beetles: Influence of host volatiles. *Science* (Wash.) 166:905-906.

(85)    J.A. Klun, O.L. Chapman, K.C. Mattes, P.W. Wojtlowski, M. Beroza. 1973. Insect sex pheromones: Minor amount of opposite geometrical isomer critical to attraction. *Science* (Wash.) 181:661-663.

(86)    J. Boeckh, E. Priesner, and D. Schneider. 1963. Olfactory receptor response to the cockroach sexual attractant. *Science* (Wash.) 141:716-717.

(87)    D. Schneider, B.D. Block, J. Boeckh, and E. Priesner. 1967. Die Reaktion der männlichen Seidenspinner auf Bombykol und seine Isomeren: Elektroantennogramm und Verhalten. *Z. Vgl. Physiol.* 54:192-209.

(88)    K.-E. Kaissling, and E. Priesner. 1970. Die Reichschwelle des Seidenspinners. *Naturwissenschaften* 57:23-28.

(89)    R.C. Gesteland, and W. Müller. 1974. Intensity coding in olfactory receptor cells. *Fed. Proc.* 33:1472.

(90)    R.J. Bartell, and H.H. Shorey. 1969. Pheromone concentrations required to elicit successive steps in the mating sequence of males of the light-brown apple moth, *Epiphyas postvittana. Ann. Entomol. Soc. Am.* 62:1206-1207.

(91)    R.S. Kaae, H.H. Shorey, and L.K. Gaston. 1973. Pheromone concentration as a mechanism for reproductive isolation between two lepidopterous species. *Science* (Wash.) 179:487-488.

(92)    R.M.M. Traynier. 1968. Sex attraction in the Mediterranean flour moth, *Anagasta kuhniella*; location of the female by the male. *Can. Entomol.* 100:5-10.

(93)    J.S. Kennedy and D. Marsh. 1974. Pheromone-regulated anemotaxis in flying moths. *Science* (Wash.) 184:999-1001.

(94)    S.R. Farkas and H.H. Shorey. 1972. Chemical trail-following by flying insects: A mechanism for orientation to a distant odor source. *Science* (Wash.) 178:67-68.

(95)    S.R. Farkas, and H.H. Shorey. and L.K. Gaston. 1974. Sex pheromones of Lepidoptera. Influence of pheromone concentration and visual cues on aerial odor-trail following by males of *Pectinophora gossypiella. Ann. Entomol. Soc. Am.* 67:633-638.

(96)    H.H. Shorey. 1973. Behavioral responses to insect pheromones. In *Annual Review of Entomology,* Vol. 18. Ray F. Smith, Thomas E. Mittler, and Carroll N. Smith, Eds. Annual Reviews, Palo Alto, Calif., pp. 349-380.

(97)    D. Schneider and U. Seibt. 1969. Sex pheromone of the queen butterfly: Electroantennogram responses. *Science* (Wash.) 164:1173-1174.

(98)    G.G. Grant. 1971. Electroantennogram responses to the scent brush secretions of several male moths. *Ann. Entomol. Soc. Am.* 64:1428-1431.

(99)    G.G. Grant, U.E. Bradly, and J.M. Brand. 1972. Male armyworm scent brush secretions: Identification and electroantennogram study of major components. *Ann. Entomol. Soc. Am.* 65:1224-1227.

(100)   W. Seibt, D. Schneider, and T. Eisner. 1972. Hairpencils, wing pouches and courtship of the butterfly *Danaus chrysippus. Z. Tierpsychol.* 31:513-530.

(101)   W.L. Brown, Jr., T. Eisner, and R.H. Whiiaker. 1970. Allomones and kairomones: Transspecific chemical messengers. *Bioscience* 20:21-22.

(102)   G.N. Lanier, M.C. Birch, R.F. Schmitz, and M.M. Furniss. 1972. Pheromones of *Ips pini* (Coleoptera: Scolytidae): Variation in response among three populations. *Can. Entomol.* 104:1917-1923.

(103)   J.P. Vite, and D.L. Williamson. 1970. *Thanasimus dubius*: Prey perception. *J. Insect Physiol.* 16:233-239.

(104)   P.D. Greany, and E.R. Oatman. 1972. Analysis of host discrimination in the parasite *Orgilus lepidus* (Hymenoptera: Braconidae). *Ann. Entomol. Soc. Am.* 65:377-383.

(105)   M. Sternlicht. 1973. Parasitic wasps attracted by the sex pheromone of their

coccid host. *Entomophaga* **18**:330-342.

(106)   R.L. Jones, W.J. Lewis, M. Beroza, B.A. Bierl, and A.N. Sparks. 1973. Host-seeking stimulants (kairomones) for the egg parasite, *Trichogramma evenescens. Environ. Entomol.* **2**:593-596.

(107)   W.J. Lewis, R.L. Jones, and A.N. Sparks. 1972. A host-seeking stimulant for the egg parasite *Trichogramma evanescens*: Its source and a demonstration of its laboratory and field activity. *Ann. Entomol. Soc. Am.* **65**:1087-1089.

(108)   R.L. Jones, W.J. Lewis, M.C. Bowman, M. Beroza, and B.A. Bierl. 1971. Host-seeking stimulant for parasite of corn earworm: Isolation, identification, and synthesis. *Science* (Wash.) **173**:842-843.

(109)   W.J. Lewis, and R.K. Jones. 1971. Substance that stimulates host-seeking by *Microplitis croceipes* (Hymenoptera: Braconidae), a parasite of *Heiothis* species. *Ann. Entomol. Soc. Am.* **64**:471-473.

(110)   L.B. Hendry, P.D. Greany, and R.J. Gill. 1973. Kairomone mediated host-finding behaviour in the parasitic wasp *Orgilus lepidus. Entomol. Exp. Appl.* **16**:471-477.

(111)   D.B. Hays, and S.B. Vinson. 1971. Acceptance of *Heliothis virescens* (F.) (Lepidoptera, Noctuidae) as a host by the parasite *Cardiochiles nigriceps viereck* (Hymenoptera, Braconidae). *Anim. Behav.* **19**:344-352.

(112)   B.M. Hegdekar and A.P. Arthur. 1973. Host hemolymph chemicals that induce oviposition in the parasite *Itoplectis conquisitor* (Hymenoptera: Ichneumonidae). *Can. Entomol.* **105**:787-793.

(113)   S.A. Corbet. 1971. Mandibular gland secretion of larvae of the flour moth, *Anagasta kuehniella,* contains an epideictic pheromone and elicits oviposition movements in a Hymenopteran parasite. *Nature* (Lond.) **232**:481-484.

(114)   S.A. Corbet. 1973. Concentration effects and the response of *Nemeritis canescens* to a secretion of its host. *J. Insect Physiol.* **19**:2119-2128.

(115)   S.B. Vinson, and W.J. Lewis. 1965. A method of host selection by *Cardiochiles nigriceps. J. Econ. Entomol.* **58**:869-871.

(116)   M.S. Blum. 1969. Alarm pheromones. *Ann. Rev. Entomol.* **14**:57-80.

(117)   M.S. Blum. 1971. Dimensions of chemical sociality. *Proc. 2nd Int.* IUPAC Cong. Israel, **3**:147-159.

(118)   M.S. Blum, and S.L. Warter. 1966. Chemical releasers of social behaviour. VII. Isolation of 2-heptanone from *Conomyrma pyramica* (Hymenoptera: Formicidae: Dolichoderinae) and its modus operandi as a releaser of alarm and digging behaviour. *Ann. Entomol. Soc. Am.* **59**:774-779.

(119)   M.S. Blum, S.L. Warter, and J.G. Traynham. 1966. Chemical releasers of social behaviour. VI. The relation of structure to activity of ketones as releasers of alarm for *Iridomyrmex pruinosus* (Roger). *J. Insect Physiol.* **12**:419-427.

(120)   M.S. Blum, F. Padovani, and E. Amante. 1968. Alkones and terpenes in the mandibular glands of *Atta* species (Hymenoptera: Formicidae). *Comp. Biochem. Physiol.* **26**:291-299.

(121)   M.S. Blum, F. Padovani, H.R. Hermann, Jr., and P.B. Kanowski. 1968. Chemical releasers of social behaviour. XI. Terpenes in the mandibular gland of *Lasius umbratus. Ann. Entomol. Soc. Am.* **61**:1354-1359.

(122)   F.E. Regnier, and E.O. Wilson. 1968. The alarm-defense system of the ant *Acanthomyops claviger. J. Insect Physiol.* **14**:955-970.

(123)   E.O. Wilson, and F.E. Regnier. 1971. The evolution of the alarm defense system in the formicine ant. *Am. Nat.* **105**:279-289.

(124)   F.E. Regnier, and E.O. Wilson. 1971. Chemical communication and 'propaganda' in the slave maker ants. *Science* (Wash.) **172**:267-269.

(125)  B. Tursch, D. Daloze, M. Dupont, C. Hootele, M. Kaisin, J.M. Pasteels, and D. Zimmermann. 1971. Coccinellin, the defensive alkaloid of the beetle *Coccinella septempunctata. Chimia* **25**:307-308.

(126)  B. Tursch, D. Daloze, M. Dupont, J.M. Pasteels, and M.-C. Tricot. 1971. A defense alkaloid in a carnivorous beetle. *Experientia* **27**:1380-1381.

(127)  B.J. Wallace, and M.S. Blum. 1971. Reflex bleeding: A highly refined defensive mechanism in *Diabrotica* larvae (Coleoptera: Chrysomelidae). *Ann. Entomol. Soc. Am.* **64**:1021-1024.

(128)  M.S. Blum, and A. Sannasi. 1974. Reflex bleeding in the lampyrid *Photinus pyralis:* Defensive function. *J. Insect Physiol.* **20**:451-460.

(129)  L.R. Nault, L.J. Edwards, and W.E. Styer. 1973. Aphid alarm pheromones: Secretion and reception. *Envir. Entomol.* **2**:101-105.

## CHAPTER 4

# Control of Insect Behavior via Chemoreceptor Organs

Karl-Ernst Kaissling

*Max-Planck-Institut*
*Seewiesen, West Germany*

Insects are able to perceive numerous chemical compounds via chemoreceptor cells. The effective chemicals belong, like in vertebrates, to two stimulus *modalities,* taste and smell, according to two morphologically rather distinct groups of chemoreceptor organs. Within each modality insects can distinguish stimuli, such as sugar, salt, and water, or a great variety of odor compounds, including even carbon dioxide and water vapor. Many insects discriminate complex mixtures of chemicals (e.g., the scents of flowers), others detect certain compounds (e.g., pheromones) in extremely low concentrations. Behavioral studies in recent years have shown that insects can respond selectively to a certain ratio of two or a few constituents of a mixture, as in the case of some pheromones (1)-(6).

That insects can discriminate indicates that they must have various types of receptor cells—that is, cells with different response specificities to chemicals. Indeed, all insects investigated so far have well developed gustatory and olfactory receptor organs equipped with various types of receptor cells. Even larvae or the adults of relatively small species with only a few sensory organs (sensilla) seem to have chemically specialized receptor cell types. Insect species with highly developed chemosensory systems may have several hundred thousand receptor cells and numerous cell types.

If we consider that receptor cells of various types are distributed over the antennal surface or over other body parts, we can imagine that any chemical stimulus appears to the central nervous system (CNS) as a spatial pattern of firing and silent sensory cells. Two sufficiently different stimuli—whether they are single compounds or mixtures—would produce spatially different excitation patterns. If the organism responds differently to the two stimuli we would consider them qualitatively different. It is, therefore, obvious that the recognition of spatial excitation patterns plays a major role in the discrimination of stimulus qualities in the chemosensory system.

The spatial excitation pattern of chemoreceptors could be relatively simple if one receptor cell type responds to one stimulus compound—for example, in the presence of a single-compound pheromone or with carbon dioxide or humidity. More frequently, however, the stimulus eliciting a given behavioral response is composed of two or more essential chemical components. Correspondingly, the sensory excitation pattern would be more complex and include several cell types, enabling the organism to respond only to the "right" mixture and not to other stimuli.

Chemically induced behavioral responses often can be inhibited or reinforced by certain compounds added to the stimulus (7)-(9). Such behavioral inhibitors or synergists do not elicit behavioral responses by themselves but definitely change the sensory excitation pattern. It is conceivable that a behavioral inhibitor recruits nervous activity of previously nonstimulated cells, thereby acting as a stimulatory compound on the cellular level. Similarly, a synergistic compound could act as an excitatory stimulus on a special cell type, though not necessarily eliciting a behavioral response when given alone. The few cases investigated so far show that there are at least two cell types each responding to most of these compounds but with different relative sensitivities. Therefore, adding a behavioral inhibitor or synergist to the pheromone should change the characteristic ratio of impulses delivered by the two cell types and enable the animal to distinguish the mixture from the pure pheromone (10)-(11) (Chapter 10).

Most chemicals that modify insect behavior have, in the past, been discovered by chemical analysis and behavioral assays. Now it is necessary to understand how chemoreceptor cells function and how their excitation pattern can be induced or changed by appropriate chemicals in order to elicit or suppress the particular behavioral response.

Some of the terms I shall be using in this discussion of receptor cell function—such as *excitation, inhibition,* and *adaptation*—can have different meanings depending on the general scope and to some extent on the writer's view. Therefore, brief explanations of some of the terms should clarify their meanings for this chapter.

By *chemical stimulus* we mean any compound that excites the receptor cell. Excitation comprises a sequence of alterations (transducer processes) in the receptor cell, which finally lead to a change of the spontaneous rate of nerve impulses. Usually, the word *excitation* is used in a narrower sense—only for the increase of impulse activity—whereas its decrease is called *inhibition.* This may appear confusing since both effects are elicited by a change of the cell membrane potential (a receptor potential). An increased impulse rate is produced by a lowered membrane potential or a depolarization of the receptor cell; a decreased impulse rate follows a hyperpolarization (12).

Moreover, both depolarization and hyperpolarization can be achieved by a conductance increase of the receptor-cell membrane. We know, for example, that in nerve axons an increase of $Na^+$ conductance produces depolarization, whereas an increase of $K^+$ conductance leads to hyperpolarization (13).

In several insect species, olfactory cells have been found that can be depolarized by some compounds and hyperpolarized by others, and which give intermediate responses to mixed stimuli (14)-(20). It is conceivable that the two opposed response types are mediated by two types of molecular receptors on the same cell membrane controlling the conductance of different ion species. As an alternative, one could think of conductance channels for a single-ion species that might be "open" to a certain degree in the unstimulated cell. These channels could be driven by some stimulants toward higher, and by others toward lower, conductances and in this way cause depolarization and hyperpolarization of the cell membrane. Both effects could even be mediated by the same type of molecular receptor if it is able to react in a different way to depolarizing and hyperpolarizing compounds. Thus both excitation and inhibition of impulse activity can be produced by events that can be called excitatory on the molecular level.

Information about the excitatory mechanisms can be gathered from the transfer functions between stimulus, receptor potential, and impulse activity. Receptor potentials are usually recorded extracellularly either as the summated response from many cells—the electroantennogram—or from the cells of a single sensillum. In the latter case one tries to connect the so-called "different" electrode with the receptor lymph that surrounds the outer dendrites of all cells of a sensillum and that is electrically isolated from the receptor lymph spaces of other sensilla. This receptor lymph space has, in the unstimulated state, a positive standing potential of 40 to 50 mV when compared with the hemolymph space in which the other or "indifferent" electrode is inserted. Usually, the receptor lymph space becomes less positive with an excitatory stimulus and more positive with an inhibitory stimulus—that is, with respect to increased and decreased impulse activity. These changes of the standing potential are considered as negative or positive receptor potentials. It is still not clear to what extent these extracellular receptor potentials reflect the changes of membrane potentials of the outer dendrites and of the impulse-generating region near the cell soma. The receptor potential of the impulse-generating region is also called the generator potential (12).

## I. RECEPTOR MOLECULES

In considering the mechanism of excitation in chemoreceptors, one generally assumes that the outer dendrite of the receptor cell has special molecular receptor (or acceptor) units bound to its membrane (12). These hypothetical receptor molecules are thought to become activated and induce conductance changes across the cell membrane if a stimulus molecule is bound to them in an appropriate manner. One reason for assuming particular receptor molecules is the remarkable specificity of the receptor cells. Other processes— such as adsorption to the cuticular wall of the sense organs or diffusion from the outer surface toward the cell membrane—seem to contribute very little to the specificity of the cell (21). Moreover, the early inactivation and the late enzymatic destruction of stimulus molecules is relatively unspecific (21)-(22). The molecular structures responsible for the specificity of receptor cells should be located on the receptor cell because gustatory as well as olfactory hairs often contain the outer dendrites of several receptor cells, each of which has a different specificity (6) (10) (11) (15)-(18) (23). The surrounding hair wall, which has to be penetrated by the stimulus molecules, is constructed by a special cell—the trichogen cell—and could hardly be imagined to generate different specificities for the receptor cells inside the hair (24, 25).

Membrane-bound receptor molecules have been isolated from other biologic tissues. The acetylcholine-activated receptor molecule induces a conductance increase of about $10^{-10}$ $\Omega^{-1}$ in the muscle endplate membrane (26)-(28). A similar conductance increase per bound stimulus molecule was calculated for the bombykol receptor cell (29). That is, opening a single-ion channel would be sufficient to produce the elementary receptor potential of a 200-to-300-$\mu$V amplitude and about 50 msec duration, as observed in olfactory cells (29) (30).

The acetylcholine receptor of the electroplax of the electric eel, *Torpedo californica,* is a protein molecule with molecular weight of 500,000, and seems to be composed of several polypeptide units (31). These, as well as other membrane receptor molecules, are considered to be integrated in the cell membrane and, therefore, depend strongly on the state of their lipid surrounding. It is not yet known whether the receptor molecule itself can act as an ion gate or if it mediates the excitation to other structures in the membrane. In any case, the receptor molecule itself has to become activated after binding with a stimulus molecule, which means that it must undergo conformational or other changes, like those of the rhodopsin molecule in visual cells (32). There seems to be sufficient time for all of these events leading up to the conductance change, since the average reaction time between the adsorption of an odor molecule and the elementary receptor potential is about 200 to 500 msec (29) (33). Even visual cells in insects can have remarkably long reaction times of more than 100 msec between a light flash and elementary receptor potentials (34).

Receptor proteins involved in chemoreception have been detected in bacteria. The galactose-binding protein (molecular weight 35,000) is one of the nonmembrane-bound receptor molecules of *Escherichia coli*; it is found in the periplasmic space of the cell wall (35) (36). Sugar-binding molecules have been isolated from the tongues of vertebrates (37) (38). These are proteins that are very likely to be receptor molecules, since they bind various sugars according to their threshold concentrations. Surprisingly, they have even a higher affinity for saccharin than for sucrose, which corresponds well to the lower behavioral threshold for saccharin. Both bacterial-binding protein and the sweet-sensitive proteins change their conformation if they bind stimulus molecules, as has been shown by *in vitro* experiments (37)-(39). However, they do not split the sugar ligands.

Glucosidases, which split disaccharides, have been isolated from insect taste receptors (40)-(41). Several glucosidases, each with different specificities of binding, have been found in the same insect. Only some of them are considered as likely candidates for receptor molecules because their binding preferences for various sugars are similar to the rank order of effectivenesses found in single-cell recordings. The splitting of the substrate, however, does not seem to be an essential event in the transduction process, since glucose also stimulates the receptor cell, but it is not yet completely excluded.

Some taste receptor cells seem to have more than one type of receptor molecule or, as an alternative, two sites on a single receptor molecule. Thus, taste cells of *Mamestra brassicae* larvae respond to a mixture of sinigrin and *1*-naphthyl-$\beta$-glucosid considerably better than to either of the compounds alone (42).

Several types of receptor molecules per cell have been assumed for the so-called "generalist" cells found in some insect species (17)-(19). These olfactory cells occur in large numbers on the antenna, and each have different but overlapping chemical specificities. There may be several types of receptor molecules present in different proportions in the various cells. Each stimulus induces a different excitation pattern among these cells which, in principle, could enable the insect to distinguish a large number of odors.

On the other hand, the so-called "specialist" cells are most sensitive in detecting a single key compound (12). Their chemical specificity is, in the extreme, identical for all cells of a given type which, therefore, should have the same type of receptor molecules. There exist transitional types between generalists and specialists. Thus the honeybee has, among other cell types, many cells that are most sensitive to queen substance but that vary in their sensitivity to several fatty acids (43)-(44). This variance enables the honeybee to distinguish between all of these compounds (44)-(45).

There is still no direct evidence for the existence of receptor molecules in olfactory cells. These cells, however, can be extremely specific—that is, they can detect very small changes of the molecular structure of stimulants (12). Often the most effective stimulus compound or the key compound can be as much as 100 or 1000 times more effective than any other related compound. As an example, the undecane receptor cell of the European garden ant, *Lasius fuliginosus,* needs 1000 times higher concentrations of decane or dodecane than of undecane for a standard response (46). Such extraordinary specificities, combined with extreme sensitivities, support presence of receptor molecules in olfactory cells.

Molecular interactions such as binding between odor and receptor molecule or activation of the receptor molecule could have different chemical specificities. This could explain the fact that the specificity of the response depends on the evaluated response parameter, such as the average frequency of nerve impulses or their distribution in time. Single cell recordings in *Bombyx mori* (29) (30) (47) and in saturniid moths (48) lead to the conclusion that the number and size of the elementary receptor potentials (see p. 48) varied independently in response to the stimulus compound. The *number* of elementary responses might be determined by the number of receptor molecules brought into an activated state by the odor molecule. Different *sizes* of the elementary potentials could, however, indicate different types of activation of the receptor molecule. It has been possible to suppress selectively the one response type that showed large elementary potentials by applying very weak stimuli of osmium tetroxide (29). This result shows that, in principle, it is possible to modify the chemical specificity of a chemoreceptor cell. Thus, one could alter the nerve impulse pattern and, possibly, confuse the organism.

## II. MODIFICATION OF RECEPTOR-CELL FUNCTION BY VARIOUS COMPOUNDS

There are, of course, several ways of inhibiting the function of receptor molecules, thereby reducing the general sensitivity of the cell. The metabolic activity of glucosidases isolated from taste hairs of certain insect species can, for instance, be inhibited by octylamine or octanol (49). These compounds also depress the sensitivity of the sugar receptor cell and inhibit the nerve impulses of the salt and water cell in the same taste bristle (50). Therefore, these compounds seem to have several sites of action-namely, at the putative receptor molecules of the sugar cells and at unknown membrane components of other taste cells. Treatment with sulfhydryl reagents and other biochemical inhibitors blocks chemoreceptor function (51)-(57). Acetazolamide, designed as a carbonic anhydrase inhibitor, suppresses the nerve-impulse response of $CO_2$-receptor cells in the honeybee. It has not yet been determined whether other olfactory cells in the treated bees remained unaffected (58).

It would, of course, be ideal to find inhibitors that block chemoreceptor cells of only one type of a given chemical specificity. One would then not be dependent upon protein-denaturing inhibitors that harm biologic tissue in general. One way to find such specific inhibitors would be to modify the natural key compound until it blocks at the binding site of the receptor molecules. This would require extensive work by chemists, synthesizing large numbers of candidate inhibitors and testing electrophysiologically in single-cell recordings. In addition, general toxicants should be more extensively tested. General toxicants for receptor molecules or for membranes could perhaps be useful for control purposes if they reach the target tissue—for example, olfactory cells of insects—in sufficient amounts, but reach the tissues of other organs in doses far below the toxic level.

The cuticle of insects is extremely impermeable to water and water-soluble compounds except at the tips of taste-receptor bristles. Access of water-soluble compounds to olfactory cells is, as far as we know, excluded. For this reason, the olfactory hairs of the cuticle must be penetrated with electrodes to establish electrical contact. The normally lipophilic odor molecules penetrate the cuticle easily but then they encounter the aqueous receptor lymph surrounding the receptor-cell dendrites. Most probably the cuticular pore channels found in the wall of most olfactory hairs (12) facilitate this penetration of the stimulus molecules to the cell membrane. Compounds could be effectively passed to the olfactory cell along this pathway in order to interfere with the membrane lipids. For example, the function of olfactory receptor cells can be effectively destroyed by dipping the antenna of a saturniid moth briefly in pentane. However, the mechanoreceptor cells, attached to the taste bristles and lacking the cuticular pore channels, will not as readily permit the penetration of pentane to their dendrites and survive this treatment.

Various lipophilic compounds have been found that are chemically not related to each other nor to the key compound of a receptor cell but which nevertheless act on olfactory receptor cells. These compounds can exert several types of effects on the same cell, depending on the applied concentration. The effects of long-chain amines—like octylamine—on bombykol receptor cells (22), (Fig. 1), and on homologous receptor cells of the female, which are insensitive to bombykol but respond to linalool (59) (Fig. 2) are as follows:
1. Weak vapor concentrations of octylamine, barely detectable by the human nose, hyperpolarize the receptor cell and suppress the spontaneous impulse activity.
2. High concentrations depolarize the cell and induce nerve impulses.
3. Octylamine repolarizes the cell if given immediately after a bombykol or linalool stimulus.
4. The sensitivity of the cell to bombykol can be reduced several orders of magnitude by stimulation with octylamine but recovers within several

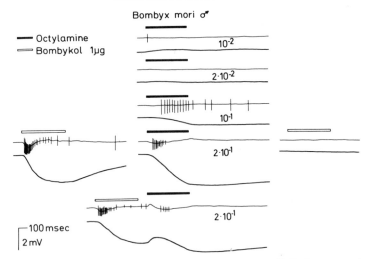

Fig. 1.   Effects of octylamine on a single bombykol receptor cell. Each response has been extracellularly recorded with **AC**-amplification (upper traces, sometimes showing nerve impulses) and with **DC**-amplification (lower traces, showing receptor potentials). The horizontal bars above each pair of traces indicate the time of stimulation (0.5 sec). Sequence of stimuli: At first, 1µg bombykol (*left*); then, four octylamine stimuli of increasing concentration (*middle*) ($10^{-2}$ means dilution 1:100 with paraffin oil, $10^0$ would mean pure octylamine on the source). The weakest stimulus shows a small but significant deflection of the receptor potential upwards, which indicates hyperpolarization. Stronger octylamine stimuli ($10^{-1}$, $2 \times 10^{-1}$) depolarize and produce nerve impulses. Later stimulation, bombykol (*right*). No response after octylamine. Final stimulation, bombykol, after recovery from octylamine inhibition, and, immediately afterwards, octylamine again (*below*). Now octylamine initially produces a deflection upwards (repolarization) and then a depolarization together with nerve impulses.

minutes, depending on the concentration of octylamine.

5. Strong vapor concentrations of octylamine depolarize the cell irreversibly when given for several seconds.

Some of these effects are similar to those observed on the impulse response of taste cells (see above).

One possible explanation of these effects is based on the observation that many lipophilic compounds, such as straight-chain alcohols, prevent hemolysis in red blood cells if applied in low concentrations but facilitate hemolysis in high doses (60). The conclusion is that these compounds can stabilize or destroy the lipid matrix of the membrane. Correspondingly, octylamine could tighten or break the membrane, according to its dose, thereby influencing the ion conductance of the membrane. Conductance changes have been observed in extracellular recordings from olfactory sensilla, stimulated with strong doses of pheromones and of octylamine (61). The repolarization effect and the

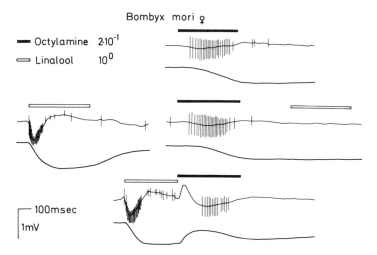

Fig. 2.     Effects of octylamine on a single linalool-sensitive receptor cell of the female. See explanation of Fig. 1. Sequence of stimuli: At first, pure linalool on the source (*left*). Then, octylamine (*above*). Later, octylamine, followed by linalool with no response to the latter (*middle*). Finally, linalool, after recovery from octylamine inhibition, and immediately afterwards, octylamine again (*below*). Octylamine repolarizes and depolarizes as in the bombykol receptor cell (Fig. 1). Weak octylamine stimuli also show hyperpolarization as in the male. Note the typical time course of the octylamine response, which might indicate competing de- and repolarizing effects. Responses to linalool or bombykol usually show a much steeper increase of the receptor potential after the onset of the stimulus and the highest frequency of impulses at the beginning.

decrease of sensitivity shows that the amine probably also affects the action of the receptor molecules or of the ion gates, possibly by way of the alteration of their lipid milieu. Octylamine probably favors the closed state of ion gates. Such a mechanism would seem likely since the state of membrane proteins is known to depend essentially on the state of the surrounding lipid (62) (63).

It is astonishing that the receptor cell is able to recover from the octylamine effect after some time and that the octylamine-induced receptor potential decreases after the end of stimulation: octylamine, like bombykol and linalool, undergoes an early inactivation. This relatively unspecific inactivation effect has been considered to be more likely a physical displacement than a chemical decomposition of the stimulus molecules. Strong octylamine concentrations, however, damage the cell irreversibly.

Other long-chain amines exert effects similar to those of octylamine on the olfactory cells (22). The same is probably true for a large number of other compounds. For instance, geranoli (17) and linalool act similarly on pheromone receptor cells. Linalool itself excites the bombykol receptor cell of the male much less than the homologous female receptor. It does not produce hyperpolarization, but repolarizes the cell if given on top of a receptor potential induced by bombykol (Fig. 3).

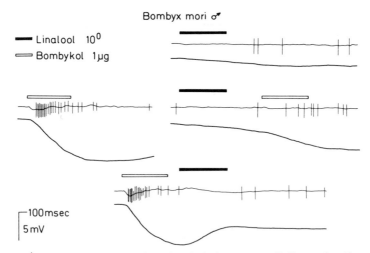

Fig. 3.   Effects of linalool on a single bombykol receptor cell. See explanation of Fig. 1. Sequence of stimuli as in Fig. 2. Linalool does not stimulate the cell (*above*), but inhibits the response to bombykol (*middle*) and repolarizes if given after bombykol (*below*).

Vapor of pentane or chloroform hyperpolarizes in strong concentrations, suppresses the response to bombykol effectively, and repolarizes if given after bombykol (Fig. 4). These effects are suggested from electroantennogram measurements, but remain to be confirmed by single-cell recordings. Thus, in principle, it is possible to modify the action of chemoreceptor cells by compounds that probably do not act directly on the receptor molecules but rather on other components of the cell membrane. Generally, these compounds induce narcotic effects of various degrees and types. Therefore, extensive screening studies would be required to find compounds that selectively affect insect chemoreceptors and that could be used for control purposes.

## III. MODIFICATION OF RECEPTOR-CELL FUNCTION BY NATURAL STIMULI

So far we have discussed possible modification of receptor-cell function by chemicals other than the natural key compounds. But these, too, if given over long time intervals or in high doses, can change the sensitivity of the cell and in this way affect the chemosensory excitation pattern.

Bombyx mori ♂, EAG

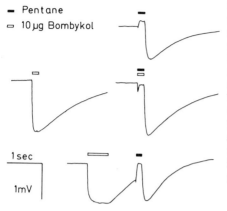

- Pentane
□ 10 μg Bombykol

1 sec

1 mV

Fig. 4. Effects of pentane on the electroantennogram (EAG). Sequence of stimuli: First, pentane, in high vapor concentration, produces a positive EAG (probably hyperpolarization) followed by a negative EAG (probably depolarization) after the end of stimulation (*above*). Then, bombykol. Later, pentane, simultaneously with bombykol (*middle*). Complete inhibition of the bombykol response. The EAG after the end of the stimulation is larger than after pentane alone. Finally, pentane, immediately after bombykol, produces a pronounced repolarization (*below*).

Absolute and incremental detection thresholds of chemoreceptors can increase owing to the fact that most chemosensory organs accumulate active stimulus molecules. For example, odor molecules can strongly adsorb on the surface of olfactory hairs and continue to excite the receptor-cell dendrites after exposure to the outer stimulus has ceased (12) (21) (29). Such excitation amounts to an increased "noise level"; consequently, additional stimuli must be stronger to be detected as a signal. This accumulation effect becomes especially obvious after strong stimuli, which produce long-lasting receptor potentials or impulse firing (12) (22) (29). Obviously the postulated mechanisms of early inactivation, causing the decrease of the receptor potential at the end of stimulation, can be overloaded (cf. Fig. 8).

Other mechanisms clean the antenna of stimulus molecules but act more slowly. In *Bombyx* it has been shown that within minutes the odor molecules diffuse from the hairs toward the epithelium (33) and are later enzymatically destroyed (21). Finally, products appear in the hemolymph of the antenna. Enzymatic decomposition of odor molecules also was found on parts of the body other than the antenna and in other species (21) (64) (65).

Threshold changes due to accumulation should be distinguished from reversible changes of receptor-cell sensitivity induced by adequate stimuli. These latter changes involve alterations in the receptor cell and are called adaptation. The recovery, sometimes called disadaptation, can take many minutes—as in visual cells—or even hours (22) (65) (Figs. 5 and 6).

Adaptation of receptor cells is a complex phenomenon and could, theoretically, have many origins. It could be caused by a lowered affinity of binding to the receptor molecule, by a decreased rate of its activation, by a reduced increase of membrane conductance per activated receptor mole-

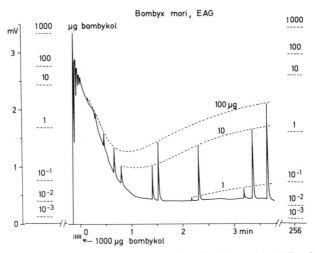

Fig. 5.    Adaptation of the electroantennogram (EAG) of a male of *Bombyx mori*. Deflection upwards means depolarization. Continuous recording over several hours. Sequence of stimulation: First, Bombykol stimuli of increasing strength ($10^{-3}$ up to 1000 μg). The maximum amplitudes are shown as broken lines. Then, at time zero, five stimuli of 1000 μg of bombykol have been given as adapting or conditioning stimuli (at the five vertical bars). The EAG declines relatively slowly after the fifth stimulus. Several test stimuli have been applied in the following 3 min. The responses (connected by *broken lines*) are considerably decreased and start to recover, indicating adaptation and disadaptation of the cells. Complete recovery is reached after several hours as the amplitudes of test stimuli indicate on the right side of the figure. Same experiment as evaluated earlier (23) (66).

cule (66) (67), by a change of properties of the membrane lipids, or by many other effects. Various parts of the receptor cell can adapt differently because the nerve impulses often adapt much sooner than the receptor potential. Two cells of the same sensillum can adapt independently with respect to the impulse response (Fig. 7). This shows that the receptor cell itself has been altered. Otherwise, it has yet to be clarified whether adaptation could also be caused by changes in the receptor lymph and in the auxiliary cells, which contribute to the function of all receptor cells of the same sensillum (68). Extensive studies of single cells would be necessary to understand the mechanisms of adaptation.

A progressive decrease of sensitivity to repeated stimuli, observed in behavioral assays, is called habituation. Habituation may, in principle, be caused by the above-discussed accumulation, by sensitivity changes of receptor cells, or by alterations in the CNS. The latter could occur even at very low stimulus concentrations, where accumulation at or adaptation of the receptor cells has not occurred. This means that the sensory excitation pattern is unchanged but evaluated by the brain in a different manner.

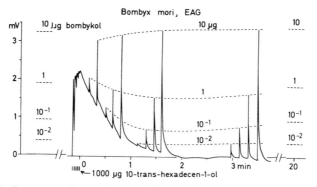

Fig. 6.   Similar to experiment shown in Fig. 5. The conditioning stimulus, however, was 10-*trans*-hexadecen-1-ol instead of bombykol (10-*trans*, 12-*cis* hexadecadienol). The decay after conditioning is relatively slow and similar to the decay after a bombykol stimulus of the same strength in Fig. 5. The test stimuli with bombykol indicate only a very small adaptation of the EAG. The suggestion is that the prolonged decay indicates accumulation of active molecules on the antenna and that the adaptation of the EAG depends on the degree of excitation (here of depolarization).

Increased sensitivity of behavioral responses can be called sensitization. Dogs (69) and honeybees (70) showed considerably lower thresholds to odor compounds after their ingestion or injection. This effect, curiously, appeared 4 days after application in both species. Initially, however, the sensitivity was decreased, which could have been due to the accumulation of the odor substances in the body, increasing the background activity of the receptor cells. The increased sensitivity after 4 days might indicate alterations in the receptor cells or, more likely, in the CNS.

Habituation or *decreased* sensitivity is considered as a possible basis for the so-called "confusion technique" used to control pest insects (71). Habituation can be produced by treating large areas with high concentrations of pheromone for long time intervals. The maintained concentration of an attractant alone, acting as an increased background, could also produce disorientation, increasing the incremental detection threshold and thereby reducing the "active space" of natural odor sources. Also, large numbers of artificial-stimulus sources could compete with the natural sources, reducing their effectiveness. Finally, different concentrations of the same olfactory stimulus have been shown to have a disorienting effect on behavior responses (72) (73).

## IV. INTERNAL FACTORS INFLUENCING RECEPTOR-CELL FUNCTION

It has been suggested that the sensitivity of receptor cells of insects can be internally controlled even though they are not innervated by efferent nerve

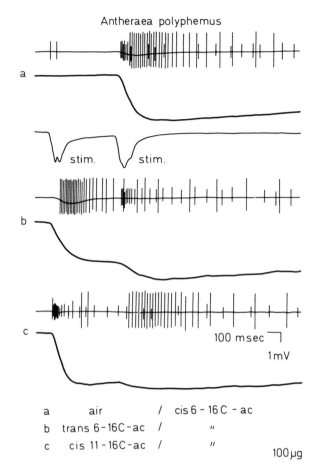

Antheraea polyphemus

stim.     stim.

a

b

c

100 msec

1mV

a        air          /   cis 6 - 16 C - ac
b    trans 6-16C-ac  /         "
c    cis 11 - 16C-ac /         "
                                    100 μg

Fig. 7. Adaptation of nerve impulses without adaptation of the receptor potential. Large and small nerve impulses are from two cells of the same olfactory hair. As in Fig. 1, nerve impulses and receptor potentials are shown in separate traces. In addition the registration of airstream velocity is given in the third line from the top. First two stimuli (*above*): no response to air, both cells responding to 6-*cis*-hexadecenyl acetate. Next two stimuli (*middle*): the cell with large impulses responds to 6-*trans*-hexadecenyl acetate. The following stimulus increases the receptor potential and elicits the small impulses but fails to induce additional large ones. Obviously the large impulses have been adapted by the preceding stimulation. Below: the reciprocal experiment is shown. 11-*cis*-hexadecenyl acetate excites and adapts the cell which fires the small impulses. [The cells with the large impulses would respond most sensitively to 6-*trans*, 11-*cis* hexadecadienyl acetate, a pheromone constituent of *Antheraea polyphemus* (6).]

fibers (74). Therefore, there is interest in controlling insects by the use of hormones. It is assumed that the distal openings of taste hairs will close after feeding, thereby reducing the access of stimulants to the receptor dendrites. This assumption is based on an increased electrical resistance measured across the entire maxillary palps with many taste bristles in locusts (75) (76). Cockroach antennae showed only small resistance changes—but a significant increase in the trans-epithelial voltage—after perfusion with serotonin. Ion-transport systems in the epithelium have been considered to be the site of action of serotonin, and might also be influenced by natural insect hormones (68). These and other mechanisms might serve to modulate receptor-cell function.

## V. CONCLUSION AND SUMMARY

The chemosensory system of an insect consists of an array of receptor cells of several types with different chemical specificities. Every chemical stimulus appears to the CNS as a characteristic spatial pattern of excited and nonexcited cells.

Several possibilities for controlling insect behavior by way of the chemosensory system have been discussed. One could elicit or facilitate specific behavioral responses by inducing the adequate sensory-excitation pattern. This has been tried often by using natural stimuli like pheromones and food stimulants and sometimes by artificial attractants—as with fruit flies (Chapters 9 and 19)—or by using repellents (Chapter 18). Repellents obviously induce meaningful sensory-excitation patterns, thereby possibly mimicking unknown natural olfactory cues. It is, of course, important to consider other environmental and physiological conditions which modify these responses (77).

Other methods can be used to change or suppress the natural behavior. The sensory excitation pattern can be distorted by *modifying the natural stimulus situation*. Ratios between stimulus components could be changed by increasing the concentration of one of several essential compounds. Natural inhibitors could be applied. High concentrations of natural stimuli, either maintained in the surrounding environment or accumulated on the sense organs, can be used to decrease the incremental sensitivity of the receptor cells.

*Alterations in the receptor cells* could be brought about by applying appropriate compounds affecting specific receptor molecules or the receptor-cell membrane itself. Correctly diluted, they would act only on insect chemoreceptors, and not on other organs. Perhaps one should consider whether nature has developed special mechanisms to modify chemoreceptor

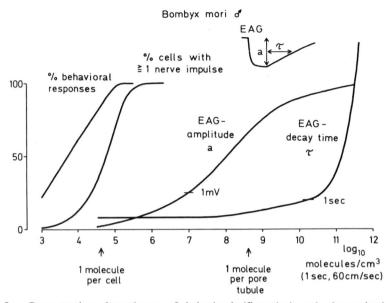

Fig. 8. Concentration dependence of behavioral (fluttering) and electrophysiologic responses in the male silkmoth *Bombyx mori*. The abscissa gives the strength of stimulus: molecules/cm³ air, given for 1 sec at a wind velocity of 60 cm per sec (13). Percent of behavioral responses = relative number of males responding to the stimulus (84). Percent of cells with ≥ 1 nerve impulse = relative number of cells responding in the electrophysiologic test within 2 sec after the onset of the stimulus (84). *a* = maximum EAG-amplitude (13). $\tau$ = half-time of decay of the EAG(23). The stimulus $10^3$ molecules/cm³ was produced by a load of 3 x $10^{-6}$ µg of bombykol on the stimulus source.

function in insects: defensive secretions of insects or, more important, compounds from plants (deterrents) should be tested (Chapter 2). Moreover, natural stimuli could change the properties of receptor cells—that is, cause adaptation; or receptor-cell properties could be changed indirectly by way of the hormonal system. In this connection, the use of insectistatic agents deserves consideration (78).

Finally, sensory excitation could induce *habituation* effects in the CNS. This might be accomplished even with weak stimulation, which would not change the sensory excitation patterns but, instead modify their evaluation by the CNS.

The various behavior-modifying strategies require rather different concentrations of the stimulants acting on chemoreceptors (Fig. 8). Inducing nerve-impulse firing and behavioral responses might require only very low concentrations. The fluttering response of *Bombyx mori* can be elicited with $10^3$ molecules/cc of air applied for 1 sec in an airstream of about 60 cm/sec (79). Habituation effects in the CNS could, in principle, occur at very low concentrations—in fact, as soon as an animal has reacted once. Adaptation of the receptor cells would definitely need more than one molecule per receptor cell. One molecule per cell is reached in *Bombyx* with about $10^5$ molecules/cc. Transient decreases of sensitivity of the impulse response become obvious at about $10^7$ molecules/cc. The prolongation of the receptor potential, probably due to the discussed accumulation, is first evident at about $10^9$ molecules/cc. This stimulus corresponds to a load of about one molecule of bombykol per pore tubule in the hair wall. Finally, strong adaptation effects like those shown in Fig. 5 were obtained with a conditioning stimulus of about $3 \times 10^{11}$ molecules per cc for about 4 sec (22). This is equivalent to about $10^3$ molecules per pore tubule, or about $10^7$ molecules per receptor cell.

Any effective inhibition of a receptor cell would require at least one molecule of a very powerful compound per cell. Usually hyperpolarization needs stronger stimuli than an equivalent depolarization (15). Especially high concentrations may be necessary if one wants to block receptor molecules. If we set the minimum number of receptor molecules per cell equal to the number of pore tubules (about $10^4$ per cell in *Bombyx*) we would need at least $10^3$ inhibitor molecules to block only 10% of the receptor molecules. This is equivalent to a stimulus concentration of $10^8$ molecules of inhibitor per cubic centimeter of air, which is $10^5$ times more than needed to elicit the fluttering response by bombykol. These considerations, of course, are very preliminary and do not exclude the practical value of any of these methods.

Several of the possible control methods discussed in this chapter seem to be promising, and have been developed through the use of behavioral assays, including field tests. However, extensive studies on chemoreceptor organs might help in finding specific chemical stimulants and in designing appropriate stimulus programs which could be used to control behavioral responses most efficiently.

## REFERENCES

(1)  Y. Tamaki, H. Noguchi, T. Yushima, and C. Hirano. 1971. Two sex pheromones of the smaller tea tortrix: Isolation, identification and synthesis. *App. Entomol. Zool.* 6:139-141.

(2)  Y. Tamaki, H. Noguchi, T. Yushima, C. Hirano, K. Honma, and H. Sugawara. 1971. Sex pheromone of the summer fruit tortrix: Isolation and identification. *Kontyu* 39:338-140.

(3)  G.M. Meijer, F.J. Ritter, C.J. Persoons, A.K. Minks, and S. Voerman. 1972. Sex pheromones of summer fruit tortrix moth *Adoxophyes orana*: Two synergistic isomers. *Science* 175:1469-1470.

(4)  J.A. Klun, and J.F. Robinson. 1972. Olfactory discrimination in the European corn borer and several pheromonally analogous moths. *Ann. Entomol. Soc. Amer.* 65:1337-1340.

(5) H.E. Hummel, L.K. Gaston, H.H. Shorey, R.S. Kaae, K.J. Byrne, and R.M. Silverstein. 1973. Clarification of the chemical status of the pink bollworm sex pheromone. *Science* **181**:873-875.

(6) J. Kochansky, J. Tette, E.F. Taschenberg, R.T. Cardé, K.-E. Kaissling, and W.L. Roelofs. 1975. Sex pheromone of the moth *Antheraea polyphemus*. *J. Insect Physiol.* **21**:1977-1983.

(7) W.L. Roelofs and A. Comeau. 1971. Sex pheromone perception: Synergists and inhibitors for the red-banded leaf roller attractant. *J. Insect Physiol.* **17**:435-448.

(8) J.A. Rudinsky, M.M. Furniss, L.N. Kline, and R.F. Schmitz. 1972. Attraction and repression of *Dendroctonus pseudotsugae* (Coleoptera: Scolytidae) by three synthetic pheromones in traps in Oregon and Idaho. *Can. Entomol.* **104**:815-822.

(9) W.L. Roelofs and R.T. Cardé. 1974. Sex pheromones in the reproductive isolation of lepidopterous species. In *Pheromones*. M.C. Birch, Ed. North-Holland, Amsterdam, pp. 96-114.

(10) R.J. O'Connell. 1975. Olfactory receptor responses to sex pheromone components in the red-banded leaf roller moth. *J. Gen. Physiol.* **65**:179-205.

(11) C.J. den Otter. 1974. Electrophysiology of sex pheromone sensitive cells on antennae of the summerfruit tortrix moth *Adoxophyes orana*. Abstr. 1st Congr. Eur. Chemoreception Res. Org. Orsay/Paris, p. 45.

(12) K.-E. Kaissling 1971. Insect olfaction. In *Handbook of Sensory Physiology* Vol. IV, 1st ed. L.M. Beidler, Ed. Springer-Verlag, Berlin, pp. 351-431.

(13) B. Katz, 1966. *Nerve, Muscle, and Synapse*. McGraw-Hill, New York.

(14) J. Boeckh. 1962. Elektrophysiologische Untersuchungen an einzelnen Geruchsrezeptoren auf den Antennen des Totengräbers *(Necrophorous*, Coleoptera). *Z. Vgl. Physiol.* **46**:212-248.

(15) J. Boeckh. 1967. Inhibition and excitation of single olfactory receptors, and their role as a primary sensory code. *Int. Symp. Olfaction and Taste II*. T. Hayashi, Ed. Pergamon, Oxford, pp. 721-735.

(16) V. Lacher. 1964. Elektrophysiologische Untersuchungen an einzelnen Rezeptoren für Geruch, Kohlendioxyd, Luftfeuchtigkeit und Temperatur auf den Antennen der Arbeitsbiene und der Drohne (*Apis mellifica* L.). *Z. vgl. Physiol.* **48**:587-623.

(17) D. Schneider, V. Lacher, and K.-E. Kaissling. 1964. Die Reaktionsweise und das Reaktionsspektrum von Riechzellen bei *Antheraea pernyi* (Lepidoptera, Saturniidae). *Z. Vgl. Physiol.* **48**:632-662.

(18) W.A. Kafka. 1970. Analyse der molekularen Wechselwirkung bei der Erregung einzekner Riechzellen. *Z. Vgl. Physiol.* **70**:105-143.

(19) H. Mustaparta. 1975. Response of single olfactory cells in the pine weevil *Hylobius abietis* L. *J. Comp. Physiol.* **97**:271-290.

(20) U. Waldow. 1970. Elektrophysiologische Untersuchungen an Feuchte-, Trocken-, und Temperaturrezeptoren auf der Antenne der Wanderheuschrecke. *Z. Vgl. Physiol.* **69**:249-283.

(21) G. Kasang and K.-E. Kaissling. 1972. Specificity of primary and secondary processes in *Bombyx* antennae. *Int. Symp. Olfaction & Taste* IV. D. Schneider, Ed. Wiss. Verlagsgesellschaft Stuttgart, pp. 200-206.

(22) K.-E. Kaissling. 1972. Kinetic studies in olfactory receptors of *Bombyx mori. Int. Symp. Olfaction & Taste,* IV. D.Schneider, Ed. Wiss.Verlagsgesellschaft, Stuttgart, pp. 207-213.

(23) K.-E. Kaissling. 1969. Kinetics of olfactory receptor potentials. C. Pfaffmann, Ed. *Int. Symp. Olfaction & Taste.* III. Rockefeller University Press, New York, pp. 52-70.

(24)   K.D. Ernst. 1969. Die Feinstruktur von Riechsensillen auf der Antenne des Aaskäfers *Necrophorus* (Coleoptera). *Z. Zellforsch.* **94**:72-102.

(25)   K.D. Ernst. 1972. Die Ontogenie der basiconischen Riechsensillen auf der Antenne von *Necrophorus* (Coleoptera). *Z. Zellforsch.* **129**:217-236.

(26)   J.C. Meunier, R. Sealock, R. Olsen, and J.P. Changeux. 1974. Purification and properties of the cholinergic receptor protein from *Electrophorus electricus* electric tissue. *Eur. J. Biochem.* **45**:371-394.

(27)   G.G. Hammes, P.B. Molinoff, and F.E. Bloom. 1973. Receptor biophysics and biochemistry. *Neurosci. Res. Progr. Bull.* **11**:3.

(28)   B. Katz and R. Miledi. 1972. The statistical nature of the acetylcholine potential and its molecular components. *J. Physiol.* **224**:665-699.

(29)   K.-E. Kaissling. 1974. Sensory transduction in insect olfactory receptors. In *Biochemistry of Sensory Functions*. L. Jaenicke, Ed. 25th Colloq. Ges. Biol. Chemie, Mosbach. Springer-Verlag, Heidelberg, pp. 243-273.

(30)   K.-E. Kaissling. 1974. Riechphysiologische Untersuchungen an Insekten. Mitt. Max-Planck-Gesellsch, pp. 400-423.

(31)   M.A. Raftery, J. Bode, R. Vandlen, Y. Chao, J. Deutsch, J.R. Dugnid, K. Reed, and T. Moody. 1974. Characterization of an acetylcholine receptor. In *Biochemistry of Sensory Functions,* L. Jaenicke, Ed. 25th Coll. Ges. Biol. Chemie, Mosbach. Springer-Verlag, Berlin, pp. 541-564.

(32)   G. Wald. 1973. Visual pigments and photoreceptor physiology. In *Biochemistry and Physiology of Visual Pigments.* M. Langer, Ed. Springer-Verlag, Berlin, pp. 1-13.

(33)   R.A. Steinbrecht and G. Kasang. 1972. Capture and conveyance of odour molecules in an insect olfactory receptor. *Int. Symp. Olfaction & Taste, IV.* D. Schneider, Ed. Wiss. Verlagsgesellschaft, Stuttgart, pp. 193-199.

(34)   J. Scholes. 1965. Discontinuity of the excitation process in locust visual cells. *Cold Spring Harbor Symp. Quant. Biol.* **30**:517-527.

(35)   Y. Anraku. 1968. Transport of sugars and amino acids. I. Purification and specificity of the galactose- and leucine-binding proteins. *J. Biol. Chem.* **243**:3114-3122.

(36)   J. Adler. 1974. Chemotaxis in bacteria. In *Biochemistry of Sensory Functions.* L. Jaenicke, Ed. 25th Coll. Ges. Biol. Chemie, Mosbach. Springer-Verlag, Berlin, pp. 107-131.

(37)   F.R. Dastoli and S. Price. 1966. "Sweet sensitive protein" from bovine taste buds: Isolation and assay. *Science* **154**:905-907.

(38)   F.R. Dastoli. 1974. Taste receptor proteins. *Life Sci.* **14**:1417-1426.

(39)   T.J. Silhavy, W. Boos, and H.M. Kalckar. 1974. The role of the *Escherichia coli* galactose-binding protein in galactose transport and chemotaxis. In *Biochemistry of Sensory Functions.* L. Jaenicke, Ed. 25th Coll. Ges. Giol. Chemie, Mosbach. Springer-Verlag, Berlin, pp. 165-205.

(40)   K. Hansen and J. Kühner. 1972. Properties of a possible acceptor protein. *Int. Symp. Olfaction & Taste* IV. D. Schneider, Ed. Wiss.Verlagsgesellschaft, Stuttgart, pp. 350-356.

(41)   K. Hansen. 1974. α-Glucosidases as sugar receptor proteins in flies. In *Biochemistry of Sensory Functions.* L. Jaenicke, Ed. 25th Coll. Ges. Biol. Chemie, Mosbach. Springer-Verlag, Berlin, pp. 207-233.

(42)   H. Wieczorek. 1976. The Glycoside Receptor of the Larvae of *Mamestra brassicae* L. (Lepidoptera, Noctuidae). *J. Comp. Physiol.* **106**, 153-176.

(43)   K.-E. Kaissling and M. Renner. 1968. Antennale Rezeptoren für Queen Substance

und Sterzelduft bei der Honigbiene. *Z. Vgl. Physiol.* **59**:357-361.

(44)  E. Vareschi. 1971. Duftunterscheidung bei der Honigbiene–Einzelzell-Ableit-ungen und Verhaltensreaktionen. *Z. Vgl. Physiol.* **75**:143-173.

(45)  E. Vareschi and K.-E. Kaissling. 1970. Dressur von Bienenarbeiterinnen und Drohnen auf Pheromone und andere Duftstoffe. *Z. Vgl. Physiol.* **66**:22-26.

(46)  K. Dumpert. 1972. Alarmstoffrezeptoren auf der Antenne von *Lasius fuliginosus* (Latr.) (Hymenoptera, Formicidae). *Z. Vgl. Physiol.* **76**:403-425.

(47)  K.-E. Kaissling. 1976. The problem of specificity in olfactory cells. In *Structure-Activity Relationships in Chemoreception.* ECRO Symposium Wädens-wil, Ed.: G. Benz, Information Retrieval, London 1976, 137-148.

(48)  K.-E. Kaissling. Unpublished.

(49)  K. Hansen. 1969. The mechanism of insect sugar reception, a biochemical investigation. In: *Int. Symp. Olfaction & Taste*, III. G. Pfaffman, Ed. Rockefeller University Press, New York, pp. 382-391.

(50)  R.A. Steinhardt, H. Morita, and E.S. Hodgson. 1966. Mode of action of straight chain hydrocarbons on primary chemoreceptors of the blowfly, *Phormia regina*. *J. Cell Physiol.* **67**:53-62.

(51)  D.M. Norris, S.M. Ferkovich, J.E. Baker, J.M. Rozental, and T.K. Borg. 1971. Energy transduction in quinone inhibition of insect feeding. *J. Insect Physiol.* **17**:85-97.

(52)  N. Koyama and K. Kurihara. 1971. Modification by chemical reagents of proteins in the gustatory and olfactory organs of the flesh fly and cockroach. *J. Insect Physiol.* **17**:2435-2440.

(53)  I. Shimada, A. Shiraishi, J. Kijima, and H. Morita. 1972. Effects of sulfhydryl reagents on the labellar sugar receptor of the flesh fly. *J. Insect Physiol.* **18**:1845-1855.

(54)  H.Z. Levinson, K.-E. Kaissling, and A.R. Levinson. 1973. Olfaction and cyanide sensitivity in the six-spot burnet moth *Zygaena filipendulae* and the silkmoth *Bombyx mori. J. Comp. Physiol.* **86**:209-214.

(55)  R.H. Villet. 1974. Involvement of amino and sulfhydryl groups in olfactory transduction in silk moths. *Nature.* **248**:707-709.

(56)  E.F. Block, IV, and W.J. Bell. 1974. Ethometric analysis of pheromone receptor function in cockroaches. *J. Insect Physiol.* **20**:993-1003.

(57)  J.L. Frazier and J.R. Heitz. 1975. Inhibition of olfaction in the moth *Heliothis virescens* by the sulfhydryl reagent Fluorescein mercuric acetate. *Chem. Senses & Flavor* **1**:271-281.

(58)  G. Stange. 1974. The influence of a carbonic anhydrase inhibitor on the function of the honey bee antennal $CO_2$-receptors. *J. Comp. Physiol.* **91**:147-159.

(59)  Found by E. Priesner in J. Boeckh, K.-E. Kaissling, and D. Schneider. 1965. Insect olfactory receptors. *Cold Spring Harbor Symp. Quant. Biol.* **30**:263-280.

(60)  P. Seemann. 1970. Membrane expansion and stabilization by anaesthetics and other drugs. In *Permeability and Function of Biological Membranes.* L. Bolis, A. Katchalsky, R.D. Keynes, W.R. Loewenstein, and B.A. Pethica, Eds. North-Hol-land, Amsterdam, pp. 40-56.

(61)  K.-E. Kaissling and J. Thorson. Unpublished.

(62)  S.J. Singer. 1975. The molecular organization of membranes. In *Functional Linkage in Biomolecular Systems.* F.O. Schmitt, D.M. Schneider, D.M. Crothers, Eds. Raven, New York, pp. 103-106.

(63)  G. Guidotto. 1975. The specificity of lipid-protein interactions. In *Functional Linkage in Biomolecular Systems,* F.O. Schmitt, D.M. Schneider, and D.M. Crothers, Eds. Raven, New York, pp. 112-115.

(64)  S.M. Ferkovich, M.S. Mayer, and R.R. Rutter. 1972. Conversion of the sex pheromone of the cabbage looper. *Nature* (Lond.) *New Biol.* **242**:53-55.

(65)  S.M. Ferkovich, M.S. Mayer, and R.R. Rutter. 1973. Sex pheromone of the cabbage looper: Reactions with antennal proteins in vitro. *J. Insect Physiol.* **19**:2231-2243.

(66)  K.-E. Kaissling. 1974. Sensorische Transduktion bei Riechzellen von Insekten. *Verh. Dtsch. Zool. Ges.* **67**:1-11.

(67)  J. Thorson and M. Biederman-Thorson. 1974. Distributed relaxation processes in sensory adaptation. *Science* **183**:161-172.

(68)  J. Küppers and U. Thurm. 1975. Humorale Steuerung eines Ionentransports an epithelialen Rezeptoren von Insekten. *Verh. Dtsch. Zool. Ges.* **67**:46-50.

(69)  W. Neuhaus. 1958. Die Veränderung der Riechschärfe des Hundes durch orale Duftstoffgaben. *Z. Vgl. Physiol.* **41**:221-241.

(70)  H. Martin. 1974. Zur Frage der Primärprozesse am Geruchsrezptor. *J. Comp. Physiol.* **90**:339-366.

(71)  M.C. Birch, K. Trammel, H.H. Shorey, L.K. Gaston, D.D. Hardee, E.A. Cameron, C.J. Sanders, W.D. Bedard, D.L. Wood, W.E. Burkholder, and D. Müller-Schwarze. 1974. Programs utilizing pheromones in survey or control. In *Pheromones*. M.C. Birch, Ed. North-Holland, Amsterdam, pp. 411-461.

(72)  R.S. Kaae, H.H. Shorey, and L.K. Gaston. 1973. Pheromone concentration as a mechanism for reproductive isolation between two lepidopterous species. *Science* **179**:487-488.

(73)  W. Müller. 1968. Die Distanz und Kontaktorientierung der Stechmücken (*Aedes aegypti*) Wirtsfindung, Stechverhalten und Blutmahlzeit. *Z. Vgl. Physiol.* **58**:241-303.

(74)  E. Omand. 1971. A peripheral sensory basis for behavioral regulation. *Comp. Biochem. Physiol.* **A38**:265-278.

(75)  F.A. Bernays, W.M. Blaney, and R.F. Chapman. 1972. Changes in chemoreceptor sensilla on the maxillary palps of *Locusta migratoria* in relation to feeding. *J. Exp. Physiol.* **57**:745-753.

(76)  E.A. Bernays and R.F. Chapman. 1972. The control of changes in peripheral sensilla associated with feeding in *Locusta migratoria* (L.) *J. Exp. Biol.* **57**:755-763.

(77)  H.H. Shorey. 1974. Environmental and physiological control of insect sex pheromone behavior. In *Pheromones*. M.C. Birch, Ed. North-Holland, Amsterdam, pp. 62-80.

(78)  H.Z. Levinson. 1975. Possibilities of using insectistatics and pheromones in pest control. *Naturwissenschaften* **62**:272-282.

(79)  K.-E. Kaissling and E. Priesner. 1970. Die Riechschwelle des Seidenspinners. *Naturwissenschaften* **57**:23-28.

CHAPTER 5

# Olfactory Responses to Distant Plants and Other Odor Sources

J. S. Kennedy*

*Agricultural Research Council Insect Physiology Group*
*Department of Zoology and Applied Entomology*
*Imperial College, London, England*

This chapter deals mainly with responses of insects to plant odors, but it is focused upon the behavioral mechanisms and will therefore refer also to responses to pheromones and other odors where they help to illustrate a point. This is a confused and confusing subject, not because it is especially difficult but because of general neglect. The neglect has meant that the study of behavior has relied on untested and even unrecognized assumptions to a greater extent than any of the other scientific disciplines—chemistry, sensory physiology, ecology, and taxonomy—that are brought to bear in the joint endeavor to control insect behavior (1).

In the first place we have little idea how widespread distant olfactory responses are in nature, and their occurrence cannot be taken for granted. When counts have been made of flying "pioneer" barkbeetles (2) and aphids (3) (4) as they actually come in to land on host plants and nonhost

*Address: Imperial College Field Station, Silwood Park, Ascot Berks, SL5 7PY. England.

67

plants in comparable situations, the landing rates have turned out to be much the same. Here it would appear that any chemical discrimination between host and nonhost begins only after a relatively indiscriminate, visual, landing response, and takes the simple form of either staying or leaving.

However, such counts must be interpreted with caution because the fliers may have to pass through a migratory ("dispersal") phase before they become ready to colonize a plant or respond to its airborne odor (5) (Chapter 11). Host-finding and mate-finding behavior each involve several different responses, and the migrant may not become ready to respond in all these ways simultaneously. A migrant *Aphis fabae* becomes ready to make a visual landing response to a plant before it is ready to settle down and feed on it (6) (7); it might be that a flying bark beetle becomes ready to make visual landing responses to trees before it has become responsive to the specific airborne odors from host trees or other beetles.

The literature contains enough records of olfactory responses before contact with the plant to make it at least likely that these are common. There are enough records to suggest, further, that odors often somehow generate oriented movement toward plants, casting doubt on Thorsteinson's (8) suggestion that the effect of host-plant odors is generally restricted to arrestment after arrival at the plant.

The chemosensory distinction between olfaction and taste (9) is inadequate at the behavioral level, and we must make a further operational distinction between close-range and distant olfactory responses. Conventionally, the "distant" category excludes points a few millimeters or even a few centimeters away from a source, but includes perhaps those at decimeters and certainly those meters away. This arbitrary definition will be followed here, although it would tend to exclude walkers in favor of fliers, and the walkers must be included when discussing mechanisms because so much more work has been done on them. It will become apparent from what follows that a less arbitrary definition of distant olfactory responses may eventually become acceptable—namely, those occurring far enough from the source to preclude a chemotactic approach to it in response to directional cues provided by the odor itself.

Indeed, there has been general agreement since the 1930s—when von Buddenbrock (10) and Wigglesworth (11) wrote their textbooks—that chemotaxis cannot operate much more than millimeters from a source, because farther away the concentration gradients become too shallow and disrupted, especially in outdoor conditions. We shall come back to this point, as it is now being disputed; but it has not been disputed that there are oriented movements toward plant odor sources, from decimeters or more away, that are cued by stimuli other than the odor. The odor acts in these cases by switching on, or off, or otherwise regulating some nonchemotactic movement toward the source: orientation into wind (positive anemotaxis) is most often quoted. In addition there are distant responses to odors that are not taxes at all. The limited aim of this contribution is therefore to categorize the varieties of distant olfactory response that have been considered so far and to indicate which of them seem to be well established in at least one case.

## I. CLASSIFICATION

Dethier, Browne, and Smith's well-known 1960 paper (12) on the designation of chemicals in terms of insect responses brought some order into this subject for applied entomologists when the authors listed the various locomotory ways that an already-moving insect may react to chemical stimuli. It may, they wrote:

1. Continue without change of rate of linear progression, rate of turning or direction.
2. Stop
3. Slow its rate of linear progression
4. Increase its rate of turning
5. Increase its rate of linear progression
6. Decrease its rate of turning
7. Orient toward a source
8. Orient away from a source

Responses 2 to 6 were grouped together as nondirected responses; to this group we must add the initiation of movement when the insect was previously at rest. The technical term (13) for any such nondirected response is *kinesis* (plural, *kineses*), a category subdivided again into *orthokinesis,* where the response is a change in the amount or speed of linear movement—starting, stopping (item 2), slowing down (item 3) or speeding up (item 5); and *klinokinesis,* where the response is a change in the rate or frequency of turning, but at random with respect to the direction of the source—either turning more (item 4) or turning less (item 6). Kineses may be qualified as *direct* when the movement or the turning increases with stimulus strength, and *inverse* when they decrease with stimulus strength (14) (15).

Dethier et al. (12) placed items 7 and 8 in a second major group of responses, those actually directed with reference to the source; they are termed *taxes* (singular, *taxis*). In item 7 the taxis is positive, directed toward the source ("attraction"), while in item 8 it is negative, directed away from it ("repulsion"). Subdivisions of taxis are many but, the two relevant ones are described in Section IV.A. Thus for present purposes the classification of reactions is reduced to:

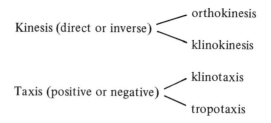

All these are types of locomotory response elicited directly by the chemical stimulus itself (Sections III and IV). To them we must add the important class of responses to nonchemical sensory cues for orientation, the responsiveness to which is regulated by chemical stimuli from distant sources. The latter are not therefore direct responses to chemical stimuli; they are orientation responses that are modulated by chemical stimuli but made to the wind (Sections II and VII) or some visual target (Section VI).

The known or suspected distant responses to odor sources can thus be arranged, first, according to whether the chemical stimulus elicits them directly or only modulates them; second, by their mechanism (kinesis or taxis); and third, according to whether the responding insect is on the ground (or other fixed substrate) or in the air. The third dichotomy is important because walking and flying insects are in different situations with respect to the directional information they receive (Sections II and IV). Tables 1 and 2 provide spaces for all these categories. A *plus* sign entered in one of the spaces means it is thought there is good evidence that the given type of response does occur, whereas a *minus* sign means there is not. The grounds for each of these verdicts are given in subsequent sections.

## II. POSITIVE ANEMOTAXIS

### A. Demonstration

Most of the information we have on insect behavioral responses to odorous chemicals comes from olfactometer tests; regrettably, it is impossible to extract the evidence required to identify distant responses from the results obtained with most types of olfactometer with or without airflow. As the name implies, such devices are focused on the receptor end of the process, the object being to assay the chemicals. Rarely is it possible to say that the performance of the insects in these devices definitely does include reactions that could help to bring them to a distant odor source in the field. One reason for this is that an insect may not respond to the same stimulus in the same way when walking as it does when flying, and in most olfactometers the

TABLE I

Responses Elicited Directly by Chemical Stimuli

| Locomotory | Kinesis | | Taxis |
| Mode | Orthokinesis | Klinokinesis | Klino- or Tropo- |
|---|---|---|---|
| Walking | + | (+) | (+) |
| Flying | + | + | − |

insects are stimulated while walking. A more important reason is the presence of steep gradients of odor concentration which permit reactions that would be possible in nature only close to an odor source. These difficulties are discussed further in Chapter 13.

A known mechanism of distant orientation that could work in the field is odor-conditioned, but nonchemotactic, positive anemotaxis. The existence of this mechanism has been demonstrated in the laboratory by using a wide, uniform airstream across the full width of a tunnel and permeating it uniformly with plant odor, thus eliminating any orientation by odor gradients or the edges of airstreams.

In this unambiguous situation upwind-oriented movement toward a plant odor source has been obtained in walking locusts (16) (17) and also in free-flying locusts (18), walking Colorado potato beetles (19), and free-flying *Drosophila* (20,21). It has been obtained likewise in male moths both walking (22)-(24) and flying (25) (26) in wind uniformly permeated with female sex pheromone. This evidence justifies *plus* signs for "case established" in two of the anemotaxis spaces in Table 2. For walking insects, thus in contact with a fixed substrate as a reference point, the *plus* appears under the heading of *mechanical stimuli* since these are known to be involved (27), although other sensory modalities (temperature, 17) may also be.

TABLE II

Chemically Modulated Responses to Nonchemical Stimuli

| Locomotory mode | Mechanical Stimuli | Visual Stimuli | |
| | From Wind (anemotaxis) | From Single Features | From Wind (optomotor anemotaxis) |
|---|---|---|---|
| Walking | + | + | − |
| Flying | − | + | + |

## B. Flying Optomotor Reactions

On the other hand, anemotactically steered flying insects cannot be entered under the same heading, for reasons that have been elaborated many times [e.g., (28)]. A flying insect is supported by a fluid medium—the air—and the movement of the air relative to the insect is mainly due to the insect's own active movement through the air. The mechanical stimuli it receives from the air are due to its own motions as it flaps, lifts, thrusts, rolls, pitches, and yaws its way along; turbulent motion of the air from other causes may also contribute. Wind is something else—movement of the air relative to the ground. The insect supported by the air out of contact with the ground can receive no mechanical stimulation from that overall movement of the air over the ground, so far as is known. A *minus* sign is therefore entered in the mechanically stimulated anemotaxis space in Table 2.

Nevertheless, a low-flying insect does receive sensory information about its motion relative to the ground, information derived from the apparent motion of the ground pattern it sees unrolling beneath it as it flies. Any movement relative to the ground of the air supporting the insect must contribute as drift to that apparent movement of the ground pattern as seen by the insect. It has been established by work on mosquitoes (29)-(31), honeybees (32) (33), *Drosophila* (20), aphids (34) and moths (35), that it is by orienting to this "optomotor" stimulation, insofar as it is caused by wind drift, that a flying insect orients to wind.

The insect reacts to the visible wind drift by "compensatory" turning and by adjustment of its flight speed (more precisely, its airspeed) and the result is that it faces into the wind and, if it makes headway upwind, it keeps the images of the ground pattern moving approximately from fore to aft over its eyes. The entraining and orienting effect of this optomotor stimulation was demonstrated experimentally by the overriding effect of artificially moving a floor pattern in still and moving air (20) (29)-(31) (34) (35); while the need for a visible ground pattern has been confirmed by the disappearance of wind orientation in darkness or red light (20) (26) (29) or outdoors over a featureless substrate, calm water (32) (33) [cf. (36): pp. 192-193, 208-220]. The *plus* sign for flying anemotaxis therefore appears in Table 2 under the heading *optomotor anemotaxis.*

## III. CHEMOKINESES

Returning to a discussion of the directly chemical responses in Table 1, there are some well-established cases of orthokinetic reactions to plant odors by both walkers and fliers, for instance in *Drosophila* (37) (38) and mosquitoes (39). They include both direct orthokinesis—that is, starting to move or speeding up—which could occur at some distance from a source; and inverse orthokinesis, such as the arrest of flying sweet clover weevils (40) and grass grub beetles (41) within decimeters of a strong odor source. Both direct orthokinetic responses to pheromones (notably at low concentrations) and inverse ones (at high concentrations) are cited in other chapters of this book.

Klinokinesis—whether in the direct form of more random turning when the odor stimulation increases, or in the classical and more efficient inverse form of more random turning when it decreases, followed by adaptation (13) (42)— is known to occur in walking insects in response to plant odors in small laboratory arenas with steep odor gradients (43) (44), but it is unlikely by itself to bring them to a field source more than a few decimeters away. It might help to keep these insects in an air current carrying odor from a nearby plant, but this has not been studied. At present, pedestrian klinokinesis can therefore be given only a parenthetical *plus* sign in Table 1.

On the other hand, klinokinesis probably does help flying insects, whose higher speeds render odor gradients effectively steeper in terms of sensory input, to find plants from perhaps a meter away in relatively still air. Deserving of special mention here are the little-known but simple and ingenious field experiments of Douwes (45) on the behavior of females of a Geometrid moth, *Cidaria albulata,* when they are ready to oviposit.

Douwes plotted the flight paths and landings of these moths among and around their host plants with the aid of a grid of stakes 0.5 m apart. He showed that they turned more often and apparently at random, thus making a more tortuous path, within patches of the host plants rather than outside them; they did this even when they did not alight. They also alighted more often, preferentially on host plants, but also on neighboring nonhosts. When a clump of host plants was removed and replaced by nonhosts, the turnings decreased in the area; but they increased again when host plants were restored to the same area. Finally, Douwes confirmed that the increased turnings and landings among host plants were indeed due to the plant odor; he blew air over a clump of host plants and along a 2-m plastic duct into an area of nonhost plants, and recorded increased landing and turning behavior there.

In these field experiments Douwes could not rigorously exclude the indirect effect of host plant odor on visual responses, which is certainly known, over centimeters at least, in other Lepidoptera (46). But his work strongly suggests a significant chemoklinokinetic effect over decimeters. There is a need for more field experiments on this kind of response mechanism because it could be involved in many cases where insects flit about near host plants before landing [e.g., (47) (48)]. It perhaps offers a new target for behavior-modifying control measures.

Rigorous laboratory experiments on chemoklinokinesis in flying insects are equally desirable. None seems to have been done with plant odors or pheromones, and the only relevant work is that of Daykin et al. (49) on mosquitoes flying horizontally in two contiguous, but nonmixing, vertical airstreams differing in temperature, humidity, or $CO_2$ content. When a mosquito crossed the sharp boundary between the two airstreams into more "favorable" air, it usually kept going without reacting; but when it crossed the other way, into the unfavorable airstream, it usually made a turn. If the turn was sharp and immediate, it took the insect promptly back into the favorable stream; but if the first turn was insufficient to do that, more turns followed (presumably owing mainly to the confined space) until the insect happened to reenter the favorable stream. The result was that flying mosquitoes aggregated within the favorable airstream as a result of spending most of their time there, and would have quickly encountered a host had one been there.

There was no evidence given or suggestion made in the description of these experiments that the turns at the boundary were biased in the direction of the favorable zone. On the contrary, Daykin et al. [(49): p. 250] stressed that it was the "klino-kinetic response to a perceived change of condition at the boundary, rather than any response within the test stream itself, or well outside the test stream, that is responsible for the aggregation," in accordance with the "pattern of klino-kinesis with adaptation observed in many other invertebrates" (13). Unfortunately their summary described this behavior loosely as turning "back" on leaving the favorable airstream, which could be taken to mean a directed response: chemotaxis. Indeed they strengthened this misunderstanding by figuring a hypothetical "attack program" in which the mosquito, each time it emerges from an odor plume downwind of a host, then turns not at random but promptly back into the plume, thus zigzagging along it in the manner reported for many insects in pheromone plumes (50). The misunderstanding was soon removed in Wright's book (21), but subsequent authors have continued to cite this mosquito work inappropriately in the context of plume-following [e.g., (51): p. 262, (28): p. 88]. Natural odor plumes at a distance from a source do not have boundaries as sharp as those in the twin-stream mosquito chamber; and klinokinesis, of course, cannot produce directed turns. Douwes's work suggests that, in theory, klinokinesis could help an insect to regain a lost odor plume eventually, by undirected turns. But the zigzagging observed along such plumes shows that some other, more efficient, directed response is at work, the nature of which is discussed in Section VII.A.

## IV. CHEMOTAXES

In a subject where thinking begins and ends with the action of chemicals, there is a tendency to assume the chemicals act to modify behavior as they modify other functions of the insect, directly. If a volatile chemical induces oriented movement toward its source, then one calls it a lure or attractant, tacitly assuming chemotaxis (1). In fact, the evidence now accumulated (Section II) has established that chemotaxis is not necessary to guide an insect toward a distant, upwind, odor source because odor-conditioned anemotaxis can also do so. But over the last few years the proposition that anemotaxis is not always necessary—because chemotaxis can also do that—has been elaborated afresh by Shorey and his co-workers (28) (50)-(54).

### A. Klinotaxis and Tropotaxis

With diffuse stimuli like odors there are two recognized mechanisms of chemotaxis: klinotaxis and tropotaxis (13). They are both illustrated in Martin's (55) (56) work on walking honeybees orienting to steep transverse concentration gradients of plant volatiles. Martin used a Y-tube without airflow but slightly inclined, so that the heavy odor molecules would flow along it from boxes containing deposits of odorant solution at the far ends of the two arms of the Y. Bees with their antennae free, or fixed normally in a splayed position, oriented tropotactically at the junction of the Y—that is, by turning in response to the difference of stimulus intensity detected between the two antennae at one moment in time.

If the two antennae were fixed close together, or if one was cut off, then tropotaxis disappeared. But the bees could still orient klinotactically to similar differences of odor concentration between their left and right sides at the junction of the Y-tube. They did this by visibly altering their behavior to the classical klinotactic mode: they now swung their antennal olfactory receptors to left and right as they walked, thereby detecting the left-right difference of stimulus intensity at successive moments in time. A bee with a single free antenna, swung that; a bee with the two immovable and closely apposed antennae swung its whole body from side to side and thus walked on a finely zigzagging path, like an ant moving along a pheromone trail or a wingless male moth walking along a line of female sex pheromone streaked across paper (53). Intact bees used both tropo- and klinotaxis in Martin's Y-tube, but even so they needed a 2:1 difference in the concentration of odorant solution at the ends of the two arms of the Y.

No satisfactory data are yet available on the minimum difference in the actual odor concentration between the two sides of an insect that is required for tropotactic or klinotactic reactions: it is probably much less than 2:1 (24). But some indications can be obtained from the behavior of ants and termites following their pheromone trails on the ground. They do this by chemotactic responses to the very steep odor gradients across the line of deposited material (57) (58), and they keep very close to it indeed: a one-inch gap in the trail can disorient them (58). Ritter and Coenen-Saraber (59) found that their termites were able to follow the trail (which included unchanged plant-derived material among the active ingredients) when it was 3 mm away from them under a mesh screen, but failed to do so at 9 mm. A natural situation where honeybees are guided chemotactically to a source of plant odor, according to von Frisch (33), is on a flower bearing a small scented and colored patch which leads the bee to the nectar, after alightment.

Such observations support the traditional view (58) that gradients steep enough to permit chemotactic orientation toward an odor source are likely to exist only very close to the source. Observations of carrion beetles (60) and moths (35) flying normally upwind toward a distant odor source, even after one antenna was removed, seem to rule out tropotaxis. For a given insect, klinotaxis might be expected to require a rather less steep gradient than tropotaxis because a wider band may be sampled as the insect swings its receptors alternately to right and left of its path, thus making spatial comparisons over time instead of between its two antennae at one time. There is no evidence for this, however, and there has been no suggestion that the sampled band would be wide enough for distant orientation toward an odor source by a walking insect. But it is conceivable that in nature an odor plume coming from a well-defined source, and lying close to the ground where the airflow is gentle, may retain well-defined boundaries presenting steep transverse gradients for several decimeters downwind. In that case chemotactic reactions could help to bring a walking insect into the plume and keep it moving along the plume; but without the aid of anemotaxis the reactions would not necessarily guide the insect toward the source. So again only a parenthetical *plus* sign is entered in the *walking chemotaxis* space in Table 1.

## B. Flying Chemotaxis

In flying insects chemotaxis is now being postulated (26)(50) as a mechanism of orientation to an odor source a meter or more away; the higher speed of fliers means that in principle they could respond chemotactically to a shallower gradient than could walkers. On this theory, wind is essential to carry the odor to a distance by drawing out the odor cloud into a long plume from the source. But the wind need not provide the sensory cues either for the oriented upwind movement toward the source, or for the transverse reorientations toward the plume's axis, which prevent the flier from straying away from it. The evidence for this theory consists so far of one wind-tunnel experiment (51)(52), and it does not apply to up- and downwind orientation along the plume (the authors conceded this point when queried (52) but it is sometimes lost sight of [(50): p. 362; (28): p. 86]).

Male *Pectinophora gossypiella* were stimulated to take flight one by one from the downwind end of a tunnel by exposing them to a wind-borne plume of female sex pheromone carried down the center of the tunnel from a small source near the upwind end. After takeoff, the wind was stopped by shutters, leaving the odor plume hanging in the stationary air. Sixteen (60%) out of 27 moths so released flew waveringly for about 1˙5 m along the tunnel and passed within 15 to 20 cm of the source within 20 sec. This performance was comparable (full figures were not given) with that of moths in a control set of trials when the wind carrying the plume was kept blowing throughout. In a third set of trials the pheromone was released for only 5 sec, and the wind was again stopped after takeoff, leaving the moths flying in stationary air with no pheromone plume; in this situation a much smaller proportion of them continued along the center of the tunnel.

These results are not regarded (35) as conclusive evidence that moths can stay with an odor plume chemotactically because the trials with the stationary plume did not exclude the possibility that the moths had set course anemotactically on takeoff, when the wind was still blowing (at exactly what point in the moths' progress the wind had stopped, is not clear). If 60% of them did not change course much, in the remaining seconds of flight, over a meter or so, then the observed performance can be explained without invoking chemotaxis. Until the results can be repeated with fliers entering an odor plume only after the air has entirely stopped moving, the case for chemotaxis as against anemotaxis remains unproven. There are many reports of moths and other insects flying upwind in an odor plume, and then changing course sharply when they come out of it so that the odor stimulus ceases (20) (25) (26) (35) (61) (62). That many of the experimental *P. gossypiella* likewise changed course when the odor stimulus ceased, in the third treatment, was therefore to be expected and is not evidence that the moths were actually guided by the stationary odor plume when it was present.

There does not appear to be any other experimental evidence for a chemotactic theory of plume-following, and until one case is established a *minus* sign must stand in the space for flying chemotaxis in Table 1. But the interest aroused by that striking report for *P. gossypiella* calls for some consideration of the chemotactic theories built extensively on it.

## V. AERIAL TRAIL-FOLLOWING

When Butler (61) (63) introduced the terms *terrestrial trail* and *aerial trail* in discussing marker pheromones, he was referring only to the physical location of the odor trails and not at all to the responses used by insects in following them. He argued, as is done here, that orientation was chemotactic with terrestrial trails but anemotactic with aerial ones. By contrast, Shorey and co-workers (28) (50) (51) (53) suggest that the wind-borne odor plume from a distant source is literally a scent trail along which a flying insect is guided chemotactically in the manner of a walking insect on a terrestrial trail.

There are two sorts of difficulties for this theory. The first is a matter of scale, odor concentration, and gradient steepness; it has already been touched on in Section IV. The second difficulty, connected with the first, concerns the proposed behavioral mechanisms of aerial trail-following (Section V.B).

## A. Stimulus Situation

The aerial "trail" is equivalent to what Wright (63) and then Bossert and Wilson (64) calculated as the "active space" within which the odor is above threshold for appropriate responses by a flier. A glance at the well-known diagrams of Bossert and Wilson (64) (65) based on those calculations shows that the terrestrial trail is by orders of magnitude narrower and more concentrated than the aerial one. The fundamental distinction between an aerial odor plume for a flying insect and a terrestrial odor trail for a walking one is that the insect on the terrestrial trail is continuously within millimeters of the volatilizing material along the whole of the route, whereas the flying insect never finds itself in a similar situation until it arrives at the extreme upwind end of the plume—its destination. For the trail-walker the odor source is a line source; for the plume-flier it is effectively a point source, which does not extend back along the route.

An intermediate case, where volatilizing material deposited along a chain of separate spots on the ground creates an aerial odor trail that is used by flying insects, has been demonstrated in Meliponine bees (33) (66). Nothing is known of the reactions they use to follow this trail. An aerial trail in the full sense of the term could exist if a large number of point sources of odor were shuttled continuously along an air line. Indeed, this is exactly what Wenner (67) speculates may be formed when there is a heavy traffic of flying honeybees between a food source and the hive. The mechanism by which he suggests they would find and stay with this trail is odor-induced anemotaxis.

Chemotactic following of the aerial odor plume from a single point source is difficult to conceive because of the filamentous, turbulent structure of such plumes (21) (50). Transverse concentration gradients, with a peak at the central axis of the plume, are assumed to exist; but they could only be detected by averaging the concentration over time at each of a number of points across the width of the plume (50). The insect does not stay for more than a fraction of a second at any point in the plume, so it is inconceivable how it could do the averaging needed for it to detect and respond, tropotactically or klinotactically, to the higher average concentration on the side of it toward the central plume axis.

The filaments in a plume are large relative to the width of the plume. Even if the flier could average, over time, the whole succession of instant differences of concentration that it detected between its two sides as it flew along an obliquely upwind path out toward the edge of the plume, it is scarcely conceivable that it would often register a significantly higher average concentration on the side toward the plume axis. Yet it would have to detect such a difference routinely, for this to explain the repeatedly inward turns observed. They can hardly be chemotactic, therefore, and so are not comparable in mechanism with the inward chemotactic turns of an insect walking along a terrestrial odor trail (53). Just the same problems arise if a flier detects filaments as pulses of odor (63) and could average the pulse frequency on its two sides.

Terrestrial trails are not polarized—but wind-borne odor plumes evidently are—for flying insects; and the above difficulties for the chemotactic theory are still greater when it comes to explaining why an insect flies toward rather than away from the source, because along this axis the odor gradients are even flatter.

## B. Response Mechanisms

The second kind of difficulty concerns the mechanism of olfactory guidance and refers equally to orientation in both axes of a wind-borne odor plume—that is, to staying with it and to moving up it. Hypotheses concerning this have varied and become quite entangled. Wright (63) once put forward a modified chemoklinokinetic hypothesis which he later abandoned (21), but which is now being revived (28) (50). So long as the insect encounters an increasing or steady odor concentration (or frequency of odor pulses, as Wright proposed), it flies straight; if it encounters a decrease in concentration it breaks into a programmed sequence of flights between alternate left and right turns, and so zigzags. The difficulty with this theory is that a decrease of odor concentration would be experienced by a flier moving out from the odor plume in any direction. The resulting zigzags could thus occur in any direction relative to the plume and its source. Yet observers agree that the zigzags of a flier, although irregular, nevertheless do not occur in all directions. They occur specifically across the plume's axis, and along it toward the source. The theory cannot therefore encompass the facts of orientation it sets out to explain: it is not in fact a chemotactic theory.

The "programming" part of Wright's (63) hypothesis will be considered further in Section VII.A. His idea of odor pulse-frequency detection was used by J.E. Moorhouse [(68), c.f. (17), (69): pp. 102-103] to construct an ingenious hypothesis of "osmo-motor" orientation to a wind-borne odor plume, but this referred to an insect on the ground and is not applicable to a flying one that cannot "feel" the wind (Section II).

Invoking chemoklinotaxis proper raises special difficulty. Some of the experimental evidence cited in this connection referred in reality to undirected, klinokinetic responses, as mentioned in Section III. The theoretical constructs (28) (50), too, endow klinotaxis with properties it is not known to have. Any temporal comparison of stimulus intensities, in any direction, is not enough for klinotaxis. It must be left-right comparison, as described in Section IV. Thus it is overrating the capabilities of klinotaxis to suggest that, if an insect "could integrate and compare within its nervous system the various intensities of odor through which it has passed," then "steering reactions (chemoklinotaxis) would be made to turn the insect back toward those positions in space where it previously obtained highest pheromone stimulation" [(50): p. 360]. Unless the insect has sampled the stimulus intensities specifically on its two sides, it has no cue for orientation about its vertical axis in that horizontal plane.

If, on the other hand, the flier turns in response to a difference detected not during left and right sampling but simply between the successive intensities through which it has passed, then the turn can only be at random. This is exactly what insects often do—for instance, the flying *Cidaria* described in Section III—but it is undirected klinokinesis, not klinotaxis, and it cannot therefore steer an insect back into an odorous zone left behind it.

Another new proposal (28) is a variation on Wright's (63) original theme: the flier keeps going straight ahead so long as it experiences an increase of odor concentration (or pulse frequency), as it will when flying up the plume; but when it experiences a decrease, as it will when flying down the plume away from the source, it is stimulated to turn round and fly the other way. The two precedents cited (28) for this hypothetical mechanism are the work of Koshland and others (70)-(72) on bacteria and of Wensler (39) on mosquitoes. The latter's experiments did not establish the mechanism of the mosquitoes' approach to an odor source; but the mechanism by which bacteria move up a chemical diffusion gradient is, for those trying to understand insect behavior, enviably clear-cut and rigorously established. It is, in fact, klinokinesis again, quite unlike the above hypothetical mechanism. Wherever a bacterium is in the concentration gradient, and whichever way it is going at the time, when it turns it does so at random with respect to the gradient. It turns frequently; and it moves up the gradient (most untidily) only because it travels somewhat further, before turning at random again, when it happens to meet a higher concentration than when it meets a lower one. This bears no resemblance to the way insects fly upwind toward an odor source, which is a far more orderly operation evidently utilizing some good directional cues.

In view of the strong appeal of chemotactic theories of aerial trail-following, it is worth underlining that there are no known means by which an insect can respond to a difference of odor concentration, which it has detected simply between successive positions along its path, by making a turn related to that direction in which it has come. A flier detecting such a change of concentration over time as it moves across or along an odor plume gets no information about the direction of the concentration gradients in space. Unless it has some directional information as well, from the sun, say, or from visible ground features, any turning response can only be at random, and this would be chemoklinokinesis, not chemotaxis.

In conclusion, there is no established case or circumstantially plausible hypothesis of chemically oriented flight toward a distant, upwind odor source. Since a walking moth will orient chemotactically to a terrestrial trail of volatilizing material from an artificial line source and move zigzagging along it (53), the presumption is that it has occasion to use chemotaxis when it is equally close to some natural source—say, a mate. But its chemotactic movement along the line source throws no light on the orientation mechanism used by the same insect when flying in zigzags up a wind-borne odor trail from a point source, for it does not then receive comparable odor cues. It would help to avoid confusion about mechanisms if the term *trail* were restricted to terrestrial trails and some other, more appropriate term such as *plume* were used for wind-borne odor clouds.

## VI. ODOR-CONDITIONED VISUAL ORIENTATION TO OBJECTS

Most of the many records of upwind flight toward plant or animal odor sources provide no rigorous evidence of anemotaxis, as Shorey (50) has pointed out. If the odor "turns on" a visual orientation response to some object at the source, say a tree trunk for a bark beetle, then, even if there is no anemotaxis, we must expect to see most insects coming in from the downwind side of the object because the wind restricts the odor to that side and carries it furthest that way. Odor-conditioned visual orientation responses to objects have indeed been demonstrated in various walking and flying insects when stimulated by plant odors (11) (46) (73)-(75), including "pioneer" bark-beetles (76) (77), and also in flying bark beetles and other flying insects stimulated by pheromones (Chapter 11) (28) (50) (54) or odors from host animals (78)-(80). Such a visual response to a single conspicuous feature goes under various names such as *hypsotaxis, horotaxis, telotaxis,* or *skototaxis,* but its known occurrence under the influence of plant odor justifies the *plus* sign in both spaces for single visual features in Table 2.

At the same time this evidence does not of course exclude the simultaneous occurrence of odor-conditioned anemotaxis, which could help at greater distances and in obstructed visual fields. Some outdoor experiments have been done using permeable screens to create a uniform visual field; the insects still flew upwind toward concealed sources of odor from host plants (81) (82), host animals (83), or conspecifics (84) some meters away.

Thus flying optomotor anemotaxis (Section II.B) remains the most generally applicable mechanism of guidance to a distant odor source and demands closer examination.

## VII. THE ANEMOTACTIC SYSTEM

There is now known to be much more to the system of odor-conditioned, optomotor anemotaxis in flying insects (Section II.B) than the mere turning on of positive anemotaxis by the odor stimulus, although that has been the general idea (13) (28) (29) (42) (58). Anemotaxis has been thought of as orienting the longitudinal body axis more or less parallel to the wind direction and so facing upwind or downwind; but oblique or transverse orientations to wind also are well documented. Moreover, only the ground track of the flier is usually recorded, and in a wind this may diverge widely from the body orientation or course of the insect. The flier's course must also be known, together with its airspeed and its height, and the wind speed at its height, in order to describe its optomotor responses (36) (85). Analysis of this system has still a long way to go, but two key points are already established.

### A. Reversing Anemomenotaxis

First, one of the elements in Wright's (63) hypothesis outlined in Section V.B has been confirmed. This is his perceptive idea that a decrease of odor stimulation actuates a built-in, central nervous "program" of zigzagging. There are a number of clear accounts of flying insects advancing upwind toward an odor source in typical fashion but then overshooting it; thereupon they switch to much wider, exaggerated zigzagging across wind, and no longer gain ground against the wind but do not lose ground either. After a short period of this activity they go downwind or in some other direction (20) (21) (25) (62). This has been called *casting* or *searching,* and has not been identified explicitly as anemotaxis because the insects were not flying upwind or downwind. But the fact that casting is oriented across wind implies that it is in reality a kind of anemotactic behavior. There can be no question of chemotactic orientation after overshooting the odor source because no odor is present.

The same behavior has been reproduced in male moths in the laboratory by suddenly removing a female pheromone source from the airstream in which the insect is flying (35). What follows is a rapid but progressive transition from narrow, advancing zigzagging to wide "stationary" zigzagging due to a sequence of changes in the optomotor responses. Again, this description refers only to the readily observed ground tracks and ground speeds of the insects, not to their courses (orientations) and airspeeds: the process has yet to be analyzed and quantified in terms of the actual responses.

Thus there appears indeed to be a central program, and it is initiated in this case not by the onset of an odor stimulus but by its termination, as an "off-effect." It is not a program of zigzags *in vacuo* (Wright's modified klinokinetic hypothesis) but a program of zigzags resulting from optomotor orientation at an angle to the wind which is reversed between left and right of the wind-line at lengthening intervals (35).

Orientation at an angle to the wind is by no means unique to this case. When a flying insect orients directly into wind it makes correcting turns in response to any oblique movement of the ground pattern images over its eyes due to "slide-slip" (29) (50). But it is known from observations on both walkers (24) (86) and fliers (33) (89) that an insect may also adopt a track at some definite angle across wind and correct any deviations from that, as in casting after overshooting an odor source. This type of orientation is called anemomenotaxis [(90); cf. photomenotaxis, orientation at an angle to a light source—e.g., the sun in the "sun-compass reaction"].

Since losing the scent by coming out of the end of an odor plume initiates a program of reversing anemomenotaxis, it would be surprising if losing the scent by coming out of the side of the plume did not do the same. In the latter case the first reversal would often bring the flier promptly back into the plume; but when a moth comes out of the side of a plume and fails to reenter it quickly, then the full "casting" program may ensue (25). Further evidence that zigzagging along a plume could be due to anemomenotactic reversals when the flier reaches the margins of the plume, comes from the second key point that has been established experimentally: if no margins are reached (because the entire wind tunnel is permeated with odor), then no zigzagging occurs (20) (25).

At this point it is illuminating to digress into the behavior of certain walking insects in a similar situation. Schwink (22) (23) observed zigzagging by silkmoth *(Bombyx mori)* males walking upwind in a room generally contaminated with female sex pheromone. Kramer (24) confirmed and extended this observation in some elegant experiments where the walking moth was kept artificially at one spot on top of a rotatable sphere. The sphere was rotated by feedback from the walking insect in such a way as to compensate exactly the insect's progression in any direction, while allowing it to orient itself in the air stream. When the airstream was uniformly permeated with "bombykol" the insect walked consistently at an angle of 30 to 50° to the wind and, at irregular short intervals, switched over spontaneously from one side of the wind-line to the other. Here again there is a program of reversing anemomenotaxis—now under constant odor stimulation instead of after it has ceased—but again with no possibility of chemotaxis.

If, now, one of the spontaneous turns in an odorous airstream took the insect's olfactory receptors into an odorless airstream, the decrease of stimulation immediately provoked an extra anemomenotactic reversal over to the other side of the wind-line—and thus back into the odorous stream. A free insect walking upwind in an odor plume would also experience a decrease of odor stimulation when its path took it over one margin of the plume into odorless wind. If it reacted every time this happened in the same way as it did when captive on the rotating sphere, the result would be a zigzagging path up the plume unerringly to its source; and free silkmoth males in an ordinary wind tunnel actually followed such paths up a bombykol plume.

However, the male silkmoth's behavior in uniform odor is, so far, unique. No such reversing anemomenotactic behavior was seen in locusts walking (or flying) in a wind tunnel uniformly permeated with grass odor (nor, indeed, did they "cast" when the odor was suddenly removed): the uniform odor induced only positive anemotaxis (17). The same was true of other insects observed flying in wind tunnels uniformly permeated with stimulating odor. *Drosophila melanogaster* flying in a wind permeated with fermenting banana odor (20) (21) and male moths (*Anagasta kühniella*) flying in a wind uniformly permeated with female sex pheromone (25) showed positive anemotaxis, flying more or less straight upwind. It was only when the same insects were given the same odors confined to a narrow plume that they flew in zigzags upwind.

## B. Odor-Modulated Anemotactic Behavior

In those insects, therefore, we know that the odor stimulus, at the given concentrations, "turns on" only positive anemotaxis; it does not turn on reversing anemomenotaxis, zigzagging. Decrease of the odor stimulus turns on zigzagging when the flier overshoots the odor source [or the source is removed (35)]; and it does so also, we have inferred, when the flier reaches the margin of the plume, on its way upwind toward the source. In this latter situation one anemomenotactic reversal is usually enough to bring the insect back into the plume to be stimulated by the odor again very quickly. Thus the simplest hypothesis is that in this short interval the casting program, seen in its entirety when the odor loss is prolonged, is initiated, but interrupted again at an early stage. Repetition of this truncated sequence each time the flier reached the plume margin would result in the familiar, obliquely zigzagging upwind progression along the plume toward the source. An insect coming to the right margin of the plume will have been flying on a track angled to the right of the wind-line, so its first reversal will be over to the left of it, as in the walking silkmoth. Successive turns would then tend automatically to be toward the plume's central axis, and so give a false impression of chemotaxis (35).

The conclusion is that crosswind flying during net upwind zigzagging along an odor plume, as well as during casting after lasting loss of contact with the plume, is all due to reversing **anemomenotactic** reactions to decreases in the odor signal strength. Such decreases could occur inside a filamentous odor plume, as well as at its edges, which could account for the observed irregularity of zigzagging. Directly upwind flight, which is sometimes observed in a plume, would occur whenever the signal strength was maintained long enough for positive anemotaxis to supervene. The combination of these optomotor orientation reactions with orthokinetic, optomotor modulation of the flier's airspeed (34) (35), all governed by odor concentrations (54) and changes in them in ways still far from clear, makes up a system of guidance to a distant odor source which at least has the virtue of unifying what information there is. It does not of course exclude the participation of other mechanisms for which there is as yet no evidence.

This anemotactic system would be a highly integrated one, but that is not too much to expect of an insect's central nervous system. The rather different anemomenotactic behavior of foraging honeybees, for instance, is even more delicately modulated by their sun compass and clock, and demonstrates how much the setting of the anemotactic angle may be adjusted in response to stimuli received. The *anemotactic angle* means the angle maintained between the wind and the insect's ground track, not that between the wind and the insect's body axis, "heading," or course, which will always be smaller.

A honeybee can keep on the right ground track from hive to food and back again in spite of variable crosswinds (33) (87). It presumably does this by turning its body axis obliquely enough into wind to compensate for the wind drift, as von Frisch and Lindauer (87) calculated it must do, and as they also observed directly. This plainly means that in adjusting its course to the varying wind direction and speed the bee is adopting a variety of angles between its body axis and the sun; and all these angles are different from the one held while moving along the same ground track in calm air. Yet that calm-air angle to the sun is the one that is communicated during dancing in the hive and used by recruits to find the food source (33) (67) (91).

Somehow the bee keeps on the right ground track relative to the sun in crosswinds regardless of all those deviations from the calm-air angle of body to sun. What it is actually doing, therefore, is preserving a constant angle between the sun and the direction of its ground track, detected as the direction of apparent movement of the ground pattern passing over its eyes. So it seems, as Johnson [(85): p. 156] has pointed out, that it is the angle between the sun and the direction of that optomotor stimulation that the bees hold constant when flying in calm and windy conditions alike, and communicate when dancing in the hive—and not the angle between the sun and the body axis.

With regard to the odor-modulated anemotactic system of guidance to a distant odor source, a final point worth noting is that although positive anemotaxis is an essential component of the system, it is not its most distinctive feature. Certainly, the receipt of an odor stimulus does in some cases "turn on" positive anemotaxis when it was absent before. Before locating an odor plume, fliers may show anemomenotaxis (88), downwind flight (92), or no directional bias (37) (93). But orienting into wind is often observed in low-flying insects without an odor stimulus, and with or without their making headway against the wind (34) (35) (94)-(97), and then an odor stimulus may merely increase the insect's airspeed (96). Orienting into wind with appropriate airspeed adjustment is a common "station-keeping" mechanism for flying insects in nature, a defense against being carried away by the wind; it is also a prerequisite for controlled landings (34). What is more distinctive about the guiding effect of a biologically significant odor signal from a distant source is its after-effect on orientation to wind—that is, the effect of a decrease of odor stimulation in initiating reversing anemomeno-taxis. It is apparently thanks to this effect of attenuation of the chemical signal that an insect chancing upon an odor plume pursues it tenaciously to the source.

## VIII. CONCLUSION

The anemotactic system appears to be the most powerful mechanism of guidance to an odor source, and on present evidence it is probably the sole mechanism from several meters away, unless the source is conspicuous enough for a visual response. But of course this does not mean it is the most vulnerable mechanism for pest-control purposes. The main obstacle at this stage in the development of pest-control methods designed to use chemicals to modify behavior is that we have no way of deciding which of the various mechanisms discussed here would prove the most vulnerable in any given pest. We do not usually know which of them are present, or how they are severally affected by chemicals, even by those chemicals already identified as active, so that single-measure bioassays are liable to misinterpretation (98).

## IX. SUMMARY

1. An attempt is made to categorize the varieties of olfactory response so far considered as likely to help bring insects to distant plants and other odor sources, and to indicate which of these responses is well established in one case at least.
2. The responses are distinguished on three main criteria: whether the insect is walking or flying; whether the chemical stimulus elicits the response directly or modulates responsiveness to other sensory cues; and the mechanism of the response—kinesis or taxis, and so on.
3. Undirected, chemoorthokinetic, and chemoklinokinetic responses have been shown to play a significant part, and in flying insects klinokinetic

responses, although little studied, are probably often important near odor sources in calm air.

4. Chemotactic responses occur in the steep concentration gradients very close to local odor sources and along chemical trails on the ground, but there is still no convincing evidence or plausible hypothesis for distant chemotactic guidance or flying insects. Existing hypotheses endow chemoklinokinesis and chemoklinotaxis with directional properties they are not known to possess.

5. Olfactory enhancement or induction of visual orientation responses to single objects, such as tree trunks, is well established.

6. At present the sole known other mechanism of distant attraction of flying insects is odor-modulated, optomotor anemotaxis, combined with chemoorthokinesis. Positive (i.e., upwind) anemotactic orientation in response to the onset and maintenance of odor stimulation, and reversing crosswind orientation (anemomenotaxis) in response to decrease of odor stimulation, are both established. There is good circumstantial evidence that they are combined with odor-modulated optomotor control of airspeed to make up an integrated system capable of generating zigzagging flight up a wind-borne odor plume as well as "casting" after losing the plume. Much remains to be clarified about the workings of this system.

## ACKNOWLEDGMENTS

I am indebted to Dr. David Marsh in particular for many informative discussions on pheromone action, to him and Dr. J.N. Brady, Dr. K.-E. Kaissling, Dr. J.E. Moorhouse, and Dr. H.H. Shorey for criticisms of this essay in draft, and to Dr. E. Kramer and Dr. J.E. Moorhouse for permission to quote their unpublished results. To Harry Shorey I am additionally indebted for the stimulus to reexamine these controversial questions.

## REFERENCES

(1)   J.S. Kennedy, 1972. The emergence of behaviour. *J. Aust. Entomol. Soc.* 11:168-176.

(2)   D.L. Wood. 1972. Selection and colonization of ponderosa pine by bark beetles. *Symp. Roy. Entomol. Soc. Lond.* 6:101-117.

(3)   J.S. Kennedy, C.O. Booth, and W.J.S. Kershaw. 1959. Host finding by aphids in the field. I. Gynoparae of *Myzus persicae* (Sulzer). *Ann. Appl. Biol.* 47:410-423.

(4)   J.S. Kennedy, C.O. Booth, and W.J.S. Kershaw. 1959. Host finding by aphids in the field. II. *Aphis fabae* Scop. (gynoparae) and *Brevicoryne brassicae* L.; with a re-appraisal of the role of host-finding behaviour in virus spread. *Ann. Appl. Biol.* 47:424-444.

(5)   J.S. Kennedy. 1975. Insect dispersal. In *Insects, Science, and Society.* D. Pimentel, Ed. Academic, London, pp. 103-119

(6)   J.S. Kennedy, and C.O. Booth. 1963. Coordination of successive activities in an

aphid. The effect of flight on the settling responses. *J. Exp. Biol.* **40**:351-369.

(7)  J.S. Kennedy, and A.R. Ludlow. 1974 Co-ordination of two kinds of flight activity in an aphid. *J. Exp. Biol.* **61**:*173-196.*

(8)  A.J. Thorsteinson. 1960. Host selection in phytophagous insects. *Ann. Rev. Entomol.* **5**:193-218.

(9)  V.G. Dethier. 1970. Some general considerations of insects' responses to the chemicals in food plants. In *Control of Insect Behavior by Natural Products.* D.L. Wood, R.M. Silverstein, and M. Nakajima, Eds. Academic, New York, pp. 21-28.

(10) W. von Buddenbrock. 1937. *Grundriss der vergleichenden Physiologie,* 2nd ed. Gebrüder Borntraeger, Berlin.

(11) V.B. Wigglesworth. 1939. *The Principles of Insect Physiology.* Methuen, London.

(12) V.G. Dethier, L.B. Browne, and C.N. Smith. 1960. The designation of chemicals in terms of the responses they elicit from insects. *J. Econ. Entomol.* **53**:134-136.

(13) G. Fraenkel and D.L. Gunn. 1961. *The Orientation of Animals. Kineses, Taxes and Compass Reactions.* Dover, New York.

(14) J.S. Kennedy 1945. Classification and nomenclature of animal behaviour. *Nature* (Lond.) **156**:754.

(15) J.D. Carthy. 1965. *The Behaviour of Arthropods.* Oliver and Boyd, Edinburgh.

(16) P.T. Haskell, M.W.J. Paskin, and J.E. Moorhouse. 1962. Laboratory observations on factors affecting the movements of hoppers of the desert locust. *J. Insect Physiol.* **8**:53-78.

(17) J.S. Kennedy, and J.E. Moorhouse. 1969. Laboratory observations on locust responses to wind-borne grass odour. *Entomol. Exp. Appl.* **12**:487-503.

(18) J.E. Moorhouse, Personal communication.

(19) J. de Wilde, K. Hille Ris Lambers-Suverkropp, and A. van Tol. 1969. Responses to air flow and airborne plant odour in the Colorado bettle. *Neth. J. Pl. Path.* **75**:53-57.

(20) F.E. Kellogg, D.E. Frizel, and R.H. Wright. 1962. The olfactory guidance of flying insects. IV. *Drosophila. Can. Entomol.* **94**:884-888.

(21) R.H. Wright. 1964. *The Science of Smell.* Allen and Unwin, London.

(22) I. Schwink. 1954. Experimentelle Untersuchungen über Geruchssinn und Strömungswahrnehmung in der Orientierung bei Nachtschmetterlingen. *Z. Vgl. Physiol.* **37**:19-56.

(23) I. Schwink. 1958. A study of olfactory stimuli in the orientation of moths. *Proc. 10th Int. Congr. Entomol.,* Montreal **2**:577-82.

(24) E. Kramer. 1975. Orientation of the male silkmoth to the sex attractant Bombykol. In *Olfaction and Taste* 5. D. Denton and J.D. Coghlan, Eds. Academic, New York, pp. 329-335.

(25) R.M.M. Traynier. 1968. Sex attraction in the Mediterranean flour moth, *Anagasta kühniella:* location of the female by the male. *Can. Entomol.* **100**:5-10.

(26) K.H. Dahm, I. Richter, D. Meyer, and H. Röller. 1971. The sex attractant of the Indian-meal moth *Plodia interpunctella* (Hübner). *Life Sci.* **10**:531-539.

(27) M. Gewecke. 1974. The antennae of insects as air-current sense organs and their relationship to the control of flight. In *Experimental Analysis of Insect Behaviour.* L.B. Browne, Ed. Springer, Berlin, pp. 100-113.

(28) S.R. Farkas, and H.H. Shorey. 1974. Mechanisms of orientation to a distant pheromone source. In *Pheromones.* M.C. Birch, Ed. North-Holland, Amsterdam, pp. 81-95.

(29) J.S. Kennedy. 1940. The visual responses of flying mosquitoes. *Proc. Zool. Soc. Lond.* **109**:221-242.

(30) R.H. Wright. 1962. The attraction and repulsion of mosquitoes. *World Rev. Pest. Contr.* **1**:2-12.

(31) P.N. Daykin. 1967. Orientation of *Aedes aegypti* in vertical air currents. *Can. Entomol.* **99**:303-308.

(32) H. Heran and M. Lindauer. 1955. Windkompensation und Seitenwindkorrektur der Bienen beim Flug über Wasser. *Z. Vgl. Physiol.* **47**:39-55.

(33) H. von Frisch. 1967. *The Dance Language and Orientation of Bees.* Harvard University Press, Cambridge, Mass.

(34) J.S. Kennedy, and A.A.G. Thomas. 1974. Behaviour of some low-flying aphids in wind. *Ann. Appl. Biol.* **76**:143-159.

(35) J.S. Kennedy, and D. Marsh. 1974. Pheromone-regulated anemotaxis in flying moths. *Science* (N.Y.) **184**:999-1001.

(36) J.S. Kennedy. 1951. The migration of the desert locust *(Schistocerca gregaria* Forsk.). I. The behaviour of swarms. II. A theory of long-range migrations. *Phil. Trans. Roy. Soc. B,* **235**:163-290.

(37) C. Flügge. 1934. Geruchliche Raumorientierung von *Drosophila melanogaster. Z. Vgl. Physiol.* **20**:463-500.

(38) G. Steiner. 1954. Uber die Geruchs-Fernorientierung von *Drosophila melanogaster* in "ruhender" Luft. *Naturwissenschaften* **41**:287.

(39) R.J.D. Wensler. 1972. The effect of odors on the behavior of adult *Aedes aegypti* and some factors limiting responsiveness. *Can. J. Zool.* **50**:415-420.

(40) H. Hans and A.J. Thorsteinson. 1961. The influence of physical factors and host plant odour on the induction and termination of dispersal flights in *Sitona cylindricollis* Fahr. *Entomol. exp. appl.* **4**:165-177.

(41) G.O. Osborne, and C.P. Hoyt. 1970. Phenolic resins as chemical attractants for males of the grass grub beetle, *Costelytra zealandica. Ann. Entomol. Soc. Am.* **63**:1145-1147.

(42) J.S. Kennedy. 1965. Mechanisms of host plant selection. *Ann. Appl. Biol.* **56**:317-322.

(43) V.G. Dethier. 1947. *Chemical Insect Attractants and Repellents.* Blakiston, Philadelphia.

(44) V. Perttunen, E. Kangas, and H. Oksanen. 1968. The mechanisms by which *Blastophagus piniperda* L. (Col. Scolytidae) reacts to the odour of an attractant fraction isolated from pine phloem. *Ann. Entomol. Fenn.* **34**:205-222.

(45) P. Douwes. 1968. Host selection and host finding in the egg-laying female *Cidaria albulata* L. (Lep. Geometridae). *Opusc. Entomol.* **33**:233-279.

(46) N. Tinbergen. 1951. *The Study of Instinct.* Clarendon, Oxford.

(47) R.T. Yamamoto, R.Y. Jenkins and R.K. McCluskey. 1969. Factors determining the selection of plants for oviposition by the tobacco hornworm *Manduca sexta. Entomol. Exp. Appl.* **12**:504-508.

(48) T.H. Coaker and S. Finch. 1972. The association of the cabbage rootfly with its food and host plants. *Symp. Roy. Entomol. Soc. Lond.* **6**:119-128.

(49) P.N. Daykin, F.E. Kellog, and R.H. Wright. 1965. Host-finding and repulsion of *Aedes aegypti. Can. Entomol.* **97**:239-263.

(50) H.H. Shorey. 1973. Behavioral responses to insect pheromones. *Ann. Rev. Entomol.* **18**:349-380.

(51) S.R. Farkas and H.H. Shorey. 1972. Chemical trail-following by flying insects. A mechanism for orientation to a distant odor source. *Science* (N.Y.) **178**:67-68.

(52) S.R. Farkas and H.H. Shorey. 1973. Odor-following and anemotaxis. *Science* (N.Y.) **180**:1302.

(53) H.H. Shorey and S.R. Farkas. 1973. Sex pheromones of Lepidoptera. 42.

Terrestrial odor-trail following by pheromone-stimulated males of *Trichoplusia ni*. *Ann. Entomol. Soc. Am.* **66**:1213-1214.

(54) S.R. Farkas, H.H. Shorey, and L.K. Gaston. 1974. Sex pheromones of Lepidoptera. Influence of pheromone concentration and visual cues on aerial odor-trail following by males of *Pectinophora gossypiella*. *Ann. Entomol. Soc. Am.* **67**:633-638.

(55) H. Martin. 1965. Zur Nahorientierung der Biene im Duftfeld zugleich ein Nachweis für die Osmotropotaxis bei Insekten. *Z. Vgl. Physiol.* **48**:481-533.

(56) H. Martin. 1965. Osmotropotaxis in the honey-bee. *Nature* (Lond.) **208**:59-63.

(57) W. Hangartner. 1967. Spezifität und Inaktivierung des Spurpheromons von *Lasius fuliginosus* und Orientierung der Arbeiterinnen im Duftfeld. *Z. Vgl. Physiol.* **57**:103-136.

(58) V.G. Dethier. 1957. Chemoreception and the behavior of insects. *Surv. Biol. Progr.* **3**:149-183.

(59) F.J. Ritter and C.M.A. Coenen-Saraber. 1969. Food attractants and a pheromone as trail-following substances for the Saintonge termite. Multiplicity of the trail-following substances in *L. trabea*-infected wood. *Entomol. Exp. Appl.* **12**:611-622.

(60) V.G. Dethier. 1947. The role of the antennae in the orientation of carrion beetles to odors. *J. N.Y. Entomol. Soc.* **55**:285-293.

(61) C.G. Butler. 1967. Insect pheromones. *Biol. Rev.* **42**:42-87.

(62) C.G. Butler. 1970. Chemical communication in insects: behavioral and ecologic aspects. *Adv. Chemorecep.* **1**:35-78.

(63) R.H. Wright. 1958. The olfactory guidance of flying insects. *Can. Entomol.* **90**:81-89.

(64) W.H. Bossert and E.O. Wilson. 1963. The analysis of olfactory communication among animals. *J. Theor. Biol.* **5**:443-469.

(65) E.O. Wilson. 1963. Pheromones. *Sci. Am.* **May**:1-11.

(66) M. Lindauer and W.E. Kerr. 1958. Die gegenseitige Verständigung bei den stachellosen Bienen. *Z. Vgl. Physiol.* **41**:405-434.

(67) A.M. Wenner. 1974. Information transfer in honey-bees: A population approach. In *Nonverbal Communication*. L. Krames, P. Pliner, and T. Alloway, Eds. Plenum, New York, pp. 113-169.

(68) J.E. Moorhouse. Unpublished.

(69) J.S. Kennedy. 1966. Some outstanding questions in insect behaviour. *Symp. Roy. Entomol. Soc. Lond.* **3**:97-112.

(70) R. Macnab and D.E. Koshland, Jr. 1972. The gradient-sensing mechanism in bacterial chemotaxis. *Proc. Natl. Acad. Sci. USA* **69**:2509-2512.

(71) D.E. Koshland, Jr. 1974. Chemotaxis as a model for sensory systems. *Fed. Eur. Biochem. Soc. Lett.* **40**(Suppl):S3-S9.

(72) N. Tsang, R. Macnab, and D.E. Koshland, Jr. 1973. Common mechanism for repellents and attractants in bacterial chemotaxis. *Science* (N.Y.) **181**:60-63.

(73) Tinbergen, N. 1958. *Curious Naturalists*. Country Life, London.

(74) G. Steiner and R. Wette. 1954. Über die Wirkung eines homogenen Duftfeldes auf optische Wahlen von *Drosphila hydei*. *Naturwissenschaften* **41**:172.

(75) S.D. Beck. 1965. Resistance of plants to insects. *Ann. Rev. Entomol.* **10**:205-232.

(76) L.H. McMullen and M.D. Atkins. 1962. On the flight and host selection of the Douglas-fir beetle, *Dendroctonus pseudotsugae* Hopk. (Coleoptera: Scolytidae). *Can. Entomol.* **94**:1309-1325.

(77) J.A. Rudinsky. 1966. Host selection and invasion by the Douglas-fir beetle,

*Dendroctonus pseudotsugae* Hopkins, in coastal Douglas-fir forests. *Can. Entomol.* 98:98-111.

(78) W.C. Bradbury and G.F. Bennett. 1974. Behaviour of adult Simuliidae (Diptera). II. Vision and olfaction in near-orientation and landing. *Can. J. Zool.* 52:1355-1364.

(79) J. Brady. 1972. The visual responsiveness of the tsetse fly *Glossina morsitans* Westw. (Glossinidae) to moving objects: the effects of hunger, sex, host odour and stimulus characteristics. *Bull. Entomol. Res.* 62:257-279.

(80) A.G. Gatehouse and C.T. Lewis. 1973. Host location behaviour of *Stomoxys calcitrans. Entomol. Exp. Appl.* 16:275-290.

(81) W.J. Mistric, Jr. and E.R. Mitchell. 1966. Attractiveness of isolated groups of cotton plants to migrating boll weevils. *J. Econ. Entomol.* 59:39-41.

(82) C. Hawkes. 1971. Behaviour of the adult cabbage root fly. Field dispersal and behaviour. *Rep. Nat. Veg. Res. Stn. 1970:*93.

(83) G.A. Vale. 1974. The responses of tsetse flies (Diptera, Glossinidae) to mobile and stationary baits. *Bull. Entomol. Res.* 64:545-588.

(84) J.P. Seybert and R.I. Gara. 1970. Notes on flight and host-selection behavior of the pine engraver, *Ips pini* (Coleoptera: Scolytidae). *Ann. Entomol. Soc. Am.* 63:947-950.

(85) C.G. Johnson. 1969. *Migration and Dispersal of Insects by Flight.* Methuen, London.

(86) K.E. Linsenmair. 1969. Anemomenotaktische Orientierung bei Tenebrioniden und Mistkäfern *(Insecta, Coleoptera). Z. Vgl. Physiol.* 60:445-449.

(87) K. von Frisch and M. Lindauer. 1955. Über die Fluggegeschwindigkeit der Bienen und über ihre Richtungsweisung bei Seitenwind. *Naturwissenschaften* 42:377-385.

(88) G. Steiner. 1953. Zur Duftorientierung fliegender Insekten. *Naturwissenschaften* 40:514-515.

(89) P.B. Kannowski and R.L. Johnson. 1969. Male patrolling behaviour and sex attraction in ants of the genus *Formica. Anim. Behav.* 17:425-429.

(90) K.E. Linsenmair. 1968. Anemomenotaktische Orientierung bei Skorpionen *(Chelicerata, Scorpiones). Z. Vgl. Physiol.* 64:154-211.

(91) J.L. Gould. 1974. Honey-bee communication. *Nature* (Lond.) 252:300-301.

(92) M.T. Gillies and T.J. Wilkes. 1974. Evidence for downwind flights by host-seeking mosquitoes. *Nature* (Lond.) 252:388-389.

(93) H. Mustaparta. 1974. Response of the pine weevil *Hylobius abietis* L. (Col. Curculionidae) to bark beetle pheromones. *J. Comp. Physiol.* 88:395-398.

(94) J.A. Chapman, and J.M. Kinghorn. 1958. Studies of flight and attack activity of the ambrosia beetle, *Trypodendron lineatum* (Oliv.), and other Scolytids. *Can. Entomol.* 90:362-372.

(95) J.A. Chapman. 1962. Field studies on attack flight and log selection by the ambrosia beetle *Trypodendron lineatum* (Oliv.) (Coleoptera: Scolytidae). *Can. Entomol.* 94:74-92.

(96) L.R. Cole. 1970. Observations on the finding of mates by male *Phaeogenes invisor* and *Apanteles medicaginis* (Hymenoptera: Ichneumonoidea). *Anim. Behav.* 18:184-189.

(97) J.A. Downes. 1969. The swarming and mating flight of Diptera. *Ann. Rev. Entomol.* 14:271-298.

(98) R.T. Cardé, T.C. Baker, and W.L. Roelofs. 1975. Behavioural role of individual components of a multichemical attractant system in the oriental fruit moth. *Nature* (Lond.) 253:348-349.

## CHAPTER 6

# Insect Feeding Deterrents in Plants

Katsura Munakata
*Laboratory of Pesticides Chemistry*
*Faculty of Agriculture*
*Nagoya University*
*Nagoya, Japan*

Continuous, heavy usage of some insecticides has created serious problems, the most obvious being direct toxicity to nontarget organisms such as parasites, predators, pollinators, and man. Certain of the chemicals may be concentrated in food chains and accumulated in the nontarget organisms. Also, the target pests frequently develop resistance to the insecticides. Considerable research is needed in addition to that which has been undertaken to provide alternative, noninsecticidal methods of insect control.

One of the most effective alternatives to the use of insecticides might involve manipulation of natural environmental chemicals, such as attractants, repellents, stimulants, antifeedants, and arrestants, which are normally encountered by the insects and act as stimuli, controlling their behavior.

This review is restricted to a consideration of the chemical substances in plants that act as feeding deterrents when they are contacted by insects. An insect feeding deterrent is defined as a chemical that inhibits feeding, although it does not kill the insect directly. The insect often remains near the leaves containing the deterrent and may die through starvation. The terms *gustatory repellent, antifeedant,* and *rejectant* are often used synonymously with *feeding deterrent.* Studies of naturally occurring antifeeding, in which no plant material is consumed by the insect, may reveal the presence of feeding deterrents in plants and lead to further studies of the correlation between the chemical structure of the deterrents and their activity in inhibiting feeding. Identified and synthesized antifeedants might be used as alternatives to insecticides for the protection of crops from noxious insects. Chapman (1) recently prepared a comprehensive review on the effect of chemicals that inhibit feeding by phytophagous insects.

## I. EXAMPLES OF INSECT ANTIFEEDANTS

A number of interesting reports of the presence of insect antifeedants in plants are reviewed below to give an indication of the current status of this field. The deterrents naturally produced by plants play a major role in determining the range of food plants selected by phytophagous insects and in conferring resistance against insect attack to many plant species. Identification of the chemical structures of the feeding deterrents provides useful knowledge to the fields of natural-product chemistry, pesticide chemistry, and insect physiology.

### A. Aphids and Phlorizin

Phlorizin, a phenolic compound present only in the apple genus, *Malus,* has been reported to function as a probing stimulant for the two apple-feeding aphid species, *Rhopalosiphum insertum* and *Aphis pomi.* The same compound acts as a deterrent, inhibiting probing by two nonapple-feeding species, *Myzus persicae* and *Amphorophora agatonica.* Phlorizin also inhibits ingestion in the two latter species, as well as in *A. pomi.* However, *A. pomi* feeds in the phloem tissue of apple, which apparently contains no phlorizin, so the ingestion of this apple-feeding species is not inhibited (2).

### B. Blister Beetles and Coumarins

Sweetclover (*Melilotus* spp.) is one of a wide range of plants fed upon by blister beetles, *Epicauta* spp. No serious damage to sweetclover was observed until low-coumarin strains of the white-flowered species, *M. alba,* were developed. Coumarin precursors in sweetclover are the glucosides of *cis-* and *trans-O*-hydroxycinnamic acid (*O*-HCA). *trans-O*-HCA glucoside is partially converted to the *cis*-isomer through the action of UV light. During tissue disruption, such as that caused by blister beetle feeding, the $\beta$-glucosidase contained within the leaves hydrolyzes *cis-0*-HCA glucoside to yield coumarinic acid, which lactonizes spontaneously to form coumarin. The *trans-0*-HCA glucoside has no deterrent effect (3). However, both *cis-0*-HCA glucoside and coumarin have been found to act as strong feeding deterrents for the beetles. Experimental variation of the amounts of these two compounds within different lines of sweetclover has resulted in proportional changes in beetle feeding activity.

## C. Locusts and Triterpenoids

The neem tree, *Azadirachta indica* (or *Melia indica*), can be used as an antifeedant. A suspension of crushed seeds or leaves applied directly to leaves of other plants or to stored products deters feeding of some insect species. Extracts of the seeds also show repellent action against the migratory locust, *Locusta migratoria*, and the desert locust, *Schistocerca gregaria*. A triterpenoid was isolated from freshly crushed neem tree fruit and identified as meliantroil ($C_{30}H_{50}O_5$). Meliantroil causes 100% antifeeding activity against desert locusts when applied at 8 $\mu g/cm^2$ to filter paper soaked with 0.25 M sucrose (4).

A systemic investigation of extracts of neem tree seeds revealed another, more potent locust antifeeding compound, which was identified (5) (6) as the triterpenoid, azadirachtin ($C_{35}H_{44}O_{16}$). Azadirachtin caused 100% inhibition of feeding when impregnated onto sucrose-soaked filter paper at 1 $ng/cm^2$.

## D. Insect Feeding Deterrents and Systemic Action

For practical application to plants, insect antifeeding substances should not only be persistent, they should also be absorbed by the plant tissues and be translocated to the actively growing points. If a deterrent can not translocate to the new plant growth, that area might be selectively attacked by insect pests.

The systemic action of the deterrent azadirachtin was studied by applying various concentrations in water to the soil around young bean plants. The efficacy and persistence of the treatments were assayed with adult desert locusts that had been starved for 8 to 10 hours. After 25 days the plants grown in soil treated with azadirachtin at concentrations of 10 ppm or higher had received only slight damage from the locusts. Similar results were obtained using older established plants up to the fruiting stage, of both tall and dwarf bean varieties. Also, bean seedlings grown from seeds soaked for 24 hr in 0.01, 0.1, and 1% solutions of azadirachtin were protected against locust damage for 1 week. However, high organic activity and alkalinity in soils reduced the systemic antifeeding activity of azadriachtin (7).

## E. Potato Leafhopper and *Solanum* Alkaloids

Potato leafhoppers, *Empoasca fabae,* exhibit differing aggregation and oviposition responses to the various varieties or crosses of *Solanum tuberosum* and other *Solanum* species (8). These differences are due in part to certain solanum alkaloids. The alkaloid, leptine-I, extracted from leaves of *Solanum chacoense,* markedly reduced both the rate of initial imbibition by the leafhoppers and their survival time. Tomatine, solanine, solanidine, and demissidine reduced initial imbibition, but did not influence survival time, and tomatidine affected neither imbibition nor survival. Thus, the alkaloids, in relation to their qualitative and quantitative distributions in the plants and their antifeeding properties, probably play decisive roles in the natural interactions of potato leafhoppers with *Solanum* species.

## F. *Scolytus multistriatus* and Naphthoquinones

The normal host plant of the bark beetle, *Scolytus multistriatus,* is the American elm, *Ulmus americana.* When extracts from hickory, *Carya ovata,* or white oak, *Quercus alba,* were added to extracts of American elm, feeding by the bark beetles was deterred. The active principle in hickory was found to be juglone (5-hydroxy-1, 4-naphthoquinone). The beetles would even feed on hickory bark after juglone had been removed. On the other hand, the bark beetle *Scolytus quadrispinosus,* which normally feeds on hickory, tolerated juglone in the extracts on which it fed. Thus, deterrent chemicals in the bark of nonhost trees appear to be important bases for determining the distribution of the host species on which the bark beetles will feed (9) (10).

## G. Sweetclover Weevil and Nitrate

The sweetclover weevil, *Sitona cylindricollis,* is a major pest of sweetclover (*Melilotus* spp.) in the northern Great Plains of the United States. Aqueous extracts of *M. officianilis,* a weevil-susceptible sweetclover species, stimulated extensive feeding by the weevil. However, extracts of *M. infesta,* a resistant species, inhibited feeding. The compound causing the deterrent effect of *M. infesta* was isolated and identified as ammonium nitrate. Sodium and potassium nitrate are very similar to ammonium nitrate in causing feeding deterrency, and nitrate assays of young *M. infesta* leaves indicate a level five to ten times higher than that found in comparable *M. officianalis* leaves. Therefore, the hypothesis is advanced that the nitrate ion is the feeding deterrent in leaves of *M. infesta* (11).

Nitrate content is lowest in the "pinched," unfolding leaves of *M. infesta,* being at a concentration of about 100 ppm. As leaf age increases, the average nitrate content correspondingly increases. The average nitrate content of the oldest fully expanded leaves is nearly twice that of the youngest fully expanded leaves, nine times that of the loosely pinched leaves, and 80 times that of the tightly pinched leaves.

Another unidentified feeding deterrent, in addition to nitrate, is also found in the leaves of *M. infesta.* The mechanisms of resistance and susceptibility of sweetclover species will be more fully understood after all deterrent factors have been identified (12).

## II *SPODOPTERA LITURA* AND FEEDING DETERRENTS

An extensive survey for chemicals which act as feeding deterrents against the tobacco cutworm, *Spodoptera litura,* has been conducted at the Laboratory of Pesticide Chemistry, Nagoya University, Nagoya, Japan. The tobacco cutworm is a noxious pest of sweet potatoes, sugar cane, crucifers, taro, and legumes, and control of this worm is very important to Japanese agriculture. However, certain plants have feeding-deterrent activity and thus resistance against this pest. Extracts of these plants were investigated in research directed toward the isolation and identification of the active principles.

### A. Screening Method

Leaf discs were punched out with a cork borer from leaves of susceptible food plants, usually sweet potato, *Ipomoea batatas batatas.* The discs were immersed in acetone solutions of test samples or in pure acetone as a control. After air-drying, the discs were placed in polyethylene dishes with test larvae, usually third instar, of the tobacco cutworm. About half the area of the control discs was usually eaten within 2 hr, at which time the consumed areas of all discs were measured by Dethier's method (13). The consumed area of treated discs expressed as a percentage of the consumed area of control discs was used as an index of the antifeeding activity of the samples. Also, in some cases the minimum concentration of test samples needed to cause 100% antifeeding activity was determined (14).

### B. Antifeedants from *Cocculus trilobus*

*Cocculus trilobus* ("Kamiebi" in Japanese), although a host plant of the Japanese fruit-piercing moths, *Oraesia excavata* and *O. emarginata,* is scarcely attacked by other insects in nature. Therefore, it was assumed that *C. trilobus* contained toxins or feeding inhibitors against most insects.

Two alkaloids were isolated as crystalline forms from the fresh leaves of this plant and identified. One alkaloid, having insecticidal properties, is cocculolidine, and the other, having deterrent properties, is isoboldine.

In sweet-potato leaf-disc tests, isoboldine showed feeding inhibitory activity against tobacco cutworms at 200 ppm. On the other hand, *O. excavata* larvae completely consumed *C. trilobus* leaf discs treated with 1000 ppm of isoboldine (15).

## C. Antifeedants from *Parabenzoin trilobum*

Leaves of *Parabenzoin trilobum* (= *Lidera triloba*) ("Shiromoji," in Japanese) are not attacked by the tobacco cutworm. Crude extracts of these leaves also show antifeeding activity in sweet-potato leaf-disc tests. Two active compounds, shiromodiol-diacetate and shiromodiol-monoacetate, were isolated from the extract and identified (16). Threshold concentrates of both compounds were about 0.125 and 0.033% when tested against tobacco cutworms and larvae of *Abraxas miranda,* respectively. However, after a number of hours of testing, the test discs were consumed by the larvae, leading to the concept of relative deterrent activity (17).

## D. Antifeedants from *Orixia japonica*

*Orixia japonica* has been used in Japan for centuries to protect books against insect attack. The active principles in the leaves were identified as the furocoumarin or furoquinoline substances: isopimpinellin, bergapten, and kokusagin. Isopimpinellin and bergapten, when assayed in leaf-disc tests, were found to possess potent antifeeding activity against the tobacco cutworm (17).

## E. Antifeedants from *Clerodendron tricotomum*

In 1962, T. Miyake investigated the antifeeding activities of extracts from fresh leaves of various plant species against the insects *Euproctis pseudoconspersa* and *Cicadella viridis* and found that extracts of the verbenaceous plant *Clerodendron tricotomum* ("Kusagi" in Japanese) were especially active (18). More recent research has shown that extracts of air-dried leaves of *C. tricotomum* possess antifeeding activity against the tobacco cutworm also. Two crystalline clerodane substances, clerodendrin-A and clerodendrin-B, were isolated from the leaves (19) (20). Both substances inhibit feeding of larvae of a number of other insect species, including *Calospilos mirands* (at a concentration of 5000 ppm), the European corn borer, *Ostrinia nubilalis* (at 5000 ppm), and the oriental tussock moth, *Euproctis subflava* (at 1000 ppm). These limited results indicate that these compounds may act as antifeedants for larvae of polyphagous rather than monophagous insects (21).

## F. Antifeedants from Verbenaceae

After the discovery of antifeeding substances from *C. tricotomum,* constituents of other verbenaceous plants were examined for activity. Three species from Japan and 13 species from Taiwan were collected and their extracts were subjected to the tobacco cutworm leaf-disc tests. The plants *Caryopteris divaricata, Callicarpa japonica, Clerodendron fragrans, Clerodendron calamitosum,* and *Clerodendron cryptophyllum* showed strong antifeeding activity. Many clerodane compounds were isolated from these plants, and their chemical structures were investigated.

Eight antifeeding diterpenes having a clerodane skeleton were identified from *Caryopteris divaricata*: clerodin, caryoptin, dihydroclerodin-I, dihydrocaryoptin, clerodin hemiacetal, caryoptin hemiacetal, caryoptinol, and dihydrocaryoptinol (22). A weak antifeedant, phytol, was identified from *Clerodendron japonica*; a new antifeedant, 3-epicaryoptin, was identified from *C. calamitosum*; and clerodendrin-A was identified as an antifeedant from *C. cryptophyllum* (23). It is interesting that compounds containing the clerodane skeleton are found both in *Clerodendron* and *Caryopteris* species. In relation to biogenesis, it is interesting that 3-epicaryoptin but not caryoptin is found in *C. calamitosum*.

Leaf-disc feeding tests of the diterpenes at concentrations causing 100% antifeeding activity were allowed to continue beyond the standard 2 hr. The test leaves were not fed upon during observations of 24 hr or more, and the larvae eventually starved to death. Therefore, the term *absolute antifeedant* is applicable to these compounds, in contrast to the *relative antifeedants* that retard feeding only for a defined time.

## G. Antifeeding Activities and Bitterness

Gymnemic acids, the triterpene saponines from leaves of the asclepiad vine, *Gymnema sylvestre,* have been classed as taste-distorting agents because they suppress the perceived sweetness of sugars in man (24). Gymnemic acids deter feeding by a broadly polyphagous insect, the southern armyworm, *Prodenia eridania.* The antifeedant effect is also demonstrable with sugar-free diets, suggesting that the acids do not act, as they do in mammals, by suppressing the sweetness of sugars (25).

Clerodendrins, diterpene derivatives having a clerodane skeleton, have a bitter taste. However, there does not seem to be a linear relationship between the degree of bitterness, as perceived by man, and antifeeding activity against insect larvae. Polyol derivatives have a slightly bitter taste and posses weak activity, whereas the methanol adduct derivatives are not bitter but show the strongest antifeeding activity. Furthermore, the larvae do not bite the treated leaves and they eventually starve to death when concentrations of the test solutions, except those of polyol and acetate derivatives, are raised above twice the concentrations which cause 100% antifeeding. While the exact mode of antifeeding action of these compounds has not been elucidated, it may be caused by both smell and taste (26).

In further tests of bitterness versus antifeeding activity, nine compounds having a bitter taste—tenulin, helenalin, parthenin, absinthin, columbin, marrubin, limonin, nomilin, and helvolic acid—were exposed in leaf-disc tests to tobacco cutworms. Only one compound, absinthin, inhibited feeding. Thus, the deterrency of compounds may be attributable to their intrinsic distastefulness, rather than to bitterness (27).

## III. CONCLUSIONS

Antifeedants appear to be a viable alternative to the use of insecticides for control of many insect pests. Many natural products of plants possess insect antifeeding activity. These compounds probably play a role as resistance factors, protecting the plants against insect attack.

Based on our experience with isolation and identification of insect antifeedants from plants, the following research methodology is suggested:

1. Conduct surveys to determine which plant species are not eaten by test insect larvae.
2. Conduct preliminary extraction of leaves with several solvents, monitored by leaf-disc tests.
3. Isolate active compounds, monitored by leaf-disc tests.
4. Identify active compounds and, if necessary, synthesize them and related compounds.

Using these procedures, we have identified many new natural products having insect antifeeding activities. We have noted several general characteristics of these compounds:

1. Some compounds show relative and others show absolute antifeeding activity.
2. The bitter taste of a compound may have no relation to its antifeeding activity.
3. Naturally occurring antifeedants may be systemic in plants and decomposable in the environment—both advantageous characteristics.
4. Compounds having insect antifeeding activity can and should be used as genetic index substances in selective breeding programs for insect-resistant plant varieties.

Thus, there are two ways in which insect control may be effected by insect antifeeding substances:

1. The development of powerful synthetic insect antifeedants
2. The development of resistant crop varieties that naturally incorporate the antifeeding substances

Research on insect feeding deterrents in plants contributes to natural-products chemistry. Although few investigators study the antifeeding properties of natural products, such research offers a good opportunity to meet with new compounds; we have discovered more than 20 new natural products during the past 10 years. Such studies also provide new knowledge to basic biology, especially ethology.

When used for insect pest management, feeding deterrents might be especially advantageous because they control the insects indirectly through starvation, and they may not be harmful to parasites, predators, and pollinators. If crops were sprayed with efficient antifeedants, perhaps the pests would turn from the crops to weeds.

## REFERENCES

(1)  R.F. Chapman. 1974. The chemical inhibition of feeding by phytophagous insects: A review. *Bull. Entomol. Res.* **64**:339-363.

(2)  M.E. Montgomery and H. Arn. 1974. Feeding response of *Aphis pomi, Myzus persicae* and *Amphorophora agathonica* to phlorizin. *J. Insect Physiol.* **20**:413-421.

(3)  H.J. Gorz, F.A. Haskins, and G.R. Manglitz. 1972. Effect of coumarin and related compounds on blister beetle feeding in sweetclover. *J. Econ. Entmol.* **65**:1632-1635.

(4)  D. Lavie, M.K. Jain, and S.R. Shpan-Gabrielith. 1967. A locust phagorepellent from two *Melia* species. *Chem. Comm.* **18**:910-911.

(5)  J.H. Butterworth, E.D. Morgan, and G.R. Percy. 1972. The structure of azadirachtin; the functional groups. *J. Chem. Soc. Perkin Trans.* **1**(19):2445-2450.

(6)  P.R. Zanno, I. Miura, and K. Nakanishi. 1975. Structure of the insect phagorepellent Azadirachtin. Application of PRFT/CWD Carbon-13 Nuclear Magnetic Resonance. *J. Am. Chem. Soc.* **97**:1975-1977.

(7)  J.S. Gill and C.T. Lewis. 1971. Systemic action of an insect feeding deterrent. *Nature* **232**:402-403.

(8)  D.L. Dahlman and E.T. Hibbs. 1967. Responses of *Empoasca fabae* (Cicadellidae: Homoptera) to tomatine, solanine, leptine-1, tomatidine, solanidine and demissidine. *Ann. Entromol. Soc. Am.* **60**:732-740.

(9)  B.L. Gilbert, J.E. Baker and D.M. Norris. 1967. Juglone (S-hydroxy-1, 4-naphthoquinone) from *Carya ovata,* a deterrent to feeding by *Scolytus multistriatus. J. Insect Physiol.* **13**:1453-1459.

(10)  B.L. Gilbert and D.M. Norris. 1968. A chemical basis for bark beetle *(Scolytus)* distinction between host and nonhost trees. *J. Insect Physiol.* **14**:1063-1068.

(11)  W.R. Akerson, F.A. Haskins and H.J. Gorz. 1969. Sweetclover-weevil feeding deterrent B: isolation and identification. *Science* **163**:293-294.

(12)  G.L. Beland, W.R. Akerson and G.R. Manglitz. 1970. Influence of plant maturity and plant part on nitrate content of the sweetclover weevil-resistant species *Melilotus infesta. J. Econ. Entomol.* **63**:1037-1039.

(13)  V.G. Dethier. 1947. *Chemical Insect Attractants and Repellants.* Blakistone, Philadelphia, p. 210.

(14) K. Munakata. 1970. Insect antifeedants in plants. In *Control of Insect Behaviour by Natural Products.* D.L. Wood, R.M. Silverstein, and M. Nakajima, Eds. New York, Academic, pp. 179-187.

(15) K. Wada and K. Munakata. 1968. Naturally occurring insect control chemicals. Isoboldine, a feeding inhibitor, and cocculolidine, an insecticide in the leaves of *Cocculus trilobus* DC. *J. Agric. Fd. Chem.* **16**:471-474.

(16) K. Wada, K. Matsui, Y. Enomoto, O. Ogiso and K. Munakata. 1970. Insect feeding inhibitors in plants. I. Isolation of three new sesquiterpenoids in *Parabenzoin trilobum* Nakai. *Agric. Biol. Chem.* **34**:941-945.

(17) N. Yazima, K. Tsuzuki, N. Kato and K. Munakata. 1969. Insect antifeeding substances in *Orixa japonica.* Abstract Ann. Meeting Agr. Chem. Soc., Japan. p. 145.

(18) T. Miyake. 1962. Private communication.

(19) N. Kato, M. Shibayama and K. Munakata. 1973. Structure of the diterpene Clerodendrin-A. *J. Chem. Soc. Perkin* **I**:712-719.

(20) N. Kato, K. Munakata and C. Katayama. 1973. Crystal and molecular structure of the *p*-Bromobenzoate chlorohydrin of Clerodendrin-A. *J. Chem. Soc. Perkin* **II**:69-73.

(21) N. Kato, M. Takahashi, M. Shibayama and K. Munakata. 1972. Antifeeding active substances for insects in *Clerodendron tricotomum* Thunb. *Agric. Biol. Chem.* **36**:2579-2582.

(22) S. Hosozawa, N. Kato and K. Munakata. 1974. Antifeeding active substances for insects in *Caryopteris divaricata* Maxim. *Agric. Biol. Chem.* **38**:823-826.

(23) S. Hosozawa, N. Kato, K. Munakata and Yuk-Lin-Chen. 1974. Antifeeding active substances for insects in plants. *Agric. Biol. Chem.* **38**:1045-1048.

(24) T. Eisner and B.P. Halpern. 1971. Taste distortion and plant palatability. *Science* **172**:1362.

(25) M.S. Granich, B.P. Halpern and T. Eisner. 1974. Gymnemic acids: Secondary plant substances of dual defensive action. *J. Insect Physiol.* **20**:435-439.

(26) N. Kato, M. Takahashi, M. Shibayama and K. Munakata. 1972. Antifeeding active substances for insects in *Clerodendron tricotomun* **Thumb.** *Agric. Biol. Chem.* **36**:2579-2582.

(27) K. Wada and K. Munakata. 1971. Insect feeding inhibitors in plants. III. Feeding inhibitory activity of terpenoids in plants. *Agric. Biol. Chem.* **35**:115-118.

## CHAPTER 7

# Responses of Blood-Sucking Arthropods to Vertebrate Hosts

Rachel Galun
*Israel Institute for Biological Research*
*Ness-Ziona, Israel*

Behavioral responses of hematophagous arthropods to the hosts that provide them their food is one manifestation of feeding behavior. Like feeding in any other organism, hematophagy is accomplished through a series of behavioral patterns. Though these patterns are related and perhaps interdependent, they can be considered not only as parts of a single act but also as separate phenomena, each under the control of a particular set of physical and chemical conditions (1).

Blood-sucking behavior of arthropods has been thoroughly reviewed in recent years (2)-(4). The present chapter is presented as a critique of this field, and will refer to the reviews, rather than to the original papers, when possible.

## I. HOST-FINDING

The blood-sucking process can be schematically divided into four successive steps: attraction to the host and settling (orientation); probing and tasting (initiation of feeding); sucking or gorging (continuation of feeding); and withdrawal of mouthparts (termination of feeding).

First, the animal must orient itself positively with respect to the food source. The animal accomplishes this through active movement toward the warm-blooded host, guided by visual and thermal stimuli and a complex of airborne chemical stimuli emanating from the host. Though variable in detail, the chemical stimuli conform to a common general pattern and are dominated by carbon dioxide, water vapor, fatty acids and their derivatives (especially lactic acid), ammonia, and amines.

The stimulus which evokes approach in daylight seems usually to be the visual one of host movement, perhaps reinforced by host form. There is little specificity at this stage, as witnessed by the readiness with which tsetse flies and tabanids pursue vehicles of many kinds: all that seems to be needed is an appropriate combination of size, distance, and speed. Vision is of great importance in host-finding in blood-sucking Diptera. It is of much less importance in blood-sucking bugs, fleas, lice, mites, and ticks. At night, a positive anemotaxis triggered and maintained by olfactory stimuli, seems to be more important than visual stimuli in guiding certain insects to their vertebrate hosts.

Thermal stimuli have been shown to be important in attracting ticks, bugs, lice, mosquitoes, and tsetse flies to their hosts. The convection currents created in the air surrounding warm-blooded animals are well within the range of sensitivity of various blood-sucking arthropods. The importance of a warm upward current of air for host-finding has been demonstrated for ticks, blackflies, fleas, and mosquitoes (3).

Carbon dioxide plays an important role in host location. Low concentrations of $CO_2$ seem to accelerate spontaneous takeoff behavior of mosquitoes (5) and to activate mites and fleas (3). However, many blood-sucking arthropods also orient to a $CO_2$ source, and it has been utilized in traps as a bait for mosquitoes, simuliids (6), *Culicoides* (7), tabanid flies (8), and ticks (9).

Neither visual, nor thermal, nor $CO_2$ stimuli could account for any host specificity. And indeed the phenomenon of blood-sucking is not always a specific one. A single blood-sucking species may be polyphagous and feed on several species of reptiles, birds, or mammals.

## II. HOST DISCRIMINATION

Askew (10) states that because parasitism demands a high degree of specialization, the host ranges of parasitic insects are limited. A few are polyphagous, but these are nevertheless ecologically able to parasitize only those hosts that share some common attribute, such as a similar habitat, pelage, or nesting site. Many parasites are oligophagous, with a host range embracing a few genera, belonging to one or two families; and in some groups, lice in particular, the majority are monophagous. An evolutionary trend in the direction of monophagy is perhaps inevitable. In the face of competition from other parasitic species, the polyphagous parasite will have a higher survival rate on those hosts on which it is a more successful competitor, and it will therefore become more and more adapted to parasitizing in an increasingly restricted host range.

Host selection can sometimes be understood by studying very carefully the ecology and behavior of the host in relation to that of the blood-sucking arthropod. It has been known since early in this century that many tsetse fly species show a distinct hierarchy of host preference. Certain mammals that are very abundant are seldom or never used as food by the flies. Animals avoided include duiker, waterbuck, Grant's gazelle, impala, hartebeest, zebra, baboon, and dikdik.

A very distinct host discrimination can be demonstrated in the classical observation of Lamprey et al. (11) (Table 1). The impala accounted for about 70% of the mammals in the area, yet they contributed only 1% of the meals of *Glossina swynnertoni,* while the relatively scarce warthog contributed 77%.

Hosts favored by many species of *Glossina* belong mainly to the families Suidae and Bovidae. In spite of the fact that the host discrimination of tsetse flies has been studied for over 50 yr, the explanation for this discrimination remains obscure.

Glasgow (12) suggests that the discrimination of tsetse flies in favor of certain hosts is a reflection of a coincidence of the habits of the fly and the host. According to this theory, the flies habitually rest in certain trees and warthogs habitually rest underneath the same trees; thus the fly will encounter a warthog more often than any other host.

Nash (13) thinks that host selection depends on the "reliability" of the host species. Provided the species is reasonably attractive, the tsetse flies feed more often on one which has reliable habits. The warthog is such a host, with its burrow and mud wallow. The bushbuck is a most reliable individual; it lives in the depth of its selected thicket and comes out to graze in early mornings and evenings. The same can be said for the buffalo and the rhinoceros, which deposit dung in the same spot day after day.

On the other hand, unfavored hosts such as zebra, wildebeest, and gazelle

TABLE I.

Game Observations and Identifications of the Gut Contents of
*Glossina swynnertoni* (11)

| Species | Relative Abundance, (%) | Tsetse Fly Gut Contents, (%) |
|---|---|---|
| Warthog | 2.68 | 77 |
| Rhinoceros | 0.21 | 2 |
| Buffalo | 0.02 | 14 |
| Dikdik | 7.89 | 0 |
| Grant's gazelle | 2.61 | 0 |
| Giraffe | 6.63 | 0 |
| Hartebeest | 3.51 | 0 |
| Impala | 69.91 | 1 |
| Lesser kudu | 0.68 | 0 |
| Waterbuck | 4.38 | 0 |
| Other herbivores | 0.46 | 0 |
| Carnivores | 1.01 | 0 |
| Unidentified Bovidae | — | 6 |
| | 99.99 | 100 |

are migratory by nature and tend to be frightened easily and thunder about. However, differences in tsetse fly host preference cannot always be explained on the basis of different host habits, as many hosts have similar habits and share the same habitat. Thus, the hartebeest that share the woodland with *Glossina morsitans* may have some unpleasant smell for the tsetse fly (13). The impala repels attack with its constantly rippling skin. Fatness of the hosts and thickness of their fur may also play a role in causing the flies to be reluctant to feed. Fastness of the hosts and their evasiveness may also protect them from tsetse flies. These may be protective factors for monkeys and baboons.

Nash (13) describes Simpson's observations about caged monkeys. Even when the monkeys appeared to be asleep when a fly settled on them, they either caught and ate it, or they frightened it away. Nash never saw a fly steal a meal. However, in order to find out whether the flies are repelled by the monkeys, feeding experiments with anesthetized or restrained monkeys would have to be carried out.

Edman and Kale (14) studied the effect of host behavior on the feeding success of mosquitoes. Their preliminary observations showed that rodents display defensive (antimosquito) behavior, and that the cotton mouse, cotton rat, and gray squirrel effectively prevent mosquitoes from feeding. In a more quantitative way, they compared the feeding of *Culex nigripalpus* on various wild birds, mainly heron species, which were found to be their hosts in nature. As can be seen from Table 2, when the birds were restrained during the exposure, the mosquitoes fed avidly on all six species, and all mosquitoes managed to complete their blood meals. No damage was done to the mosquitoes, and over 90% of the population was recovered after one night of exposure. When the mosquitoes were placed in cages with unrestrained birds, the picture was quite different. A high percentage of feeding was successful only on the night heron and the green heron, while the little blue heron,

TABLE II.

Feeding of Mosquitoes on Adult Herons Physically Restrained and Unrestrained (14)

| Host | Restrained | | | Unrestrained | | |
|------|-----------|---------|----------|--------------|---------|----------|
| | Mosquitoes Recovered (%) | Engorged (%) | Feeding Incomplete (%) | Mosquitoes Recovered (%) | Engorged (%) | Feeding Incomplete (%) |
| Night heron | 85 | 69 | 0 | 87 | 98 | 1 |
| Green heron | 96 | 89 | 0 | 94 | 92 | 0 |
| Little blue heron | 91 | 76 | 0 | 75 | 8 | 6 |
| White ibis | 93 | 96 | 0 | 66 | 12 | 8 |
| Cattle egret | 93 | 66 | 1 | 71 | 15 | 13 |
| Snowy egret | 92 | 78 | 1 | 76 | 13 | 14 |

white ibis, cattle egret, and snowy egret protected themselves quite effectively from mosquito bites. Recovery of the mosquitoes was also much lower when they were exposed to hosts upon which they could not feed effectively.

Separation of the bird species according to behavioral criteria employed in Table 3 revealed a possible association between defensive and foraging behavior. The birds with highly developed antimosquito behavior also actively seek and chase their prey in nature. The night heron and green heron, on the other hand, wait for their food in a motionless stance.

We compared the feeding behavior of two species of ticks (15) showing a predilection for avian hosts: one polyphagous species, *Ornithodoros tholozani*, which feeds on any mammal, bird, or reptile found in its habitat, and a second species, *Argas persicus,* which could be classified as oliphagous. Experiments were designed to test host acceptance by the ticks, rather than active host selection. The ticks were confined in a bag attached to the host, and the number of feeding ticks, the size of the blood meal, and the number of eggs produced were recorded. *Ornithodoros* feed readily on rabbits, hamsters, rats, mice, guinea pigs, and chickens. The ticks engorged the same amount of blood from all the hosts tested and produced the same number of

TABLE III.

Types of Antimosquito Behavior Displayed by Each of the Seven Ciconiiformes (14)

| Antimosquito behavior | Night Heron | Green Heron | Little Blue Heron | White Ibis | Loui- siana Heron | Cattle Egret | Snowy Egret |
|---|---|---|---|---|---|---|---|
| **Using Head and Bill** | | | | | | | |
| Head shake | + | + | + | + | + | + | + |
| Head rub (body) | | | + | + | + | + | + |
| Bill snap or jab | + | + | + | + | + | + | + |
| Bill rub (body) | | | + | + | + | + | + |
| Bill rub (legs, feet) | | | + | + | + | + | + |
| Bill rub (perch) | | | | + | | | |
| Bill peck (body) | | | + | + | + | + | + |
| Bill peck (legs, feet) | | | + | + | + | + | + |
| Bill peck (perch) | | | + | | | | |
| **Using Legs and Feet** | | | | | | | |
| Foot shake | | | | + | | | |
| Foot stamp (perch) | | | + | | + | + | + |
| Foot slap (other foot) | | | + | | + | + | + |
| Head scratch | | | | + | + | + | |
| **Using Body** | | | | | | | |
| Wing flip or flap | | | + | | + | + | + |
| Body fluff | | | | + | | + | + |

eggs (with the exception of those attached to snakes, from which smaller meals were taken) (Table 4).

*Argas persicus* showed distinct host preferences. Although close to 100% fed on the avian hosts, they were very reluctant to feed on the other hosts, and none fed on rats, snakes, or toads. Furthermore, those that did feed on the less suitable hosts consumed considerably smaller blood meals, and some toxic effects of rabbit and guinea pig blood were manifested (Table 5). If we accept Askew's theory (10) that evolution leads to a specialization toward a smaller number of hosts, then we can describe the situation here as already being at a high degree of specialization. The selection of the host seems to take place at the initiation of feeding or probing. We found that in order to induce this species to feed artificially, we had to use chicken skin.

Tawfic and Guirgis (16) fed *Argas arboreus* through different membranes (Table 6) and found that the ticks refused to feed through rabbit skin and rabbit caecum, although they fed quite successfully through pigeon skin or crop, or chicken skin or crop. Their experiments suggest that the refusal to feed through rabbit-derived membranes is due to some chemical deterrent rather than to the lack of a probing stimulus. The ticks fed through inert membranes such as parafilm and also through bovine intestine that was processed in a manner that did not leave any soluble or volatile materials in it (*Baudruche* membrane). We think that the partial feeding that occurs on unfavorable hosts is also due to a deterrent which sends a continuous input for avoidance behavior. Continuation of feeding is very high in the behavioral hierarchy, but when the ticks are partially fed, the avoidance behavior dictated by the host deterrent takes over. On a suitable host, feeding is terminated presumably when the abdominal stretch receptors provide a negative feedback.

Oligophagy or monophagy exhibited by many arthropods could be mediated through *kairomones* emitted by the hosts. In arthropods only the

TABLE IV.

Effect of Various Host Species on Egg Production of *Ornithodoros tholozani*

| Host | Feeding (%) | Meal Size (mg) | Laying Eggs (%) | Average No. Eggs/Female |
|------|-------------|----------------|-----------------|--------------------------|
| Rabbit | 95 | 85 | 78 | 75 |
| Hamster | 80 | 83 | 80 | 79 |
| Rat | 95 | 81 | 73 | 84 |
| Mouse | 85 | 77 | 67 | 88 |
| Guinea Pig | 75 | 91 | 86 | 100 |
| Chicken | 95 | 86 | 62 | 92 |
| Snake | – | 60 | 71 | 26 |

TABLE V.

Effect of Various Host Species on Egg Production of *Argas persicus*

| Host | Exposure time (hr) | Feeding (%) | | Weight of Meal (mg)[a] | | Females Laying Eggs (%) | No. Eggs/ Female | Eggs/mg Ingested | Mortality of Ticks (%) |
|---|---|---|---|---|---|---|---|---|---|
| | | Female | Male | Female | Male | | | | |
| Chicken | 2 | 100 | 100 | 39 | 17 | 100 | 124 | 3.2 | 0 |
| Goose | 2 | 97 | 97 | 43 | 18 | 100 | 125 | 2.9 | 0 |
| Pigeon | 2 | 98 | 98 | 66 | 23 | 90 | 148 | 2.3 | 0 |
| Guinea pig | 5 | 44 | 16 | 26 | 13 | 12.5 | 16 | 0.6 | 37.5 |
| Rabbit | 5 | 24 | 28 | 31 | 12 | 25.0 | 33 | 1.0 | 75.0 |
| Hamster | 5 | 28 | 28 | 20 | 9 | 35 | 28 | 1.4 | 0 |
| Mouse | 5 | 16 | 16 | 30 | 14 | 35 | 28 | 2.3 | 0 |
| Rat | 5 | 0 | 0 | 0 | 0 | 0 | — | — | — |
| Snake | 24 | 0 | 0 | 0 | 0 | 0 | — | — | — |
| Toad | 24 | 0 | 0 | 0 | 0 | 0 | — | — | — |

[a] After excretion of coxal fluid.

TABLE VI.

Efficiency of Different Membranes for Feeding
*Argas arboreus* on Citrated Pigeon Blood (16)
after 1-hr Exposure

| Kind of Membrane | Nymphs | | Adults | |
|---|---|---|---|---|
| | Total Exposed | Fed (%) | Total Exposed | Fed (%) |
| Pigeon skin | 96 | 85.4 | 55 | 94.5 |
| Pigeon crop | 100 | 71.0 | 50 | 86.0 |
| Chicken skin | 100 | 75.0 | 50 | 88.0 |
| Chicken crop | 80 | 62.5 | 50 | 78.0 |
| Rabbit skin | 60 | 0 | 40 | 0 |
| Rabbit caecum | 58 | 0 | 40 | 0 |
| Parafilm | 120 | 40.0 | 80 | 35.0 |
| Cellophane | 70 | 0 | 60 | 0 |
| Baudruche skin | 200 | 81.0 | 200 | 92.0 |
| Nylon | 80 | 0 | 50 | 0 |

olfactory sense has the versatility to account for specificity. Much effort has been devoted to finding specific attractants.

Fallis and Smith (6) studied the responses of several ornithophilic simuliids to their hosts. *Simulium euryadminiculum* has the most specific feeding habits of all the black flies. It feeds exclusively on the common loon, *Gavia immer*. The flies show a very strong attraction to ether extracts of the uropygial gland and tail of the loon. Extracts of other parts of the loon's body have only a negligible effect. No isolation or identification of the active component was performed. *Simulium rugglesi* feeds readily on ducks and to a limited extent on birds that frequent forest canopies and thickets. The flies are attracted to an extract from duck gland plus $CO_2$ to a lesser extent to $CO_2$ alone, but not to the gland extract alone. *S. aurem, S. latipes, S. quebeceuse,* and *S. croxtoni* have less specific host preference, and are attracted in considerable numbers to $CO_2$ alone.

Adults of few if any ixodid ticks regard man as a host of predilection. Neither is man at the top of the host hierarchy for any *Glossina* species, yet several mosquitoes show distinct anthropophilic behavior. Careful study of *Aedes aegypti* populations has shown that light strains from Tanzania preferred man to chickens, guinea pigs, or wild rats, while the dark strains preferred these animals to man (17). Thus, host preference may be a genetic trait. Much effort has been devoted to identify the chemicals emitted by man whereby he attracts mosquitoes. Every imaginable part of the human body and its products must by now have been fractionated in every possible way in search of kairomones. As this area of research is reviewed in Chapter 18 by

Dr. Khan, who is a member of one of the most active teams exploring mosquito attractants, I will only point out some of the highlights. Certain attractive compounds have been found: lysine, alanine, methionine, carbamino compounds, lactic acid, and sex pheromones [for the various references the reader is referred to (3)]. Yet, nothing proved to be as attractive as the intact man. That seems to suggest that many blood-sucking arthropods recognize man by his complex of effluvia and not by any specific indicator.

So far we have dealt with factors that are involved in orientation of the parasite to its host and initiation of feeding. It seems that our present state of knowledge may explain these two steps for many polyphagous blood feeders. At this stage, orientation and selection of a specific host cannot be fully accounted for in chemical terms.

## III. EFFECT OF BLOOD COMPOSITION ON HOST DISCRIMINATION

Let us now examine the third phase in the feeding process—the actual imbibing of the blood meal. Using artificial feeding and offering the meal through an appropriate membrane allows us to study dependence of meal ingestion on the presence of the proper phagostimulants in the diet. The present state of our knowledge has recently been summarized in detail (4) and will be reviewed here only briefly. The chemicals identified so far as phagostimulants responsible for hematophagy are summarized in Table 7.

Several conclusions may be drawn from the information in Table 7. The range of phagostimulatory compounds tends to narrow with evolution, ending with ATP as the only phagostimulant, as observed in fleas. ATP is by far the most stimulatory compound for insects. Reduced glutathione, in the presence of glucose, seems to stimulate ticks no less than ATP or DPNH (which also require the presence of glucose). If we go further along the evolutionary line (i.e., leeches), neither ATP nor GSH exert any stimulation, and arginine seems to be the most important stimulant.

It seems that like most carnivores, hematophagous animals respond positively to molecules of relatively small size, which have wide distribution in animal tissues and blood. All the stimulatory compounds are found in concentrations of $10^{-4}$ to $3 \times 10^{-3}$ $M$ in plasma or blood cells (Table 8).

If we examine the content of phagostimulants in blood to see whether they can account for variations in feeding responses to the host, we see that, had the tick been exclusively dependent on the free amino acids in the plasma, such chemicals might have been a limiting factor. Most of the amino acids are found in the plasma at a concentration around $10^{-4}$ $M$. This is definitely a suboptimal concentration for ticks (Table 9). However, since ticks are stimulated mainly by GSH and ATP, the concentration of these compounds, even in hosts which are poor in ATP, such as cattle, is many fold higher than the ticks' stimulatory threshold, which is around $10^{-4}$ $M$ (19).

TABLE VII

Chemical Nature of Phagostimulants of Hematophagous
Animals (4)[a]

| Animal | Phagostimulants |
|--------|-----------------|
| Medical leech | Arginine > histidine > cysteine > glutamate > asparate > lysine; glucose = galactose = sorbose (only in the presence of Na) |
| Soft ticks | GSH > ATP > DPNH > ADP > ITP = GTP leucine, isoleucine, alanine, phenylalanine, proline (only in the presence of glucose) |
| *Rhodnius* bug | ATP > ADP = CTP = GTP = CDP = ITP cAMP > ADP = GDP > AMP |
| Mosquitoes and tsetse flies | ATP > ADP > AMP = cAMP |
| Rat flea | ATP |

[a]The following abbreviations are used in this table: AMP: adenosine 5'-monophosphate; cAMP: 3',5'-cyclic-adenosine monophosphate; ADP: adenosine diphosphate; ATP: adenosine triphosphate; ITP: inosine triphosphate; GTP: guanosine triphosphate; CDP: cytidine diphosphate; CTP: cytidine triphosphate; DPNH: reduced diphosphopyridine nucleotide; GSH: reduced glutathione.

The concentration of ATP required to induce 50% feeding is $6 \times 10^{-6}$ $M$ for *Rhodnius* (20), and $10^{-5}$ $M$ for *Glossina* (21) and mosquitoes (22). These threshold concentrations are at least two orders of magnitude lower than those found even in poor blood, provided the blood in ATP can gain egress to contact the insect's chemoreceptor surfaces and thereby induce feeding.

As indicated in Table 8, all blood ATP and GSH are confined to the cellular fraction of the blood, where they are firmly bound. Therefore, their availability as stimulants has been questioned. Galun and Rice (23) showed

TABLE VIII

Content of Some Phagostimulants in the Blood of Various Animals (mg/100 ml) (18)

|  | Man | | Cattle | | Rabbit | | Pigeon | |
|--------|------|--------|------|--------|------|--------|------|--------|
|  | Cell | Plasma | Cell | Plasma | Cell | Plasma | Cell | Plasma |
| ATP | 72 | 0 | 27 | 0 | 112 | 0 | 182 | 0 |
| GSH | 77 | 0 | 157 | 0 | – | – | – | – |
| Glucose | 74 | 97 | 15 | 85 | 41 | 145 | 150 | – |
| Isoleucine | 0.9 | 1.6 | – | – | – | – | – | – |
| Leucine | 1.5 | 1.9 | – | – | – | – | – | – |
| Arginine | 0.3 | 2.3 | – | – | – | – | – | – |

TABLE IX.

Effect of Amino Acids on the Feeding Response of
*Ornithodoros tholozani* (19)

| Compounds Added to Saline Containing Glucose 1 mg/ml | Concentration of Amino Acid | | |
|---|---|---|---|
| | $10^{-2}$ *M* | $10^{-3}$ *M* | $10^{-4}$ *M* |
| | Response (%) | | |
| L-Leucine | 73 | 65 | 35 |
| L-Isoleucine | 68 | 34 | 40 |
| L-Isoleucine (allo-free) | – | 3 | 11 |
| L-Valine | 30 | 33 | 35 |
| L-Phenylalanine | 5 | 50 | 25 |
| L-Tyrosine | – | 5 | – |
| L-Tryptophan | 15 | 10 | – |
| L-Aspartate | 55 | 50 | 5 |
| L-Asparagine | – | 23 | 35 |
| L-Glutamate | 0 | 10 | 15 |
| L-Glutamine | – | 30 | 35 |
| L-Arginine | 10 | 24 | 17 |
| L-Citrulline | 0 | 15 | – |
| L-Lysine | 5 | 22 | 10 |
| L-Alanine | 25 | 45 | 30 |
| Glycine | 10 | 10 | – |
| L-Proline | 23 | 40 | 40 |
| L-Hydroxyproline | – | 15 | 15 |
| L-Histidine | 15 | 10 | – |
| L-Serine | 35 | 30 | 35 |
| L-Threonine | 10 | 0 | – |
| L-Cystenie | 13 | 18 | 20 |
| L-Methionine | 10 | 5 | – |

that if red blood cells are centrifuged in such a manner that they are not accompanied by any platelets, they do not stimulate feeding. It is necessary to add blood platelets to induce feeding of mosquitoes or tsetse flies. The number of platelets varies a great deal from one animal to another (Table 10). Goats have about 1/10 of the platelet number when compared to most other animals, yet they are used as a host for breeding tsetse flies in many laboratories.

If the ATP from all the platelets were liberated, its concentration would barely reach the threshold of stimulation for mosquitoes or tsetse flies. Therefore, it has to be assumed that the necessary concentration is attained only locally at the chemoreceptor level. Either a platelet-releasing factor is located at the tip of the receptor, or platelets adhere preferentially to the receptor and release their content there. Host preference through differences

## TABLE X.

Blood Platelet Count in Various
Animals (24)

| Animal | Platelet Count (thousands/mm³) | |
|---|---|---|
| | Average | Range |
| Cattle | 684 | 542-975 |
| Pig | 403 | 296-616 |
| Goat | 50 | 8-127 |
| Rabbit | 533 | 170-1120 |
| Guinea pig | 719 | 550-880 |
| Man | 409 | 273-545 |

in host platelets is a possibility that merits investigation. Clearly, it is not merely a matter of ATP content, but many other properties of the platelets. Platelet adhesion and spreading on foreign surfaces, as well as the release reaction and aggregation of platelets from various species, should be looked into. There is evidence, for example, of quantitative and qualitative differences in the release reaction of human and rabbit platelets (25).

## REFERENCES

(1)   K.J. Lindstedt. 1971. Chemical control of feeding behavior. *Comp. Biochem. Physiol.* **39A**:553-581.

(2)   V.G. Dethier. 1957. The sensory physiology of blood-sucking arthropods. *Exp. Parasit.* **6**:68-122.

(3)   B. Hocking. 1971. Blood-sucking behavior of terrestrial arthropods. *Ann. Rev. Entomol.* **16**:1-26.

(4)   R. Galun. 1975. The role of host blood in the feeding behavior of ectoparasites. In *Dynamic Aspects of Host Parasite Relationships,* Vol. 2. A Zuckerman, Ed. Keter, Jerusalem, pp. 132-162.

(5)   P.N. Daykin, F.E. Kellogg, and R.H. Wright. 1965. Host finding and repulsion of *Aedes aegypti. Can. Entomol.* **97**:239-263.

(6)   A.M. Fallis and S.M. Smith. 1964. Ether extracts from birds and $CO_2$ as attractants for some ornithophilic simuliids. *Can. J. Zool.* **42**:723-730.

(7)   G.R. Defoliard and C.D. Morris. 1968. A dry ice baited trap for collection and field storage of hematophagous Diptera. *J. Med. Entomol.* **4**:360-362.

(8)   B.H. Wilson, N.P. Tugwell and E.C. Burns. 1966. Attraction of tabanids to traps with dry ice under field condition in Louisiana. *J. Med. Entomol.* **3**:148-149.

(9)   R. Garcia. 1962. Carbon dioxide as attractant for certain ticks. *Ann. Entomol. Soc. Am.* **55**:605-606.

(10)  R.R. Askew. 1971. *Parasitic Insects.* Heinemann, London, pp. 268-269.

(11)  H.F. Lamprey, J.P. Glasgow, F. Lee-Jones, and B. Weitz. 1962. Simultaneous

census of potential and actual food sources of the tsetse fly *Glossina swynnertoni. J. Anim. Ecol.* **31**:151-156.

(12) J.P. Glasgow. 1963. *The Distribution and Abundance of Tsetse.* Macmillan, New York.

(13) T.A.M. Nash. 1969. *African's Bane: The Tsetse Fly* Collins, London.

(14) J.D. Edman, and H.W. Kale. 1971. Host behavior: Its influence on the feeding success of mosquitoes. *Ann. Entomol. Soc. Am.* **64**:513-516.

(15) R. Galun and J. Sternberg. Unpublished.

(16) M.W. Tawfic, and S.S. Guirgis. 1969. Experimental feeding of *Argas* (persicargas) *arboreus* through membranes. *J. Med. Entomol.* **6**:191-195.

(17) H.K. Gouck. 1972. Host preference of various strains of *Aedes aegypti* and *A. simpsoni* as determined by an olfactometer. *Bull. World Health Org.* **47**:680-83.

(18) E.C. Albritton. 1952. *Standard Values in Blood.* Saunders, Philadelphia.

(19) R. Galun, and S.H. Kindler. 1968. The chemical basis of feeding in the tick *Ornithodoros tholozani. J. Insect Physiol.* **14**:1409-1421.

(20) W.G. Friend and J.J.B. Smith. 1971. Feeding in *Rhodnius prolixus*: Potencies of nucleoside phosphates in initiating gorging. *J. Insect Physiol.* **17**:1315-1325.

(21) R. Galun, and J. Margalit. 1969. Adenine nucleotides as feeding stimulants of the tsetse fly *Glossina austeni. Nature* **222**:583-584.

(22) R. Galun. 1967. Feeding stimuli and artificial feeding. *Bull. World Health Org.* **36**:590-593.

(23) R. Galun, and M.J. Rice. 1971. Role of blood platelets in haematophagy. *Nature* **233**:110-111.

(24) O.W. Schlam. 1965. *Veterinary Haematology*, 2nd ed. Lea & Febiger, Philadelphia.

(25) D.C.B. Mills, I.A. Robbs, and G.C.K. Roberts. 1968. The release of nucleotides, 5-hydroxytryptamine and enzymes from human blood platelets during aggregation. *J. Physiol.* **195**:715-729.

## CHAPTER 8

# Host-Related Responses and their Suppression: Some Behavioral Considerations

## L. Barton Browne

### *CSIRO Division of Entomology, Canberra A.C.T. Australia*

The three foregoing chapters have dealt with a number of aspects of the host-related behavior of insects, and it is my intention here to complement and extend some of that material. I have chosen for consideration three aspects that relate also to topics discussed in Chapters 17 and 18. First I discuss some aspects of the orientation behavior of insects to host odor that have not been mentioned in Chapter 5 and some possible modes of action of repellents in the light of what is now known about the response mechanisms of insects to attractants. Second, I consider, from a behavioral point of view, the possible modes of action of both feeding and oviposition deterrents and of locomotor stimulants, with special reference to the relationships that may exist between locomotion and the more specific behaviors of feeding and oviposition. Third, I consider some behavioral bases of choice in relation to host selection.

## I. ORIENTATION MECHANISMS OF INSECTS TO VOLATILE CHEMICALS

### A. Responses to Attractants

To lay a foundation for a discussion of some behavioral aspects of repellency it is necessary to mention several things that are known about orientation mechanisms involved in attraction, the discovery of which predates the recent work of Farkas and Shorey (1) (2) and Kennedy and Marsh (3), and which have received little or no mention in the writings of these authors.

An odor plume has three dimensions, and at least some insects deviate from the shortest upwind path between their starting point and the source of the attractant verically and horizontally. Kellogg et al. (4) showed this to be true for *Drosophila melanogaster* responding to odor of fementing banana emanating from a point of source. Trynier (5) demonstrated it even more clearly for males of *Ephesia kuehniella* responding to a sex pheromone. Thus an insect in a plume has not only to make left and right turns, but also to make adjustments to the vertical plane.

That an insect has to make adjustments in two planes raises a number of questions. For instance, are the casts that insects make at right angles to the wind upon losing contact with the odor, which Kennedy and Marsh (3) consider to be anemotactically guided, in only one plane or in two? Further, if it is in two planes, are the sizes of horizontal and vertical components the same when an insect loses the odor laterally as when it does so vertically? If it were found that the lateral component of casting is always substantially greater than the horizontal one, it might indicate that a major role of casting is related to loss of contact with the odor resulting from the behavior of the odor plume relative to the insect, rather than vice versa, since it seems likely that wind shifts in the horizontal plane would be more frequent than apparent shifts in the vertical plane.

Farkas and Shorey (2) have suggested that one possible basis for the determination of the direction of turns made by insects upon losing the odor stimulus is a preprogrammed response in which right and left turns are alternated, and this is substantiated somewhat by the alternation in turning direction shown by *Bombyx mori* walking in a uniform field of pheromone (6). Again, it has to be pointed out that Farkas and Shorey (2) considered turning in only the horizontal plane and that Kramer's (6) findings with walking insects necessarily apply only in one plane and might not, therefore, be applicable to flying insects. Further reason for doubting the applicability of the one-plane model to flying insects is that Kellogg et al. (4) with *D. melanogaster* and Traynier (5) with *E. kuehniella* showed that insects flew fairly directly upwind in air uniformly permeated with attractant and that these insects showed little or no sign of zigzag flight.

Another characteristic of orientation behavior that has received no recent comment, but which may be relevant to current controversies surrounding orientation mechanisms in attraction, is that some insects make downwind casts. This is clearly seen in the flight tracks of *D. melanogaster* photographed by Kellogg et al. (4) and in those of *E. kuehniella* traced by Traynier (5), who concluded that the insect made downwind casts if it did not again receive stimulation after making a number of cross-wind casts. This seems to be a mechanism by which insects overshooting the source of the attractant reestablish contact with that source. It suggests that the behavior of insects upon losing contact with the active space of the plume might be a more complex piece of programming than the results of Kennedy and Marsh (3) suggest.

Serious consideration has never been given to the possible consequences of any adaptation that may take place during the insects' upwind movement within an odor plume. Such movement may take several minutes, during which the insect is in more or less constant contact with the volatile stimulus, and it seems almost certain that some adaptation would occur in the receptors and perhaps also in the central nervous system. Since any substantial degree of adaptation would reduce the size of the active zone (7) because of the resulting increase in threshold, and perhaps render much of that part of the odor plume already experienced by the insect effectively nonstimulating, adaptation might play an important role in determining an insect's exact behavior in an odor plume. In the extreme case, for instance, in which adaptation was very rapid and virtually complete, the insect would need to move progressively into areas of higher and higher odor concentration to maintain effective stimulation and, therefore, to recognize the continuing existence of the odor. Under these admittedly hypothetical and perhaps unlikely circumstances, therefore, the odor plume itself might provide the insect with directional information, since any movement other than that towards the source would be associated with loss of stimulation by the insect.

Although the extreme situation I have just considered may never exist, adaptation may well be an integral part of the complex olfactory guidance system of insects.

The available evidence suggests that adaptation may not be very rapid or complete. *D. melanogaster* (4) and *E. kuehniella* (5) have been shown to fly upwind in air uniformly permeated with attractant odors. Therefore, these insects are clearly capable of moving upwind in the absence of any progressive increase in odor; but these flights were only of short duration—being over approximately 1 m. Haskell et al. (8), Moorhours (9), and Kennedy and Moorhouse (10) showed that *Locusta migratoria* larvae walked upwind for a distance of about 1 m in response to a wind that was fairly uniformly permeated with grass odor; on arrival at the upwind end they perched on the gauze that formed the end of the tunnel, but gradually dispersed even in the continuing presence of odor (11). These results indicate that considerable adaptation may ultimately occur but that it is not particularly rapid. Also, the departure of the insects from the upwind screen might be due to mechanisms other than adaptation. Kramer (6) obtained evidence that suggests that rapid adaptation does not occur in males of *B. mori* responding to sex pheromone. He found that the insects would respond for some time by positive anemotaxis when stimulated with a constant pheromone concentration and that their speed of walking did not diminish with time. In other experiments he showed that walking speed was positively correlated with pheromone concentration. Finally, Kennedy and Marsh (3) have shown that male *Plodia interpunctella* would make up to 16 upwind approaches toward a pheromone source in an unbroken flight if returned repeatedly to the starting line by the manipulation of optomotor cues, suggesting either that adaptation did not occur to any great degree or that disadaptation was rapid when the insects were returned to the lower concentration at the downwind end of the plume.

## B. Responses to Repellents

Dethier et al. (12) defined an attractant as "a chemical which causes insects to make oriented movements toward its source" and a repellent as "a chemical which causes insects to make oriented movements away from its source." The contribution by Kennedy in Chapter 5 of this book and the remarks I have already made here indicate that it would perhaps be better to define an attractant as "a chemical or mixture of chemicals which, acting in the vapor phase, cause an insect to behave in ways which result in its moving toward the source of the material or toward a zone of preferred concentration." Similarly, knowledge accumulated since 1960 suggests that a repellent should be defined more broadly than was suggested by Dethier et al. (12). Even a definition equivalent to that just suggested attractant—namely, that a repellent is "a chemical or mixture of chemicals that, acting in the vapor phase, causes an insect to behave in ways which result in its movement away from the source of the material"—is not entirely satisfactory since, in practice, the performance of a repellent is almost always assessed in terms of its ability to inhibit the insect's response to chemical attractants and perhaps to attractive visual stimuli. It could be argued, therefore, that the practical definition of a repellent would be "a chemical that, acting in the vapor phase, prevents an insect from reaching a target to which it would otherwise be attracted." Perhaps the formal definition of a repellent should be an amalgamation of the two suggested here.

As with attractants, our knowledge of the behavioral mode of action of repellents is fragmentary; the available evidence suggests that insects differ in their behavioral responses. Haskell et al. (8) and Kennedy and Moorhouse (10) showed that locust larvae displayed negative anemotaxis when the airflow in a flat bed wind tunnel was permeated with iso- or $d$-valeric acid or carbon tetrachloride. Grass odor, in this situation, elicits positive anemotaxis. Wright and Rayner (13) found, on the other hand, that *Aedes aegypti* showed no sign of negative anemotaxis when in a horizontally directed airstream permeated with the mosquito repellent dimethyl phthalate. Under these conditions their flight directions were as random as those of mosquitoes in clean air.

Daykin et al. (14) made a study of the ways in which the host-finding behavior of *A. aegypti* is disrupted by the presence of repellents. Their experiments showed that mosquitoes entered a host-generated plume lacking repellent without deviating, but that they made turns upon leaving the plume, usually with the result that their new course was toward the plume's central axis. When the host-generated plume contained a repellent, they turned when they first encountered the plume and usually failed to enter it. On the occasion when they did enter the plume, they emerged from it without turning. Thus, the presence of a repellent essentially reversed the behavior seen in its absence.

Wright (15) reported that the presence of a piece of repellent-soaked paper on the floor of a cage reduced to zero the number of mosquitoes biting a human arm to which no repellent had been applied. This he ascribed to the known (16) ability of the repellent to prevent mosquitoes from being activated by carbon dioxide arising from the human arm. It seems, however, that the effect of the repellent might be due to other types of disruption of the host-finding program; Daykin et al. (14) state that "if the whole environment is permeated with repellent, convection currents [arising from the host] are ignored because there is no turn back upon entering one and no turn back upon leaving it." Presumably, therefore, the turns made by an insect on reaching a repellent-bearing odor plume is a response to encountering the repellent, but the failure to turn on leaving one is due to the mere presence of the repellent.

The available information, though fragmentary, relates to the use of repellents. Two methods have been suggested. The first is the commonly employed one of applying the repellent to the potential target, in which case the active zone of the plume of repellent corresponds more or less, depending on the relative emission rates and insect sensitivities, with that of the host attractants. In the second, an area containing potential hosts is permeated with the repellent. Permeation would be effective if the repellent elicits negative anemotaxis or if it causes the insect to ignore host stimuli, and if a major degree of adaptation to the repellent does not occur. Permeation would be ineffective if the insects behavior was modified only upon entering a zone of repellent or under any circumstances in which the insect adapted substantially to its presence.

Finally, attraction to hosts might involve visual as well as chemical stimuli, [e.g., (17)] and it could be that an insect rendered indifferent to host chemicals by the action of a repellent might still reach its target by using visual cues. The only relevant data I know of that suggest that a repellent might cause an insect to become unresponsive to visual stimulation is that of Kahn (18), who showed that mosquitoes that had had their antennae painted with repellent were no longer attracted to dark stripes.

In summary, we know very little about how repellants affect behavioral responses and about how they prevent insects from finding their hosts. There has been little work specifically related to the behavior of insects in various situations in which repellents are present, and even less on the mechanisms involved in attraction to hosts. Only when a great deal more information is available will we be in a position to exploit fully the possibilities of disrupting, by the use of repellents, the responses of insects to their hosts.

## II. LOCOMOTION AND ITS RELATIONSHIP TO THE ACTION OF FEEDING AND OVIPOSITION DETERRENTS

Jermy (19) discussed the behavior of insects that had contacted a feeding deterrent. He reported that larvae and adults of the Colorado potato beetle that had tasted a feeding deterrent showed a high level of locomotor activity, which continued for some time, during which the insects would not feed on material containing no deterrent. He considered this behavior to be the result of a central inhibitory state generated by the deterrent. Further, he suggested that insects that normally range widely between periods of feeding show more intense and prolonged locomotor activity when they experience the effect of feeding deterrents than do insects that normally do not leave the host between periods of feeding.

Jermy implies that the central inhibitory state is widespread within the central nervous system, causing the insect to fail to feed for a time after contacting the deterrent and simultaneously causing it to show a level of locomotor activity appropriate to the kind of locomotory behavior normally shown by the insect between periods of feeding.

In the remainder of this section I consider more broadly and more speculatively relationships that might exist between locomotion and both feeding and oviposition, and make special reference to the situation in which the level of locomotor activity is negatively correlated with the performance of either of the other two activities.

Some evidence suggests that feeding and locomotion are, in some insects, antagonistic behaviors: the performance of one inhibits the other. Ibbotson and Kennedy (20) demonstrated this in the aphid, *Aphis fabae*; their experiments showed that a variety of treatments causing a reduction in activity also caused the insects to probe. Dethier and Gelperin (21) showed that adults of the blowfly, *Phormia regina*, that had had the ventral nerve cord cut between the subesophageal and the thoracic ganglion mass fed indefinitely and finally burst, whereas those that had had all the abdominal nerves severed behind the ganglionic mass took only two to four times the amounts taken by controls—suggesting that locomotion and feeding might be antagonistic in this insect. The two kinds of operation are equivalent in terms of sensory input—or rather lack of it—from the abdomen, and it seems possible that, as suggested by Dethier and Gelperin (21), and argued in more detail by Barton Browne (22), the onset of walking and resulting input to the brain concerning the activity of relevant motor centers in the thoracic ganglionic mass might, in the final analysis, be responsible for bringing about the cessation of feeding.

However, for both aphids and blowflies, feeding is an akinetic activity; locomotion and feeding would probably not be antagonistic in the same way in insects such as caterpillars, which progress while feeding and are quiescent between meals.

Information regarding the relationship between oviposition and locomotion is even more sketchy. Barton Browne et al. (23), in their study of the pheromonally mediated group oviposition in the blowfly, *Lucilia cuprina*, observed that probing with the ovipositor was always associated with a low level of locomotor activity brought about, at least in part, by the responses of the flies to tarsal contact with one another. Barton Browne (24) reported that strong illumination from which the flies could not escape inhibited oviposition by *L. cuprina*, presumably because it caused the insects to remain in a highly locomotor state. It seems likely that oviposition in this insect, which lays a large mass of eggs at one time and for which egg-laying is an akinetic activity, is antagonistically related to locomotion. This kind of antagonism presumably would not exist in insects, such as some butterflies, that lay a series of single eggs while pausing only briefly between periods of flight.

Two possible bases may account for the existence of an inverse relationship between two kinds of behavior in an insect receiving stimulation from a behavior-modifying chemical. The first is that sensory input elicited by the chemical has access only to the part of the central nervous system concerned with one of the inversely related behaviors and that the state of excitation or inhibition generated in the part of the central nervous system concerned with this behavior causes the opposite state to be generated in the part of the nervous system concerned with determining the level of the other activity. The second possibility, which might be termed *pseudoantagonism*, is that in which the input has access to the parts of the central nervous system concerned with determining the levels of both the activities, generating an excitatory state in one and an inhibitory state in the other.

The two theoretically possible situations just outlined suggest that chemicals that have the effect of deterring insects from feeding or ovipositing might do so in two distinct ways. First, the input elicited by the chemical might have direct access to the feeding center or the oviposition center in the central nervous system, generating in it an inhibitory state. In this case, any accompanying increase in locomotor activity could be due either to antagonism or to the fact that the input also has direct access to the locomotor center, generating in it a state of excitation. The second situation is that in which the input has access only to the locomotor center, generating in it an excitation state, with an inhibitory state being developed in the feeding or oviposition center as a result of central nervous antagonism.

It is difficult to know, at this time, what practical implications the possibilities of direct and indirect effects of chemicals on specific behaviors might have, or even whether the two could reliably be distinquished on the basis of behavioral experiments. The concept of antagonistic relationships between behaviors, especially when one of them is locomotion, which in many insects forms the connecting link between more specific activities, should be borne in mind by workers during the development of bioassays and in their consideration of methods for the practical use of chemicals for modifying host-related behavior.

## III. CHOICE BEHAVIOR

When selecting a host for oviposition or feeding, insects are often confronted with a complex choice situation. Contact stimulation probably plays a part in the host-selection behavior of all phytophagous insects, and determines finally whether an insect, in the correct physiologic state, will oviposit or feed. In some cases, however, an initial choice of plant to be visited may be made on the basis of characteristics—such as odor (25) or color or shape (26)-(28)—that the insect can perceive at a distance. Clearly, the behavioral bases for choice will differ according to the part played by these cues.

Two extreme situations illustrate how perception at a distance of host characteristics might influence insect host-selection behavior. In the first, in which the insect makes no selection on the basis of plant characteristics perceived at a distance but makes it entirely on the basis of stimulation received when in contact with the plant, the frequency with which different types of plants would be visited would depend only on their relative abundance. The likelihood that an insect would feed or oviposit on a particular kind of plant once visited would vary according to its contact-perceived characteristics; those plants that elicit a greater amount of excitatory input and little or no inhibitory input would be fed or laid upon more readily than those for which the excess of excitatory over inhibitory input is less. Under these circumstances, the greater amount of feeding or egg-laying would take place on "preferred" plants. Plants eliciting a relatively small absolute amount of excitatory input, or an amount of inhibitory input that nearly cancels out the amount of excitatory input elicited, would be fed or laid upon only by insects that were very ready to do so—for example, those that have experienced unusually long periods of food deprivation or have been unable to find an oviposition site after maturation.

Preferences based on this mechanism might be reinforced by another factor in insects that take their food as discrete meals or for which the laying of each egg is a separate behavioural act, and that recognize hosts only by contact stimulation. Some insects take relatively small meals of materials that are less than optimally stimulating (22), and they would feed less on hosts eliciting a smaller net amount of input excitatory to feeding. A similar mechanism might operate in oviposition. Thus this type of insect would depart from less than optimally stimulating hosts in a physiologic state such that it would still be ready to feed or oviposit on hosts providing greater net amounts of excitatory input.

Where stimulation is provided by contact perception only, the insect has no method of simultaneously comparing the other plant types in its environment. In some circumstances (e.g., when the most "preferred" plant is relatively scarce) choice based on contact, stimulation only would be relatively less satisfactory.

In the second extreme situation, in which the insect can differentiate accurately between different kinds of plants on the basis of characteristics perceived at a distance, the frequency with which it visits the different kinds of plants would depend on the type of distance stimulation provided by the plants and not merely on their availability; some plant types would receive fewer visits than would be expected, or perhaps none at all.

When the distance cues to which the insect is responding are visual ones, the insect can make simultaneous comparisons of the different amounts of stimulation provided. If the cues are of an olfactory nature, however, the extent of comparison is less clear, but the insects would have the opportunity for making selections before alighting on the plant and, therefore, before they are committed to feeding or ovipositing. Host selection involving distance perception of host characteristics is likely to be more satisfactory than that depending only upon contact-perceived cues.

An insect that responds basically by way of contact stimulation could increase its host-selection efficiency by learning to associate the appearance of plant types with the levels of stimulation elicited when contact is made. In this way the insect would "convert" itself into an insect that uses visual stimulation perceived at a distance. There is no known evidence of this in phytophagous insects, but Arthur (29) (30) has shown that associative learning plays a part in the host-finding behavior of two ichneumonid parasitoids. Alternatively, an insect might "file away," as a standard of comparison, the net level of excitatory stimulation elicited by the most stimulating plant it has yet contacted, after which it would not feed or oviposit readily on any plant providing a smaller net amount of input excitatory to feeding or oviposition. A third possibility is that an insect, once having fed or oviposited upon a host eliciting a satisfactory level of excitatory input from contact chemoreceptors, might then localize its searching. Since there is a tendency for plants of the same kind to grow or to be grown in clumps such behaviour would increase an insect's chances of again landing on the same host species. It is probable, for instance, that behaviour of this kind is exhibited by alates of *Aphis fabae*, Kennedy (31, 32) having shown that aphids which had just taken off from leaves of a non-host plant flew more strongly and for longer than ones which had taken off from host leaves.

Finally, the possibility that an insect might use learned behavior in its host selection has important implications for laboratory bioassays designed to assess the relative acceptabilities of hosts or to determine the chemical bases for different degrees of acceptability. Clearly, the choices made by "educated" insects would differ from those made by "naive" insects. In this event, greater difficulty than usual would be experienced in predicting an insect's field behavior on the basis of its behavior in the laboratory.

## IV. CONCLUDING REMARKS

The three behavioral aspects of the host-related response of insects discussed in this chapter—orientation mechanisms of the insects to volatile chemicals, locomotion and its relation to the action of deterrents to feeding and oviposition, and choice behavior—illustrate some of the complexities workers are faced with in seeking to facilitate the modification of host-related behavior. Clearly, it is highly important to have a detailed knowledge of the target insect's behavior.

## REFERENCES

(1)     S.R. Farkas and H.H. Shorey. 1972. Chemical trail-following by flying insects: A mechanism for orientation to a distance odor source. *Science* **178**:67-68.

(2)     S.R. Farkas and H.H. Shorey. 1974. Mechanisms of orientation to a distant pheromone source. In *Pheromones*. M.C. Birch, Ed. North-Holland, Amsterdam, pp. 81-95.

(3)     J.S. Kennedy and D. Marsh. 1974. Pheromone-regulated anemotaxis in flying moths. *Science* **184**:999-1001.

(4)     F.E. Kellogg, D.E. Frizel, and R.H. Wright. 1962. The olfactory guidance of flying insects. IV. *Drosophila. Can. Entomol.* **94**:884-888.

(5)     R.M.M. Traynier. 1968. Sex attraction in the Mediterranean flour moth *Anagaster kühniella*: Location of the female by the male. *Can. Entomol.* **100**:5-10.

(6)     E. Kramer, 1975. Orientation of the male silkworm to the sex attractant Bombykol. In *Olfaction and Taste, V.* D. Denton and J.D. Coghlan, Eds. Academic Press, New York. pp. 329-325.

(7)     W.H. Bossert and E.O. Wilson. 1963. An analysis of olfactory communication among animals. *J. Theoret. Biol.* **5**:443-469.

(8)     P.T. Haskell, M.W.J. Paskin, and J.E. Moorhouse. 1962. Laboratory observations on factors affecting movements of hoppers of the desert locust. *J. Insect Physiol.* **8**:53-78.

(9)     J.E. Moorhouse. 1971. Experimental analysis of the locomotor behaviour of *Schistocerca gregaria* induced by odour. *J. Insect Physiol.* 17:913-920.

(10)    J.S. Kennedy and J.E. Moorhouse. 1969. Laboratory observations on locust responses to wind-borne odours. *Entomol. exp. appl.* 12:487-503.

(11)    J.E. Moorhouse. Personal communication.

(12)    V.G. Dethier, L. Barton Browne, and C.N. Smith. 1960. The designation of chemicals in terms of the responses they elicit from insects. *J. Econ. Entomol.* 53:134-136.

(13)    R.H. Wright, and H.B. Rayner. 1960. The olfactory guidance of flying insects. II. Mosquito repulsion. *Can Entomol.* 92:812-817.

(14)    P.N. Daykin, F.E. Kellogg, and R.H. Wright. 1965. Host finding and repulsion of *Aedes aegypti*. *Can. Entomol.* 92:239-263.

(15)    R.H. Wright. 1968. Tunes to which mosquitoes dance. *New Sci.* 37:694-697.

(16)    J.E. Simpson and R.H. Wright. 1967. Area treatment to combat mosquitoes. *Nature* 214:113-114.

(17)    G.A. Vale. 1974. The responses of tsetse flies (Diptera, Glossinidae) to mobile and stationary baits. *Bull. Entomol. Res.* 64:545-588.

(18)    A.A. Khan. 1965. Effects of repellents on mosquito behavior. *Quaest. entomol.* 1:1-35.

(19)    T. Jermy. 1971. Biological background and outlook of the antifeedant approach to insect control. *Acta Phytopathol. Acad. Sci. Hung.* 6:253-260.

(20)    A. Ibbotson, and J.S. Kennedy. 1959. Interaction between walking and probing in *Aphis fabae* Scop. *J. Exp. Biol.* 36:377-390.

(21)    V.G. Dethier, and A. Gelperin. 1967. Hyperphagia in the blowfly. *J. Exp. Biol.* 47:191-200.

(22)    L. Barton Browne. 1975. Regulatory mechanisms in insect feeding. *Adv. Insect Physiol.* 11:1-116.

(23)    L. Barton Browne, R.J. Bartell, and H.H. Shorey. 1969. Pheromone-mediated behavior leading to group oviposition in the blowfly *Lucilia cuprina*. *J. Insect Physiol.* 15:1003-1014.

(24)    L. Barton Browne. 1962. The relationship between oviposition in the blowfly *Lucilia cuprina* and the presence of water. *J. Insect Physiol.* 8:383-390.

(25)    C. Hawkes. 1974. Dispersal of adult cabbage root fly *(Erioischia brassicae* (Boche)) in relation to a brassica crop. *J. Appl. Ecol.* 11:83-93.

(26)    R.J. Prokopy. 1968. Visual responses of apple maggot flies, *Rhagoletis pomonella* (Diptera: Tephritidae): Orchard studies. *Entomol. exp. appl.* 11:403-422.

(27)    V. Moeriche. 1969. Host plant specific colour behaviour by *Hyalopterus pruni* (Aphididae) *Entomol. exp. appl.* 12:524-534.

(28)    J.R. Meyer. 1975. Effective range and species specificity of host recognition in adult alfalfa weevils, *Hypera postica*. *Ann. Entomol. Soc. Am.* 68:1-3.

(29)    A.P. Arthur. 1966. Associative learning in *Itoplectis conquistor* (Say) (Hymenoptera: Ichneumonidae). *Can. Entomol.* 98:213-227.

(30)    A.P. Arthur. 1971. Associative learning by *Nemeritis canescens* (Hymenoptera: Ichneumonidae). *Can. Entomol.* 103:1137-1141.

(31)    Kennedy, J.S. 1965. Co-ordination of successive activities in an aphid. Reciprocal effects of settling on flight. *J. Exp. Biol.* 43:489-509.

(32)    Kennedy J.S. 1966. The balance between antagonistic induction and depression of flight activity in *Aphis fabae* Scopoli. *J. Exp. Biol.* 45:215-228.

CHAPTER 9

# Behavioral Responses of Diptera to Pheromones, Allomones, and Kairomones

Brian S. Fletcher

*CSIRO Division of Entomology, University of Sydney,*
*N.S.W., Australia*

Insect-produced chemicals that are known to modify behavior in Diptera include primer and releaser pheromones, male accessory gland secretions (which may themselves be considered as a special class of pheromones), kairomones, and overcrowding factors, which act not only intraspecifically but also function as allomones for related species.

Until now, however, very few attempts have been made to analyze the precise roles these compounds play in controlling the behavior of the species concerned. Even studies on pheromones have not had the same impetus in Diptera as they have had in some of the other economically important groups—notably Lepidoptera, Coleoptera, and Hymenoptera—with the result that very few dipteran pheromones have been characterized and identified. This in turn has held back studies on the role the components of the pheromone play in controlling the hierarchy of behavioral steps involved in a response, particularly as most of the pheromones seem to be fairly complex blends of chemicals. Moreover, even when a species uses a pheromone, its

129

release is frequently accompanied by other types of signaling (e.g., auditory or visual) or it may facilitate some aspect of behavior (e.g., orientation to a visual stimulus) that is also performed readily in the absence of a pheromonal stimulus.

What follows is a summary of the present available information on the behavioral responses of Diptera to pheromones and other insect-produced chemicals, and those situations that appear to have prospects of application in control programs are discussed briefly. It should be emphasized, however, that much of the information available at present is based on preliminary laboratory studies, and both quantitative laboratory experiments and detailed field investigations are urgently needed.

## I. THE ROLES OF PHEROMONES IN SEXUAL BEHAVIOR

For mating to occur, one sex has to locate a receptive member of the opposite sex. In Diptera, this process of pair formation frequently occurs either at a visual marker, which is usually a conspicuous feature of the habitat (e.g., hill top, treetop, dung pad, or animal) where the males either swarm or sit waiting for the females to approach in response to the visual stimulus of the marker, or at an emergence, feeding, or oviposition site where the two sexes initially aggregate in response to nonsexual stimuli. Because of this, the importance of pheromones for intraspecific communication during mating varies considerably from group to group and even between members of the same family, depending upon their particular mating habits. In a few species they are used as attractants, facilitating orientation of one sex to the other from a distance. Frequently, however, they play a role in sexual behavior only at close range or when the two sexes come into contact.

### B. Orientation from a Distance

The use of pheromones as distance attractants for the opposite sex is most frequently found in those species that have developed specialized mating habits. In Nematocera, for example, they are used by the females of those species in which the swarming habit has been replaced by mating at emergence, as in *Deinocerites cancer* (1) and *Culiseta inornata* (2), although some evidence from olfactometer experiments indicates (3) that males of some species of *Culex* mosquitoes produce a pheromone when swarming that, at high concentrations, acts as a short-range attractant for females. *Microdon cothurnatus,* the only syrphid that is known to use an attractant pheromone, is also a species that mates on emergence (4).

In *Drosophila,* distance attractants occur in those species that do not mate on the food mass, but instead on surrounding vegetation—for example, the endemic Hawaiian species (5), where the males form aggregations and release pheromones to attract the females. This change in behavior was possibly due to predation pressure, as suggested by Speith (5); similar selection pressures may have been responsible for the development of attractant pheromones in many of the tephritid fruit flies, particularly the tropical and subtropical species.

The species of Diptera that have been shown to produce pheromones that facilitate orientation of the opposite sex are listed in Table 1. In some species it is the female that attracts the male—for example, *Culiseta inornata* (2), *Hippelates collusor* (6), and *Dacus oleae* (7)—but in others it is the male which attracts the female—for example, *Dacus tryoni* (8), *Ceratitis capitata* (9), and *Anastrepha suspensa* (10). In most cases the studies were carried out in laboratory olfactometer tests where the distances involved were in the order of only a few centimeters, so at present it is not possible to distinguish between species that use pheromones for long-range orientation and those that use them only for short-range orientation. None of the species appear to use pheromones that are effective over the comparatively large distances reported in some Lepidoptera and Coleoptera.

The mechanisms that Diptera use to orientate to a pheromone source have not been investigated, but in view of the work by Wright (11) (see also Kennedy, Chapter 5) on the responses of *Drosophila* to food odors plus the current theories about orientation to pheromone sources in Lepidop-

TABLE I. Species of Diptera in which a Pheromone
is Used as a Sex Attractant

| Released by the Female | Reference |
| --- | --- |
| *Culiseta inornata* | 2 |
| *Culex pipiens* | 3 |
| *C. tarsalis* | 3 |
| *C. quinquefasciatus* | 3 |
| *Deinocerties cancer* | 1 |
| *Microdon cothurnatus* | 4 |
| *Musca domestica* | 20, 25, 26, 28-31 |
| *M. autumnalis* | 18, 25 |
| *Hippelates collusor* | 6 |
| *Dacus oleae* | 7, 77 |
| Released by the Male | |
| *Ceratitis capitata* | 9, 21 |
| *Dacus tryoni* | 8, 14, 17 |
| *Anastrepha suspensa* | 10, 24, 33, 75 |
| *Rioxia pornia* | 23 |
| Hawaiian *Drosophilia* | 5 |

tera (12) (13), it is reasonable to suppose that pheromone-stimulated anemotaxis plays an important role.

Most of the quantitative data that are at present available on the behavioral responses of Diptera to sex pheromones come from studies on *Dacus tryoni,* a species in which the male produces a multicomponent pheromone (14).

The pheromone aids both long- and short-range orientation of the female to the male and also induces sexual excitation, depending upon the concentration and possibly the composition of the pheromone the female experiences. In large field cages at dusk, the normal time of mating, mature virgin females will orientate upwind to pieces of filter paper impregnated with the pheromone. On reaching the filter paper, they land, locate the area containing a high concentration of pheromone, and probe this with their ovipositor. This probing response appears to be a form of pseudocopulatory behavior elicited by a high concentration of pheromone, which causes the female to extend her ovipositor and direct it toward the pheromone source. The response thus facilitates mating, as the tip of the male abdomen would normally be covered by the greatest amount of pheromone.

The field observations were supported by laboratory bioassays (16) in which three of the responses shown by females—that is, orientation to the pheromone source and, on arrival, extension of the mouthparts and probing with the ovipositor—were measured separately when females were exposed to concentrations of pheromone ranging between $10^{-1}$ and $10^{-4}$ male equivalents at source. There was a linear relationship between the probit of percent response and concentration for each kind of behavior but orientation of a given percentage of females to the source required a lower concentration of pheromone than that required to produce extension of the mouthparts or probing with the ovipositor. This situation, therefore, in which increasing concentrations of pheromone stimulate a hierarchy of responses, is similar to that observed in some Lepidoptera (15).

In *D. tryoni* (16) the effect of prior exposure of the females to the pheromone on their subsequent response does not appear to be so marked as that observed in Lepidoptera (see Bartell, Chapter 12). One hour or more before dusk, exposure to a relatively high concentration of pheromone (i.e., several male equivalents) was required to reduce significantly the response shown at dusk, the normal time of mating. Closer to the onset of dusk, the effects of the initial exposure to pheromone were greater, particularly when the initial exposure was carried out under a low light intensity similar to that the flies experience at dusk, rather than the higher light intensity the flies normally experience at that time in their cycle.

Both their physiologic condition and various environmental factors also strongly influence the response of females to the pheromone (17). The females do not start to become responsive to the pheromone until they reach sexual maturity, which normally coincides with the completion of vitellogenesis in their ovaries. Similar types of relationship between either response to or release of a sex pheromone and the attainment of sexual maturity have been found in other female Diptera, including *Hippelates collusor* (6), *Anastrepha suspensa* (10), *Dacus oleae* (7), and *Musca autumnalis* (18), as well as in other insect species.

Sexually mature *D. tryoni* also show a strong diurnal rhythm in their response to the pheromone, and under normal conditions respond only during dusk. As in the case of mating (19), this is because of the interaction of light intensity and a circadian rhythm of responsiveness to the pheromone. The optimal light intensity for response to the pheromone is just below 10 lux and the response decreases sharply at intensities above or below this value (17). For most of the 24-hr cycle, however, the females are unresponsive to the pheromone even under the optimal light intensity because of their internal rhythm. This interaction effectively synchronizes the response under normal light-dark cycles to the dusk period.

Mating also causes females of *D. tryoni* to become unresponsive to the male pheromone, at least for several weeks (17). Similarly, in *D. oleae* (7), a related species in which it is the female that produces the principal sex pheromone, mating stops the production or release of the pheromone for a time.

## B. Close-Range Orientation and Sexual Stimulation

It might be expected that species that use pheromones for long-range orientation would also use the same or different compounds for short-range orientation and sexual stimulation. In *Musca domestica* the female sex pheromone is a single compound (20) and this affects both orientation and sexual stimulation of males. The primary function of the housefly pheromone, however, seems to be sexual stimulation of the males, and its orientating role is only of secondary importance, so it is perhaps not surprising to find that a single compound is involved.

In the few other species for which there is any information—and these are all species in which the pheromone is released by males—long-range orientation appears to be in response to a different blend of chemicals in the pheromone than that which will stimulate short-range orientation and sexual excitation—for example, *Ceratitis capitata* (21) and *Dacus tryoni* (22). In another species of Australian tephritid fruitfly, *Rioxia pornia* (23), the male uses a volatile pheromone to attract females but stops releasing it once one arrives in his vicinity. The male then produces a mound of foam from his proboscis on which the female is induced to feed, thereby immobilizing her and enabling copulation to be initiated. The male does not produce a large foam mass at the beginning, in case the female should prove unreceptive or otherwise indisposed. Instead, he initially produces a small mound and then breaks off copulation to produce a larger mound when the female has consumed the first and starts to get restless.

The use of multicomponent pheromones and feeding lures might be related to the special problems that males face, because when a male has attracted a female to him he has to switch from the relatively passive role of pheromone-release to the active role of mounting the female and initiating copulation. At the same time, the female has to be prevented from taking flight, and brought to a level of sexual excitement where she will accept the male. However, it is possible that for both males and females short-range pheromonal communication might be best effected by different compounds from those used for distance communication.

Most of the species that produce sex pheromones that are active over a distance synthesize them in specialized glands and release them at fairly specific times of day—for example, *Ceratitis capitata* (9), *Dacus tryoni* (8) (14) and *Anastrepha suspensa* (10) (24). In contrast, the species that produce only short-range or contact pheromones seem to incorporate them into the cuticular waxes so that when they reach maturity they have a general body odor that is more or less continuously stimulating to the opposite sex.

*Musca domestica* (25) (26) and *Musca autumnalis* (27), for example, produce species-specific volatile compounds that influence the behavior of males. The active compound in *Musca domestica* can be isolated from the cuticular waxes and feces of both virgin and mated females from 1 to 2 days after emergence (28) (29); it has been identified as a C23 monoolefin, (Z)-9-tricosene, commonly called muscalure (20). In tests with pseudoflies made from knots of black shoestring, both female extracts and the pheromone caused not only increased sexual excitement in the males with more attempted copulations toward each other, but also an increased number of copulations directed toward the treated pseudoflies (25) (30). In olfactometer tests, musalure attracted up to 18% of the males compared with 4% for the blank controls, thus possessing the same attractant properties as crude extracts (20).

That males seem to respond equally well to several compounds related to musalure (31), suggests that the structural requirements of the molecule are less specific than for most pheromones in other groups. As there is also a good correlation between the activity of the compounds as attractants in olfactometer tests and stimulants in pseudofly tests, it seems that the receptor sites controlling both behavioral responses are the same.

During field evaluation in large buildings containing large housefly populations, in which a number of different traps including sticky panels, sugar-baited pans, and electric strips were baited with quantities of musalure ranging from 0.5 to 100 mg, the catches of both males and females were increased (32). This was in marked contrast to the laboratory studies in which females showed no response. In the field, therefore, muscalure may act as an aggregation compound for both sexes. This would be a similar situation to that found with the dipteran sex pheromones of *Ceratitis capitata* (Chapter 19) and *Anastrepha suspensa* (33).

In some other species, notably those that form mixed aggregations at feeding or oviposition sites, one sex, usually the female, releases a volatile pheromone that does not play any direct role in orientation but instead causes the responding sex to make sexual advances to nearby objects with suitable visual characteristics.

In many species of *Drosophila*, which mate on their food masses, the males do not seem to be able to distinguish visually females from males, or even from individuals of other species, and they approach other flies fairly randomly. However, studies on *D. melanogaster* (34) and on *D. pseudoobscura* (35) indicate that a volatile pheromone released by females stimulates the rate of orientation of males toward other individuals. Even in the presence of the female pheromone, however, they are unable to identify the sex of another fly until they actually make contact with it, usually with the tarsi or mouthparts, and at this stage additional cues, probably in the form of chemotactile or gustatory pheromones, are obtained.

Also in *Lucilia cuprina* (36), both liver-fed and nonliver-fed mature virgin females release a volatile pheromone that increases the sexual excitement and copulatory attempts of males. As in *Drosophila,* but unlike the housefly, there does not appear to be any sex selective orientation component in the pheromone stimulus, and in experiments where a female was enclosed with the males there was no significant difference in the number of copulatory attempts directed toward either sex. In another calliphorid, the screw-worm fly, *Cochliomyia hominivorax,* it is the male that releases a pheromone of this type (37) and the mature virgin females that respond. At low concentrations the females show an increase in wing fluttering and preening, and at higher concentrations searching behavior and sexual strikes toward each other, similar to those performed by males in many species of Diptera.

## C. Arrestants of Locomotion

A different type of pheromone is employed by some of the temperate species of tephritid fruitfly, which use part of the host plant (e.g., the fruit or flower-heads) as a rendezvous site for the two sexes. Either the males or females that visit this site deposit a pheromone there that acts as a locomotory arrestant for the other sex, or sometimes both sexes, thereby increasing the chance of sexual encounters. In *Rhagoletis pomonella* (38) the pheromone is deposited by mature females onto fruit they visit, and this causes arriving males to remain 2 to 5 times longer on that fruit than on unmarked fruit. A similar pheromone is also used by females of the European cherry fly, *Rhagoletis cerasi* (39). In two other species, *Rhagoletis completa* and *Urophora cardui* (39), the males mark the larval foodplant with a pheromone that acts as a locomotory arrestant for the females. These pheromones do not appear to increase orientation to the rendezvous sites and are probably nonvolatile, being perceived on contact through receptors on the tarsi.

### D. Contact Recognition

It seems likely that many of the species of Diptera that produce volatile sex pheromones that facilitate pair formation also use contact pheromones for final recognition of the opposite sex, as in *Drosophila*. In some species, however, the use of contact pheromones for sexual recognition is the only situation in which sex pheromones are used during mating.

Thus it has been demonstrated in *Aedes* mosquitoes belonging to the subgenus *Stegomyia* (40), where orientation of the male to the female is in response to sound, and the tsetse fly, *Glossina morsitans* (41), where vision plays the major role in orientation, that final recognition of the female by the male depends on contact pheromones. Although these sex-recognition pheromones do not appear to play any role in mating behavior prior to contact, studies on *Aedes* suggest that they are capable of stimulating searching behavior in the male if initial contact has been made and then lost again. Males that came into contact with females but were inhibited from mating, adopted a characteristic figure-of-eight search pattern and remained flying for much longer than males stimulated only by the sound of the female. Even when males were allowed to mate normally they subsequently remained active for longer periods than males that did not come into contact with females.

### E. Strain Recognition

Although in *Drosophila* it is the male that initiates courtship, it is the female that determines which partner she will accept, and she is frequently courted by several males before mating occurs. One result of the discrimination exercised by the females is the *rare-male effect*. This is a tendency of females, when given a choice, to mate with males from the strain with the lowest frequency in the population. This subtle type of discrimination between sexual partners has now been demonstrated to be due principally to pheromone differences between strains. Indeed, *D. melanogaster* females are apparently able to distinguish not only conspecific individuals from different strains but also individuals from different cultures, possibly by means of colony-specific odors (42).

In experiments directly related to mate selection, using the two strains, chiricahua (CH) and arrowhead (AR) of the species *D. obscura* (43), with ratios of males sufficiently different from 1:1, the males of the strain in least abundance were favored as mates by females of both strains. However, when equal numbers of CH and AR females were exposed to an ether extract of CH males in a situation where the ratio of AR:CH males was 1:1, a mating advantage occurred in favor of AR males. When the AR males outnumbered the CH males by a ratio 3:1 (thus theoretically giving the CH males an advantage) the addition of extracts of CH males removed the rare-male advantage. The mechanism involved is not known, but possibly it is related to the more rapid sensory adaptation that is likely to occur to the pheromones present in highest concentration—that is, those of the strain in greatest abundance.

This rare-male effect is thought to have evolutionary significance because it provides a mechanism whereby a species can maintain an heterogenous gene pool within a population. This phenomenon, therefore, may be of much wider occurrence among insect species than is at present realized. Possibly some of the minor components of pheromone glands of some species that, at the moment, have no known function may be involved in this effect.

### F. Swarm Cohesion

In at least one species of Nematocera, *Chironomus riparius* (44), where the males swarm over a marker, there is strong circumstantial evidence that pheromones may play a part in swarm cohesion. When fresh air was passed across a swarm in a cage, swarming stopped, whereas when the air that had passed across the swarm was recirculated, the swarming continued.

### G. Repellents

In a few species of Diptera—for example, certain Hawaiian *Drosophila* (5) and the eye gnat, *Hippelates collusor* (6)—unreceptive females appear to produce pheromones repellent to the males and that, at least in some of the *Drosophila*, may actually cause physical discomfort at close range.

## II. THE ROLE OF PHEROMONES IN OVIPOSITION BEHAVIOR

### A. Attractants and Stimulants

In some species of mosquitoes, pheromones produced by the developing stages influence the ovipositional behavior of gravid females either by attracting them to favorable oviposition sites or stimulating egg-laying on arrival. In *Culex tarsalis* two quite distinct pheromones appear to be involved.

The first is a nonvolatile compound that is released into breeding waters by fourth instar larvae, pupae, and emerging adults, and that exerts an effect only when the females come into contact with it; thus it acts as an oviposition stimulant rather than as an attractant (45). The second is a volatile compound associated with the egg rafts, which acts as an attractant rather than a stimulant because the actual proportion of the attracted females that lay eggs is no different from when they lay into distilled water (46). In *Aedes aegypti* (47) gravid females are attracted to water containing immature larvae of their own species—but not to water from which the larvae had been removed—by a volatile pheromone released by the larvae. In *Aedes atropalpus* (48), on the other hand, the larvae and pupae produce a nonvolatile pheromone that is released into the water, as in *Culex tarsalis*.

Gravid females of *Lucilia cuprina* (49) frequently oviposit in groups, and this behavior too has been shown to be mediated by a female-produced pheromone. It appears to have a weak attractant effect, which helps to create the initial aggregation, and a strong chemotactile effect, which stimulates the females to lay eggs in close proximity to each other.

## B. Ovipositional Deterrents

In several species of tephritid flies, including *Rhagoletis pomonella* (50), *R. completa* (39) and *R. fausta* (51), after a female oviposits in a piece of fruit she deposits a marking pheromone on the surface, which deters other females from ovipositing in that fruit. The amount of deterrence produced by one female is related both to the size of the fruit and to the drive level of females subsequently visiting the fruit. On larger fruits or when drive level is high, due to a shortage of fruit, a greater number of repeated ovipositions occurs. These pheromones do not appear to have any repellant effect on flies orienting to the fruit, but exert their effect once the fly actually lands on the fruit. The way they deter oviposition behavior is not known, but it has been shown in *Rhagoletis pomonella* and *R. fausta* that the pheromones do not act interspecifically (51).

## III. OVERCROWDING FACTORS AND FEEDING STIMULANTS IN MOSQUITOES

Studies on several species of mosquitoes have indicated that in overcrowded conditions the older larvae release toxic substances detrimental to younger larvae of both their own and other species. For example, third instar larvae of *Culex quinquefasciatus* raised under overcrowded conditions (5-7 larvae/ml) produce chemical factors that are highly toxic to first instar larvae of their own species and also to first instar larvae of *Culex tarsalis, Aedes aegypti,* and *Anopheles albimanus* (52). Whereas untreated larvae showed a steady growth rate and reached maturity in 7 days, treated larvae had a much slower growth rate and a much higher mortality, the effect being proportional to the dosage of overcrowding (O.C.) factor used. One unit of O.C. (i.e., the material obtained from 1500-2000 third instar larvae cultured for 4-5 days in 300 ml of water) added to 300 ml of water containing 100 test larvae caused first instar larvae to take 14 days to reach maturity and also resulted in an extremely high mortality.

Part of the effect is apparently due to the bacteriocidal properties of the overcrowding factor (53) as this causes partial starvation of the larvae, which rely on the bacteria in the water for some essential growth factors. However, some of the compounds—and the major ones have now been identified (54)— have a direct effect. One effect is interference with the secretion of the cuticle (55), making it more permeable to water and the larvae thus more susceptible to drowning.

Besides their direct effect upon larval development and survival, these substances play other important roles in the natural regulation of mosquito populations by affecting the physiology and behavior of the adults produced by surviving individuals. The average number of eggs laid in each egg raft by females of *C. quinquefasciatus* that developed from first instar larvae treated with 0.25 or 0.5 units of O.C., or from second and third instar larvae treated with 1 unit of O.C., was significantly less than for the controls. Furthermore, even after only temporary overcrowding of the larvae of several species of mosquitoes (56) the resulting adults had lower wing loadings (i.e., ratio of dry body to the cube of wing length) and this was considered to give them a greater potential for dispersal.

Under both crowded and uncrowded conditions the larvae of *Culex pipiens* (57) release a feeding stimulant into the water that stimulates both themselves and other larvae to feed more intensely. The intensive feeding presumably increases development rates and possibly also synchronizes larval development to some extent. The effect, therefore, is opposite, in some ways, to that caused by the overcrowding factors of some other species; it might be significant that although *Culex pipiens* does not itself appear to produce an overcrowding factor, it is susceptible to those produced by other mosquito species (58).

## IV. THE ROLE OF ACCESSORY SUBSTANCES IN CONTROLLING BEHAVIOR

It has now been demonstrated that the males of a number of Diptera, including various mosquitoes (59), *Musca domestica* (60), *Drosophila melanogaster* (61), *D. funebris* (62), and *Dacus tryoni* (16), transfer material from their accessory glands to the female during mating. This makes the females unreceptive to further matings, at least for a time, and in most species also stimulates oviposition (63) (64).

In *Aedes aegypti* (65) the accessory material, matrone, contains two separate proteinaceous fractions, designated α- and β-proteins. Both fractions are required to prevent insemination (66), but stimulation of oviposition is due entirely to the α-protein (67). In *Drosophila funebris* (62) the accessory material also contains two major components, but in this case one of them, PS-1 (a peptide composed of 27 amino acid residues), is solely responsible for switching off receptivity, and the other, PS-2 (probably a glycine derivative), is alone responsible for stimulating oviposition.

In contrast, in *Musca domestica* (68) at least seven distinct proteins are transferred to the female in the accessory material, which in this species originates from the male's ejaculatory duct. At present nothing is known about the roles the individual proteins subsequently play. The accessory material enters the vaginal pouches where it causes cytolysis of the cells, and from there it is transported rapidly into the hemocoele. It is thought that the active factors are then carried to the brain, where they may bind to specific receptor sites.

The switch-off of receptivity occurs immediately after the first mating, and the females are frequently monogamous for life. The effect does, however, depend to a certain extent on the quantity of accessory material passed to the female, because females made unreceptive by interrupted matings, or matings with young or depleted males, remated after an initial refractory period more frequently than females that mated normally. In mosquitoes, although the mated females are immediately refractory to a second mating, the initial effect does not appear to be due to matrone, which requires several hours, to take effect but possibly to the filling of the bursa with fluid (69).

Associated with the switch-off of receptivity, the females show behavioral changes that prevent mates from inseminating them. In the housefly, extension of the ovipositor in response to the male's courtship is inhibited (64), and in mosquitoes changes in behavior such as lowering of the abdomen and nonrelaxation of the vaginal valves occur (70). In the Amherst strain of *D. melanogaster* (61) virgins use kicking and fending off with the legs as the principal method of male rejection, whereas mated females more frequently used extrusion of the ovipositor. Accessory gland implants caused changes in the rejection responses of virgin females, resulting in decreased rates of kicking and fending off. There was no corresponding increase in ovipositor extrusion, but it was suggested that this requires additional nervous input from the ovaries or abdominal stretch receptors.

Even greater changes in behavior can be seen after mating or injection of accessory material in those species in which the females normally take a more active role in mating. In *Dacus tryoni,* for instance, mature virgin females show a peak of locomotory activity at dusk (their normal time of mating) when they also respond strongly to the male-produced sex pheromone, whereas recently mated females are not active at dusk and do not respond to the pheromone (17). Injection of accessory material into virgin females not only makes them unreceptive to males so that they try and kick them off if they attempt copulation, but also causes the other changes in behavior typical of mated females (16). In the field, of course, this would ensure that unreceptive mated females seldom come into contact with sexually active males.

## V. RESPONSES TO KAIROMONES

Studies on the tachinid fly, *Drino bohemica* (71), which parasitizes several species of sawfly larvae that feed on different plants, indicated that olfactory responses to the foodplants of the host larvae and olfactory and chemotactile responses to the larvae themselves were involved in host selection and detection by the parasite. Although there is very little information available, a similar situation would seem likely for many of the other dipteran species that parasitize or prey on insect stages associated with particular plants. In such cases the olfactory response to the host may be stimulated by chemicals (e.g., metabolic and excretory products) that the prey species does not use as pheromones for intraspecific communication. There are, however, one or two known examples of dipteran predators that actually orientate to the mating pheromones of the prey species.

The brachyceran fly, *Medetera bistriata,* is a predator during the larval stage on the young larvae of certain bark beetles. The adult flies arrive, mate, and place their eggs in the beetle galleries as soon as these are initiated. In field tests (72) it was found that *M. bistriata* responded equally well to pine billets containing *Ips grandicollis* males and *D. frontalis* females but not to uninfested billets. This behavior suggests that the pheromones the beetles used were important for attraction.

This suggestion was confirmed in experiments in which synthetic samples of the pheromones of *D. frontalis* were exposed along with $\alpha$-pinene, the major host-tree monoterpene, in sticky traps. As in the case of infested pine billets, both sexes of *M. bistriata* responded, but the females far outnumbered the males. As mating occurs at the infestation sites, the flies must rely on the bark beetle pheromones, not only as kairomones for detection of their prey but also for bringing the two sexes together.

Another species that uses the sex pheromone of its insect host as a kairomone is the tachinid fly, *Trichpoda pennipes,* a larval parasite of the green stinkbug, *Nezara viridula* (73). Both sexes were strongly attracted to unmated males (the sex that produces the pheromone) but showed little response to nymphs, females, and mating pairs. Approximately seven times as many females as males were attracted, and the females had already mated. The females therefore seem to use the stinkbug pheromone principally as a kairomone to locate their prey for oviposition, whereas the males respond because it increases their chances of meeting a female. Females were attracted to the male stinkbugs even when these were placed away from their normal habitat. In this case the response of the flies seems independent of stimuli from the host plants. However, although in natural populations of stinkbugs more males than females are parasitized, the proportion of parasitized females is fairly large; this suggests that the tachinid fly frequently uses other cues besides the kairomone for locating its host.

## VI. THE POSSIBLE USES OF INSECT-PRODUCED CHEMICALS TO CONTROL DIPTERA

Studies on Diptera have revealed a wide array of situations in which pheromones are used to modify behavior. One is not restricted, therefore, to population monitoring, male annihilation, or male confusion as possible ways of using dipteran pheromones in control programs, as is the case in Lepidoptera. Moreover, in some instances (e.g., the oviposition-deterrent pheromones of fruitflies and the overcrowding factors of mosquitoes), the pheromones play a natural role in population regulation and thus may well prove to be better candidates for use in control programs than pheromones that mediate other aspects of behavior.

Until the present, however, only one dipteran pheromone, muscalure, has been evaluated as a possible control agent (32) (74), and only two other pheromones have been characterized (21) (54). As both of these proved to have a number of components, and studies in progress suggest this is also true of other dipteran pheromones—for example, *A. suspensa* (75), *D. tryoni* (22), and *D. oleae* (76)—the formulation and dispensing of the majority of dipteran pheromones may present greater problems than those of Lepidoptera or Coleoptera. However, it is still too early to know whether this really will be the case, and the exciting possibilities that some of the pheromones present for the elimination or reduction of insecticidal control merit a concerted effort to bring them to the stage of field evaluation.

### A. Fruitfly Pheromones

Many of the tephritid sex pheromones that are used to attract the opposite sex may prove to be of use in control programs. Studies with the pheromones produced by males of the *C. capitata* (Chapter 19) and *Anastrepha suspensa* (33) have indicated that in the field they not only attract females but also males. The female-produced pheromone of *Dacus oleae* (77) may also be effective in the field, although only males are attracted. The discovery that these pheromones can attract adults in field conditions is very encouraging because at present there are no powerful synthetic attractants available for *A. suspensa* and *D. oleae*. Also, trimedlure, the best-known attractant for *C. capitata,* is much lower in efficiency than methyl eugenol for *D. dorsalis* or cuelure for *D. cucurbitae* and *D. tryoni* (see Chapter 19 for a detailed account of fruitfly attractants). Even in *D. tryoni,* field cage studies (78) suggest that the male pheromone is a better attractant for females than cuelure.

It seems unlikely, however, that the fruitfly sex pheromones could be effectively employed in a confusion-type program, not only because the distance over which they are effective appears limited but also because in most species stridulation or signalling with the wings plays a role in sexual communication. Therefore, pair formation probably does not depend so heavily on orientation to the pheromone as it does in Lepidoptera. Instead, incorporation of the pheromones into integrated control procedures seems at present to be the most rewarding approach.

Thus the olive fly pheromone (7) (77) could possibly be used to trap out most of the males in an area before the release of sterile males, a technique that is being developed to control this species. The pheromone of *D. tryoni* and maybe *D. dorsalis* and *D. cucurbitae* could feasibly be used to supplement the male lures of these species to suppress populations, particularly when the male population gets low and the majority of females are unmated and, therefore, likely to respond readily to the pheromone.

It is possible that the pheromones of the species that attract both sexes might prove effective, when combined with a suitable insecticide and distributed in dispensers or traps, in significantly reducing infestation levels. Because they attract females even in mixed populations, they could prove to be more useful than male lures in this type of control, as the latter have very little effect on female fertility or numbers until the male population has been reduced to a very low level. The male lures only work effectively on isolated populations where there is little chance of individuals migrating into the area. Even with pheromones, however, there are potential problems because, as experiments with *D. tryoni* have demonstrated (17), the females have a refractory period after mating when they are not responsive to the pheromone. Thus, fertilized females that migrated into an area would not be affected by control procedures that relied upon the pheromone. However, the powers of dispersal (39) and the mating frequency (39) of tephritids varies between species, so what may represent a barrier against the use of pheromones for control in one species may not apply in other cases. This highlights the need for a thorough investigation of the ecology and behavior of a species before attempts are made to control it by the very selective methods that are now being developed.

Another type of pheromone that appears to have exciting possibilities for application in control programs is the oviposition deterrent that is used by some of the *Rhagoletis* species (39). This seems to be tailor-made for the situation because it has been evolved naturally to regulate oviposition. As pointed out by Prokopy (50), the best method of using a deterrent pheromone for control would not be to spray the whole crop with it, as this would only cause the flies to disperse to other areas or delay the females from stinging the fruit until their drive levels become high enough to overcome the deterrent effect of the pheromone. Instead, the best approach would be to spray most of the trees with oviposition deterrent but leave the remainder—say, 10%—unprotected. The females should then congregate on the unprotected trees and there they can either be killed by insecticide treatments or caught on the visual traps, which are highly effective for these species.

## B. Mosquito Pheromones

The oviposition pheromones and overcrowding factors of mosquitoes may also prove useful in manipulating the populations of the pest species. It might be possible, for instance, to use the oviposition pheromones to direct the egg-laying of a number of mosquito species in specific pools where the immature stages could be killed by insecticides or possibly even with overcrowding factors. Alternatively, already-infested pools could be treated with high enough levels of overcrowding factor to cause practically total mortality of the immature larvae, a technique that is enhanced by the simultaneous effectiveness of the overcrowding factor of any particular species against the larvae of a number of other mosquito species. As in the case of the fruitfly oviposition deterrents, these compounds have been evolved by the insects themselves to play a role in population regulation. Therefore, one may encounter situations in which there may be fewer pitfalls when employing them for control purposes than when pheromones are used to disrupt a natural process.

## C. Other Possibilities

There are several other potential ways of using pheromones or related compounds for controlling Diptera. As has already been successfully demonstrated with muscalure (32), short-range sex pheromones, which may not by themselves be very effective, can be combined with already-existing trapping techniques to improve the traps's efficiency. Individuals used in sterile-male programs might be treated with pheromone to increase their attractiveness to the opposite sex, and thereby make them more competitive with the wild males. If problems of overcoming their introduction into the females could be solved, male accessory substances might be used to prevent many species from mating.

No doubt, as studies on Diptera continue, new pheromones will be discovered and other methods of using them in control programs will be developed.

## REFERENCES

(1) J.A. Downes. 1966. Observations on the mating behaviour of the crab hole mosquito *Deinocerites cancer* (Diptera: Culicidae). *Can. Entomol.* **98**:1169-1177.

(2) J.W. Kliewer, T. Miura, R.C. Husbands, and C.H. Hurst. 1966. Sex pheromone and mating behaviour of *Culiseta inornata* (Diptera: Culicidae). *Ann. Entomol. Soc. Am.* **59**:530-533.

(3) C.M. Gjullin, T.L. Whitfield, and J.F. Buckley. 1967. Male pheromones of *Culex quinquefasciatus, C. tarsalis* and *C. pipiens* that attract females of these species. *Mosquito News* **27**:382-387.

(4) R.D. Akre, G. Alpert, and T. Alpert. 1973. Life cycle and behaviour of *Microdon cothurnatus* in Washington (Diptera: Syrphidae). *J. Kansas Entomol. Soc.* **46**:327-338.

(5) H.T. Spieth. 1974. Courtship behaviour in *Drosophila. Ann. Rev. Entomol.* **19**:385-405.

(6) T.S. Adams and M.S. Mulla. 1968. Ovarian development, pheromone production and mating in the eye gnat, *Hippelates collusor. J. Insect Physiol.* **14**:627-635.

(7) G.E. Haniotakis. 1974. Sexual attraction in the olive fruit fly *Dacus oleae* (Gmelin). *Environ. Entomol.* **3**:82-86.

(8) B.S. Fletcher. 1968. Storage and release of a sex pheromone by the Queensland fruitfly, *Dacus tryoni* (Diptera: Trypetidae). *Nature (Lond.)* **219**:631-632.

(9) M. Feron. 1962. L'instinct de reproduction chez la mouche méditerranéene des fruits *Ceratitis capitata* Wied. (Dipt. Trypetidae). Comportement sexuel—comportement de ponte. *Revue Path. vég. Entomol. agric. Fr.* **41**:1-129.

(10) J.L. Nation. 1972. Courtship behaviour and evidence for a sex attractant in the male Caribbean fruit fly, *Anastrepha suspensa. Ann. Entomol. Soc. Am.* **65**:1364-1367.

(11) R.H. Wright. 1964. *The Science of Smell.* Allen & Unwin, London.

(12) S.R. Farkas and H.H. Shorey. 1974. Mechanisms of orientation to a distant pheromone source. In *Pheromones,* M.C. Birch, Ed. North-Holland, Amsterdam.

(13) J.S. Kennedy and D. Marsh. 1974. Pheromone regulated anemotaxis in flying moths. *Science (Wash.)* **184**:999-1001.

(14) B.S. Fletcher. 1969. The structure and function of the sex pheromone glands of the male Queensland fruit fly, *Dacus tryoni. J. Insect. Physiol.* **15**:1309-1322.

(15) R.J. Bartell and H.H. Shorey. 1969. Pheromone concentrations required to elicit successive steps in the mating sequence of males of the light-brown apple moth *Epiphyas postvittana. Ann. Entomol. Soc. Am.* **62**:1206-1207.

(16) A. Giannakakis. Personal communication.

(17) B.S. Fletcher and A. Giannakakis. 1973. Factors limiting the response of females of the Queensland fruit fly, *Dacus tryoni,* to the sex pheromone of the male. *J. Insect Physiol.* **19**:1147-1155.

(18) M.F.B. Chaudhury and H.J. Ball. 1974. Effect of age and time of day on sex attraction and mating of the face fly, *Musca autumnalis. J. Insect Physiol.* **20**:2079-2085.

(19) P.H. Tychsen and B.S. Fletcher. 1971. Studies on the rhythm of mating in the Queensland fruit fly, *Dacus tryoni. J. Insect Physiol.* **17**:2139-2156.

(20) D.A. Carlson, M.S. Mayer, D.L. Silhacek, J.D. James, M. Beroza and B.A. Bierl.

1971. Sex attractant pheromone of the house fly. Isolation, identification and synthesis. *Science* (Wash.) **174**:76-77.

(21)    M. Jacobson, K. Ohinata, D.L. Chambers, W.A. Jones and M.S. Fujimoto, 1973. Insect sex attractants. 13. Isolation, identification, and synthesis of sex pheromones of the male Mediterranean fruit fly. *J. Med. Chem.* **16**:248-251.

(22)    B.S. Fletcher and T.E. Bellas. Unpublished.

(23)    G. Pritchard. 1967. Laboratory observations on the mating behaviour of the island fruit fly, *Rioxia pornia* (Diptera: Tephritidae). *J. Aust. Entomol. Soc.* **6**:127-132.

(24)    J.L. Nation. 1974. The structure and development of the two sex specific glands in male Caribbean fruit flies. *Ann. Entomol. Soc. Am.* **67**:731-734.

(25)    W.M. Rogoff, A.D. Beltz, J.O. Johnsen and F.W. Plapp. 1964. A sex pheromone in the house fly, *Musca domestica* L. *J. Insect Physiol.* **10**:239-246.

(26)    M.S. Mayer and C.W. Thaggard. 1966. Investigations of an olfactory attractant specific for males of the house fly, *Musca domestica*. *J. Insect Physiol.* **12**:891-897.

(27)    M.F.B. Chaudhury, H.J. Ball and C.M. Jones. 1972. A sex pheromone of the female face fly, *Musca autumnalis,* and its role in sexual behaviour. *Ann. Entomol. Soc. Am.* **65**:607-612.

(28)    D.L. Silhacek, M.S. Mayer, D.A. Carlson and J.D. James. 1972. Chemical classification of a male house fly attractant. *J. Insect Physiol.* **18**:43-51.

(29)    D.L. Silhacek, D.A. Carlson, M.S. Mayer and J.D. James. 1972. Composition and sex attractancy of cuticular hydrocarbons from house flies: Effects of age, sex and mating. *J. Insect Physiol.* **18**:347-354.

(30)    W.M. Rogoff, G.H. Gretz, M. Jacobson and M. Berozoa. 1973. Confirmation of (Z)-9-tricosene as a sex pheromone of the house fly. *Ann. Entomol. Soc. Am.* **66**:739-741.

(31)    D.A. Carlson, R.E. Doolittle, M. Beroza, W.M. Rogoff and G.H. Gretz. 1974. Muscalure and related compounds. I. Response of house flies in olfactometer and pseudofly tests. *J. Agric. Food Chem.* **22**:194-197.

(32)    D.A. Carlson and M. Beroza. 1973. Field evaluations of (Z)-9-tricosene, a sex attractant pheromone of the house fly. *Environ. Entomol.* **2**:555-559.

(33)    J.A. Perdomo. 1975. Sex and aggregation pheromone bioassays and mating observations of the Caribbean fruit fly *Anastrepha suspensa* (Loew) under field conditions. Ph.D. Thesis, University of Florida.

(34)    H.H. Shorey and R.J. Bartell. 1970. Role of a volatile sex pheromone in stimulating male courtship behaviour in *Drosophila melanogaster. Anim. Behav.* **18**:159-164.

(35)    C. Sloane and E.B. Speiss. 1971. Stimulation of male courtship behaviour by female "odor" in *D. pseudoobscura. Drosophila Inform.* **46**:53-56.

(36)    R.J. Bartell, H.H. Shorey and L. Barton Browne. 1969. Pheromonal stimulation of the sexual activity of males of the sheep blow fly *Lucilia cuprina* (Calliphoridae) by the female. *Anim. Behav.* **17**:576-585.

(37)    L.W. Fletcher, J.J. O'Grady, Jr., H.V. Claborn and O.H. Graham. 1966. A pheromone from male screw-worm flies. *J. Econ. Entomol.* **59**:142-143.

(38)    R.J. Prokopy and G.L. Bush. 1972. Mating behaviour in *Rhagoletis pomonella* (Diptera: Tephritidae). III. Male aggregation in response to an arrestant. *Can. Entomol.* **104**:275-283.

(39)    M.A. Bateman. 1976. Fruit flies. In *Studies on Biological Control,* V.L. Delucchi, Ed. Cambridge University Press.

(40) H.F. Nijhout and G.B. Craig, Jr. 1971. Reproductive isolation in *Stegomyia* mosquitoes. III. Evidence for a sex pheromone. *Entomol. exp. appl.* **14**:399-412.

(41) P.A. Langley, R.W. Pimley and D.A. Carlson. 1975. Sex recognition pheromone in tsetse fly *Glossina morsitans. Nature* (Lond.) **254**:51-53.

(42) D.A. Hay. 1972. Recognition by *Drosophila melanogaster* of individuals from other strains or cultures: Support for the role of olfactory cues in selective mating. *Evolution* 26:171-176.

(43) J.E. Leonard, L. Ehrman and M. Schorsch. 1974. Bioassay of a *Drosophila* pheromone influencing sexual selection. *Nature* (Lond.) **250**:261-262.

(44) A.E.R. Downe and V.G. Caspary. 1973. The swarming behaviour of *Chironomus riparius* (Diptera: Chironomidae) in the laboratory. *Can. Entomol.* **105**:165-171.

(45) A. Hudson and J. McLintock, 1967. A chemical factor that stimulates oviposition by *Culex tarsalis* Coquillet (Diptera: Culicidae). *Anim. Behav.* **15**:336-341.

(46) C.E. Osgood. 1971. An oviposition pheromone associated with the egg rafts of *Culex tarsalis. J. Econ. Entomol.* **64**:1038-1041.

(47) R.S. Soman and R. Reuben. 1970. Studies on the preference shown by ovipositing females of *Aedes aegypti* for water containing immature stages of the same species *J. Med. Entomol.* **7**:485-489.

(48) K.S.P. Kalpage and R.A. Brust. 1973. Oviposition˙attractant produced by immature *Aedes atropalpus. Environ. Entomol.* **2**:729-730.

(49) L. Barton Browne, R.J. Bartell and H.H. Shorey. 1969. Pheromone mediated behaviour leading to group oviposition in the blowfly, *Lucilia cuprina. J. Insect Physiol.* **15**:1003-1014.

(50) R.J. Prokopy. 1972. Evidence for a marking pheromone deterring repeated oviposition in apple maggot flies. *Environ. Entomol.* **3**:326-332.

(51) R.J. Prokopy. 1975. Oviposition deterring fruit marking pheromone in *Rhagoletis fausta. Environ. Entomol.* **4**:298-300.

(52) T. Ikeshoji and M.S. Mulla. 1970. Overcrowding factors of mosquito larvae. *J. Econ. Entomol.* **63**:90-96.

(53) T. Ikeshoji and M.S. Mulla. 1970. Overcrowding factors of mosquito larvae. 2. Growth-retarding and bacteriostatic effects of the overcrowding factors of mosquito larvae. *J. Econ. Entomol.* **63**:1737-1743.

(54) T. Ikeshoji and M.S. Mulla. 1974. Overcrowding factors of mosquito larvae: Isolation and chemical identification. *Environ. Entomol.* **3**:482-486.

(55) T. Ikeshoji and M.S. Mulla. 1974. Overcrowding factors of mosquito larvae: Activity of branched fatty acids against mosquito larvae. *Environ. Entomol.* **3**:487-491.

(56) J.K. Nayar and D.M. Sauerman, Jr. 1970. A comparative study of growth and development in Florida mosquitoes. Part 3. Effects of temporary overcrowding on larval aggregation formation, pupal ecdysis and adult characteristics at emergence. *J. Med. Entomol.* **7**:521-528.

(57) R.H. Dadd. 1973. Autophagostimulation by mosquito larvae. *Entomol. exp. appl.* **16**:295-300.

(58) T.M. Peters, B.I. Chevone and R.A. Callahan. 1969. Interaction between *Aedes aegypti* (L.) and *Culex pipiens* L. in mixed experimental populations. *Mosquito News* 29:435-438.

(59) G.B. Craig, Jr. 1967. Mosquitoes: Female monogamy induced by male accessory gland substance. *Science* (Wash.) **156**:1499-1501.

(60) J.C. Riemann, D.J. Moen and B.J. Thorson. 1967. Female monogamy and its control in house flies. *J. Insect Physiol.* **13**:407-418.

(61) B. Burnet, K. Connolly, M. Kearney and R. Cook. 1973. Effects of male paragonial gland secretion on sexual receptivity and courtship behaviour of female *Drosophila melanogaster. J. Insect Physiol.* **19**:2421-2431.

(62) H. Baumann. 1974. Biological effects of paragonial substances PS-1 and PS-2 in females of *Drosophila funebris. J. Insect Physiol.* **20**:2347-2362.

(63) M.G. Leahy and J.B. Craig, Jr. 1965. Male accessory gland substance as a stimulant for oviposition in *Aedes aegypti* and *A. albopictus. Mosquito News* **25**:448-452.

(64) J.G. Riemann and B.J. Thorson. 1969. Effect of male accessory material on oviposition and mating by female house flies. *Ann. Entomol. Soc. Am.* **62**:828-834.

(65) M.S. Fuchs and E.A. Hiss. 1970. The partial purification and separation of the protein components of matrone from *Aedes aegypti. J. Insect Physiol.* **16**:931-939.

(66) M.S. Fuchs, G.B. Craig, Jr., and E.A. Hiss. 1968. The biochemical basis of female monogamy in mosquitoes. I. Extraction of the active principle from *Aedes aegypti. Life Sci.* **73**:835-839.

(67) E.A. Hiss and M.S. Fuchs. 1972. The effect of matrone on oviposition in the mosquito *Aedes aegypti. J. Insect Physiol.* **18**:2217-2227.

(68) R.A. Leopold, A.C. Terranova, B.J. Thorson and M.E. Dregrugillier. 1971. The biosynthesis of the male house fly accessory secretion and its fate in the mated female. *J. Insect Physiol.* **17**:987-1003.

(69) R.W. Gwadz and G.B. Craig, Jr. 1970. Female polygamy due to inadequate semen transfer in *Aedes aegypti. Mosquito News* **30**:354-360.

(70) R.W. Gwadz, G.B. Craig, Jr. and W.A. Hickey. 1971. Female sexual behaviour as the mechanism rendering *Aedes aegypti* refactory to insemination. *Biol. Bull. Woods Hole* **140**:201-214.

(71) L.G. Monteith. 1955. Host preferences of *Drino bohemica* (Diptera: Tachinidae), with particular reference to olfactory responses. *Can. Entomol.* **87**:509-529.

(72) D.L. Williamson. 1971. Olfactory discernment of prey by *Medetera bistriata* (Diptera: Dolichopidae). *Ann. Entomol. Soc. Am.* **64**:586-589.

(73) W.C. Mitchell and R.F.L. Mau. 1971. Response of the female southern green stink bug and its parasite, *Trichopoda pennipes* to male stink bug pheromones. *J. Econ. Entomol.* **64**:856-859.

(74) C. Djerassi, C. Shih-Colemann and J. Diekman. 1974. Insect control, of the future: Operational and policy aspects. *Science* **(Wash.)** **186**:596-607.

(75) J.L. Nation. 1975. The sex pheromone blend of Caribbean fruit fly males: Isolation, biological activity and partial chemical characterization. *Environ. Entomol.* **4**:27-30.

(76) G.E. Hanniotakis. Personal communication.

(77) G.E. Hanniotakis. 1976. Male olive fly attraction to virgin females in the field. *Ann. Zool. Ecol. Anim.* (in press).

(78) B.S. Fletcher. Unpublished.

## CHAPTER 10

# Behavioral Responses of Hymenoptera to Pheromones and Allomones

Murray S. Blum,
*Department of Entomology*
*University of Georgia*
*Athens, Georgia*

## I. THE DIMENSIONS OF CHEMISOCIALITY

Many species of social insects can indeed smell "trouble" in the form of an intraspecific chemical signal emitted by one of their conspecifics in response to a disturbance. These pheromonal stimuli—the alarm pheromones—constitute one of the many types of chemical releasers utilized by hymenopterous insects to regulate a multitude of behavioral reactions, particularly in the colonial milieu. It has become increasingly apparent that the eusocial world of ants, bees, and wasps is largely controlled by volatile compounds, and the behavior of these hymenopterans is more and more frequently being interpreted in terms of pheromonal emitters and perceivers. All levels of sociality in the Hymenoptera appear to have some chemisocial bases, and it has been suggested that the development of eusociality in advanced taxa may be correlated with selection for an increased emphasis on pheromonal stimuli vis-à-vis species in less specialized groups (1).

Alarm pheromones are typical of a multitude of behavioral releasers in that the typical alarm reaction that an insect ultimately exhibits actually represents the end product of a series of behavioral "steps" rather than a single response to the chemical stimulus. Shorey (2) has analyzed in considerable detail the sequential reactions exhibited by male moths in response to the female sex pheromone, and Kennedy (3) has provided a caveat about the interpretive pitfalls that can result from oversimplifying the nature of an insect's response to a pheromonal signal. Furthermore, the responses of social insects to volatile chemical stimuli appear also to be a function of both the concentration of the pheromones and the context in which they are utilized communicatively. Thus, a single pheromone may serve a diversity of functions because of the ability of the target insects to respond differently to it, depending upon the number of pheromonal molecules impinging on their antennae or their proximity to the nest when the signal is detected.

Notwithstanding the imprecision which currently characterizes our comprehension of the *modus operandi* of pheromones, it is evident that these chemical releasers regulate a potpourri of social responses. Intraspecific chemical signals have been demonstrated to function as behavioral releasers of aggregation, recruitment, sex attraction, flight, alarm, food exchange, territorial display, and raiding. In addition to these releaser compounds, primer pheromones play important roles in both caste determination and the suppression of ovarian development in workers. Inevitably, the dimensions of our understanding of the chemisocial world of the Hymenoptera are rapidly being expanded as an increasing number of behavioral responses are demonstrated to be controlled by pheromonal triggers.

## II. THE INEXACTITUDE OF EXOCRINE TERMINOLOGY

### A. Pheromonal and Allomonal Backfires—Kairomonal Effects

In addition to intraspecific signals—the pheromones—a host of interspecific chemical stimuli also play a key role in hymenopterous societies. Brown et al. (4) have divided these interspecific agents into two classes, depending on whether they are adaptive for the emitter individual. One class of compounds—the allomones—is regarded, like pheromones, as adaptively favorable to the producer. Allomones, as exemplified by defensive compounds, constitute the first line of defense utilized by many social insects against predators and as such are of great adaptive value.

In contrast to allomones, another class of interspecific compounds, labeled as kairomones, is regarded as adaptively favorable to the perceiver, rather than the emitter (4). These act as chemical beacons for parasites and predators, and enable the perceiver to exploit the producer individual in terms eminently unfavorable to the latter.

However, there are now substantive grounds for regarding kairomones as nothing more than pheromones or allomones that, in a few specific instances, have evolutionarily boomeranged. As such, these compounds probably possess great selective value to their producers, notwithstanding their having backfired in a few cases. In the absence of information on the roles that these compounds play in the biology of their producer species, it is totally premature, *as well as evolutionarily unsound,* to regard these compounds as maladaptive. Regarding kairomones as maladaptive for their producers (4) fails to recognize that these compounds are highly adaptive in the roles for which they have been evolved—that is, as intra- or interspecific chemical stimuli. While kairomonal effects may result from the emission of pheromones or allomones, there is little justification for regarding kairomones as a class of compounds separate from pheromones or allomones. As a consequence, the evolutionary boomerangs of pheromones and allomones will be referred to henceforth as kairomonal effects.

The results of several recent investigations militate against placing these evolutionary backfires in a separate category after recognition of their true classificatory position as important intraspecific chemical stimuli.

For example, Corbet (5) has demonstrated that in the flour moth, *Anagasta kuhniella,* regulation of larval numbers and dispersion of individuals within the system is controlled by a secretion from the mandibular glands. In crowded situations larvae of *A. kuhniella* frequently secrete mandibular-gland products, and these exocrine compounds exert a major effect on the ultimate structure of the population. The glandular exudate results in larvae emigrating from crowded situations, lengthened generation time of developing individuals, and reduced fecundity in females crowded as larvae (5). Thus the mandibular-gland products clearly regulate population densities by acting as an epideictic (dispersal) pheromone. However, the ichneumonid parasite, *Venturia canescens,* exploits this important intraspecific chemical messenger as a releaser of ovipositional-probing behavior. The amount of probing is proportional to the amount of mandibular-gland secretion, indicating that the ovipositional activity of the parasites is directly proportional to the density of the flour moth population. Significantly, this epideictic pheromone reduces crowded populations further by directly affecting larval development and adult fecundity, and by promoting parasitization of the population. Therefore, the parasites clearly respond to a critical pheromone and augment its efficacy as a population regulator.

Conceivably, hymenopterous parasites use many host pheromones as either orientation or ovipositional stimuli. Sternlicht (6) reported that two species of parasitic wasps, *Aphytis melinus* and *A. coheni,* were attracted by the sex pheromone of their coccid host, *Aonidiella aurantii.* These *Aphytis* spp. may be typical of a wide range of parasitic hymenopterans that utilize sex-specific and highly volatile pheromones of their hosts in order to locate the latter accurately. But, overall, the great selective value of sex pheromones for their producers hardly justifies considering them maladaptive because some parasitic species have exploited them.

Exocrine compounds are now known to release host-seeking and ovipositional behavior in a variety of hymenopterous parasites. Conceivably, these compounds are pheromones that may possess epideictic (5) or other functions in the biologies of the host species. Hendry et al. (7) reported that searching and ovipositional behavior in the braconid wasp, *Orgilus lepidus,* are released by two different chemical stimuli present in the frass of its host, the potato tuberworm, *Phthorimaea operculella. n*-Heptanoic acid has been identified as the stimulus for host-searching, and a contact chemical stimulus in the frass as the releaser for ovipositional-probing behavior. It will not prove surprising if these frass-derived compounds function in intraspecific population regulation, somewhat analogous to the dung piles which play such an important role in population spacing in mammals. The function of the mandibular-gland secretion produced by larvae of *Heliothis virescens* is currently unknown, but this mixture of long-chain hydrocarbons (8), which releases searching behavior in the braconid parasite, *Cardiochiles nigriceps* (9), may possess considerable value as an intraspecific chemical stimulus.

### 1. Simultaneous Intra- and Interspecific Functions of Exocrine Compounds

The products secreted from the exocrine glands of hymenopterans frequently function simultaneously as both allomones and pheromones, and, in at least a few cases, kairomonal effects have been observed additionally. These facts underscore the terminological difficulties inherent in attempting to classify these compounds as either simple intraspecific chemical stimuli or interspecific communicative agents. This may be especially true of eusocial species because of their great potential for utilizing any exocrine compounds as volatile information-bearing agents, without concern for the possible interspecific roles that these exocrine products may also possess. Probably, the simultaneous utilization of natural products as both pheromones and allomones will be shown to be a common phenomenon as more and more detailed analyses of the biology and behavior of hymenopterans are undertaken.

Ghent (10) first demonstrated that an effective defensive product could also possess an important communicative function. Attacking workers of the formicine ant, *Acanthomyops claviger,* smear intruders with a mandibular-gland secretion which is dominated by the monoterpene citronellal, a potent topical irritant when applied to the arthropod cuticle. In addition to this allomonal function, citronellal acts as a chemical tocsin that releases alarm behavior in proximate workers. Similarly, the dolichoderine, *Iridomyrmex pruinosus,* sprays aggressors with a secretion fortified with 2-heptanone (11), a compound also utilized as an alarm pheromone by this species (12). Likewise, as Maschwitz (13) has demonstrated, workers of *Formica fusca* exhibit alarm behavior when they are exposed to the allomonal product of their poison gland, formic acid.

The terminological inexactitude characterizing hymenopterous exocrine products is most evident in the investigation of the predatory behavior of *Myrmecia gulosa* undertaken by Haskins et al. (14). This archaic myrmeciine ant is specialized for preying on *Camponotus* spp. which, when attacked, eject formic acid at their predators. Formic acid is a powerful alarm pheromone for *Camponotus* spp. (15), and usually acts as a potent defensive secretion (allomone) against other arthropod species. However, instead of acting as an effective deterrent allomone against *M. gulosa* workers, formic acid drives these myrmeciines into an aggressive frenzy, resulting in the *Camponotus* workers being attacked with great vigor. Therefore, for *Camponotus* workers formic acid is simultaneously both a pheromone and an allomone, but the normal adaptive value of this allomone for the formicines is more than offset by the kairomonal effect produced on workers of *M. gulosa*.

## III. A SPECTRUM OF RESPONSES TO A SINGLE PHEROMONAL STIMULUS

An astonishing variety of responses may be encountered when the adults of many hymenopterous species are exposed to one of their intraspecific chemical stimuli. The nature and duration of the response can vary considerably, depending on the location of the perceiver when the signal is detected. By adapting their pheromones to serve multiple functions, social insects expand the dimensions of chemosociality with a finite number of chemical releasers. This phenomenon, designated as pheromonal parsimony (16), appears to be so widespread among the Hymenoptera as to be one of the key developments in the evolution of eusociality. In examining the main classes of pheromones as multifunctional chemical stimuli, some of the many faces of pheromonal parsimony that have already been described in the Hymenoptera emerge.

### A. Alarm Pheromones

More than a decade ago, Ghent (10) observed that the mandibular-gland secretion of the formicine ant *Acanthomyops claviger* possessed at least two very important functions. This secretion, which is dominated by citronellal and citral (17), appears to be released, or sprayed, only when the workers are biting an object rather than responding to a localized disturbance. In addition to being toxicants per se, citronellal and citral promote the penetration of the highly ionized toxin, formic acid, through the hydrophobic epicuticle of insects (10). However, in addition to this allomonal role, the mandibular-gland products of *A. claviger* workers function as very powerful releasers of alarm behavior and attractants.

The attack of a worker of *A. claviger* invariably results in the secretion of its mandibular-gland compounds, which release a diversity of reactions in nearby workers. Initially, workers retreat, and brood may be transported away from the general vicinity of the confrontation. The relocation of the brood is highly adaptive since it removes the immatures from situations in which they could be susceptible to immediate predation. Retreating ants, exposed to the mandibular-gland secretion, exhibit frequent cleaning behavior, particularly of the antennae. Excited workers also display a peculiar jerky movement designated as tremoring, especially when they encounter other workers (10). The tremoring reaction apparently facilitates the spread of the alarm reaction initially triggered by the monoterpenes evacuated from the capacious mandibular-glands of the attacking workers. The ambulatory rate of the stimulated ants increases considerably, and tremoring becomes more frequent as the area of the alarm reaction enlarges. These effects can be produced by the neat terpenes as well as the glandular exudate.

Eventually attack behavior is released by citronellal and citral, or mandibular-glands secretion, in workers of *A. claviger* (10). After the initial retreat, small groups of workers return to the emission source of the terpenes, waving their antennae and holding their mandibles open. Attack by a worker attracts additional workers, which quickly grab the intruder with their mandibles, at the same time evacuating the contents of their mandibular glands. Recruitment of additional workers to the site of the combat increases tremoring. Excitement and further brood relocation characterize the reactions of the ants to the increased concentration of alarm pheromone. By labeling the interloper with mandibular-gland secretion, the attacking *A. claviger* workers effectively convert it to a chemical beacon that attracts more aggressive workers.

Field observations of the release of alarm behavior in the nest population of a dolichoderine species, *Conomyrma pyramicus,* similarly demonstrate diverse releaser effects. Workers of *C. pyramicus* generate an alarm signal with 2-heptanone, a product of their well-developed anal glands (11), exocrine structures that are limited to species in this subfamily. Placement of a crushed anal gland in the entrance hole of the nest causes such an outpouring of excited workers that the ant population in the vicinity of the nest may rapidly increase by at least tenfold. The same response can be produced with pure 2-heptanone. Alarm spreads as a series of enlarging waves. Excited ants, their thoraces elevated and their mandibles spread, move in zigzag lines, frequently encountering incoming ants, which immediately exhibit similar frenetic behavior. Unlike workers of the formicine *A. claviger,* those of *C. pyramicus* readily secrete their alarm pheromone after the threshold for alarm behavior has been reached.

High concentrations of 2-heptanone release a bizarre series of reactions in alarmed workers of *C. pyramicus* (11). An exaggerated type of motion described as lurching occurs when the workers are attracted to a heptanone emission source. This behavior is most pronounced when the workers approach a particularly high concentration of the pheromone. Lurching is characterized by short, jerky, in-and-out movements as the ants move toward the pheromonal emission source. It is particularly pronounced in workers proximate to a nest orifice into which a very high concentration of 2-heptanone has been placed. These excited, lurching workers will accumulate on the downwind side of the nest, but will not enter the nest; workers that are already in the nest will rarely emerge as long as the pheromonal stimulus is present. In a few instances callow workers, which are ordinarily never seen on the surface, boil out of the nest orifice after the introduction of 2-heptanone.

High concentrations of their alarm pheromone have also resulted in *C. pyramicus* workers transporting pupae to the surface and abandoning them. The pupae were eventually carried back into the nest (11). Since Wilson and Pavan (18) have observed that emigration resulted when colonies of the dolichoderine *Tapinoma sessile* were exposed to prolonged doses of their alarm pheromone, the abortive pupal transport noted with workers of *C. pyramicus* may similarly reflect premature emigratory behavior.

Alarm pheromones have been demonstrated to release digging behavior in several formicid species, especially when the workers are exposed to high concentrations of the releaser for prolonged periods. Wilson (19) observed that workers of the myrmicine, *Pogonomyrmex badius,* exhibited digging behavior when exposed repeatedly to the alarm pheromone produced in the mandibular glands. Digging behavior did not involve interactions among the affected workers and ceased when the alarm stimulus was removed. Wilson suggested that such behavior may not reflect simple "displacement activity," but rather may be functional under certain conditions—for example, serving to excavate workers that had been buried as a result of a cave-in: McGurk et al. (20) were unable to release digging behavior in workers of *P. barbatus* after injecting the pure pheromone, 4-methyl-3-heptanone, below the soil surface although, as in the case of *P. badius*, workers of this species may exhibit digging behavior when exposed to their alarm pheromone while foraging on the surface. Digging behavior is also manifested by workers of the dolichoderine, *Conomyrma pyramicus* (11), which kick the excavated particles behind them and seldom use their mandibles for transporting the excavated soil, as occurs in nest-enlargement excavations. Workers of *C. pyramicus* will bury paper squares impregnated with their alarm pheromone, effectively removing these pheromonal-emission sources from the surface. Recently, Crewe and Fletcher (21) demonstrated that the mandibular-gland secretion of the ponerine, *Paltothyreus tarsatus,* which does not function as an alarm pheromone, releases directional digging behavior only when the pheromonal source is buried. Indeed, buried ants, as well as filter papers impregnated with dimethyl disulphide—one of the mandibular gland components—are quickly excavated by attracted workers. The results with *P. tarsatus* constitute the only unequivocal demonstration of a directed and functional releaser of digging behavior in ants.

A contextual variation in alarm behavior has been demonstrated for the myrmicine, *Crematogaster scutellaris* (22). Workers exhibit different types of alarm reactions, depending on whether the alarm stimulus is encountered in the nest, in clustered groups at feeding sites, or on the trail. Alarm behavior in the nest or at feeding sites is characterized by both the secretion of venom from the elevated gasters of the ants and a marked increase in ambulatory activity. Under these circumstances, the attack threshold is very low. On the one hand, this type of alarm behavior is triggered by abdominal secretions that are probably identical to venom constituents. On the other hand, alarm behavior on the trail is primarily released by volatile constituents of the mandibular-gland secretion.

Workers of *C. scutellaris* that perceive the mandibular-gland secretion while following a chemical trail elevate their heads and begin antennal fanning (22). Short, accelerated runs, with the gasters raised, usually follow perception of the pheromone, and feigned attacks on nestmates usually occur. Alarmed workers will often stop and retreat slightly, after which an accelerated run may occur. The mandibles are often bared, and loops or turns characterize the movements of the excited workers. Wobbling movements of the head and gaster are pronounced, and if the stimulus is strong enough, sting extrusion and venom secretion occur. In this case, the induction of one type of alarm behavior can result in the release of a second type of alarm behavior. Significantly, the threshold for the release of alarm was found to be too high when the ants were in an excited recruiting phase while feeding, and too low several hours after the termination of feeding. Also, food-laden ants, as well as certain other workers, were observed to be nonresponsive to the alarm stimuli encountered on the trail, whereas some very small workers seemed to possess such a low threshold that virtually any stimulus would result in the release of alarm behavior.

Leuthold and Schlunegger (22) concluded that the mandibular-gland products of *C. scutellaris* workers are never secreted spontaneously, but only during the act of biting by an attacking worker. Once an attack occurred on the trail, the labeled intruder became the focal center for additional attacks by other workers. Thus the mandibular-gland secretion functions purely as part of the ants' pheromonal-allomonal system in either defense or as a predator-alert signal. This is in marked contrast to the general alarm frenzy resulting from the secretion of venom when nest workers perceive a disturbance. Alarm in or near the nest is characterized by a general scattering of the workers, their gasters elevated, and their attack thresholds low. Abdominal secretions only release these reactions in groups of ants in the nest or at feeding sites, but *never* on the trail. Trail alarm is released only by the mandibular-gland products, which trigger attraction to the pheromonal-emission source, rather than the scattering that occurs in nest-induced alarm. It may be significant that a low concentration of the mandibular-gland components is omnipresent in the nest of *C. scutellaris*. Presumably, workers of this species are normally habituated to the presence of low levels of these volatiles, and the release of alarm by these compounds only occurs when workers suddenly encounter them in high concentrations under field conditions.

Moser et al. (23) quantified the response of workers of the myrmicine, *Atta texana,* to their alarm pheromone, 4-methyl-3-heptanone, and observed three levels of response. At subthreshold concentrations of pheromone, the workers elevated their heads and vibrated their antennae (detection). At slightly higher concentrations, the ketone appeared to function as a simple attractant—ants moving toward the pheromonal emission source did not manifest an aggressive posture. At higher concentrations typical alarm behavior—attraction with spread mandibles—was evident. Alarmed ants frequently challenged their sister workers, a reaction also demonstrated by workers of *Pogonomyrmex badius* exposed to a high concentration of their alarm pheromone (24). Neither winged females nor males of *A. texana* exhibited any alarm reactions when exposed to high concentrations of 4-methyl-3-heptanone. *Atta* workers carrying leaves or detritus or tending the fungus gardens were very unresponsive in the presence of high concentrations of the alarm pheromone. These results parallel the observation by Leuthold and Schlunegger (22) that food-laden workers of *Crematogaster scutellaris* exhibited a very high threshold in the presence of alarm pheromones.

Field tests with the neat alarm pheromone demonstrated that *A. texana* workers were both repelled and alarmed by high concentrations (23). Crushed heads, containing the alarm releasers, attracted large numbers of workers, which became frenzied and proceeded to dismember them at once. The crushed heads were quickly carried off at right angles to the trail, often with a retinue of excited ants following the bearer of the head. The mutilated heads were undoubtedly leaking the pheromone, which attracted workers from the downwind side. This same phenomenon has been observed when crushed workers of *Conomyrma pyramicus* are transported away from the nest area (11).

The by-products of the two exocrine glands mediate a concourse of reactions to the alarm pheromones of workers of the formicine, *Camponotus pennsylvanicus* (15). Formic acid, a potent defensive substance (allomone), releases nonoriented alarm behavior in the excited workers. However, $n$-undecane, a product of the Dufour's gland, which is ejected in admixture with formic acid, acts as a synergist for the acidic alarm pheromone, at the same time acting as an orienting agent for the frenzied ants. The utilization of the hydrocarbon is maximally adaptive since it synergizes the releaser activity of formic acid while enabling the ants to locate easily the emission source of this pheromone. In addition, $n$-undecane functions as an aggregation pheromone for *C. pennsylvanicus* workers, thus providing a duality in functions of great social adaptiveness. In all probability, the propensity of ants to form aggregations in the nest may be mediated by low concentrations of exocrine products that are commonly utilized for other communicative functions in nonnest contexts.

A rather spectacular series of reactions characterizes the behavioral repertoire of *Odontomachus* workers after exposure to their alarm pheromones. Alarm in *Odontomachus* spp. is released by alkylpyrazines liberated from the capacious mandibular glands of excited workers (25). Alarmed workers approach pheromonal emission sources with their linear mandibles spread at an angle of nearly 90° to the long axis of the body. When they are proximate to the pheromonal source, the workers frequently snap their mandibles together and, if the mandibles close on a grain of sand, the ants are violently projected backwards. Alarmed ants will frequently attack their sister workers, and these conflicts may result in the infliction of fatal injuries to some of the combattants (25).

Maschwitz (13) has demonstrated that wasps in the genus *Vespa* generate an alarm signal with the poison-gland secretion. Excited workers direct the tip of their sting toward the source of the disturbance and eject a stream of venom containing a potent releaser of alarm behavior. Alarmed wasps respond to this chemical signal by becoming airborne, and their threshold for attack is lowered considerably. Objects that are marked with *Vespa* venom are very attractive to wasp workers, which attack them with great vigor. A disturbed queen will also signal alarm by ejecting venom but, unlike her workers, she will not participate in the attack on the labeled intruder. This lack of aggressive behavior on the part of the queen is highly adaptive since it reduces the possibility that she, the functional reproductive, will be injured or killed.

Maschwitz (13) and Boch et al. (26) have made detailed studies of the alarm behavior of the honeybee *Apis mellifera*. At the hive entrance a disturbed worker elevates its abdomen, opens the sting chamber, and extrudes the sting. The agitated worker will move quickly among the other bees, buzzing and liberating very odoriferous compounds, which are disseminated by wing fanning (13). One of the components of the sting pheromone has been identified as isopentyl acetate (26), and this compound evokes typical alarm behavior in bee workers. The bees assume a tense posture, elevating their antennae and baring their mandibles while they orient to the source of the pheromone by moving in irregular circles. Ultimately, the frenzied workers lunge toward the emission source and soon inundate it with their bodies.

Alarmed bees exhibit a similar behavioral repertoire when exposed to 2-heptanone, a mandibular-gland product (27). This ketone is not as effective as isopentyl acetate in releasing alarm in honeybee workers (28), but probably functions as both a secondary alarm substance and defensive secretion at the hive entrance. Neither isopentyl acetate nor 2-heptanone will release stinging behavior, and it appears that the stimulus of movement is required before a labeled object will be stung (29) (30). Free and Simpson (31) have demonstrated that both 2-heptanone and isopentyl acetate function as the principal alarm pheromones of the honeybee, but have emphasized that additional pheromonal components of the sting-derived pheromone probably play a major role in releasing all the components of alarm behavior.

2-Heptanone has been demonstrated to be a repellent for honeybee workers themselves (32) (33), and its role as an alarm pheromone has been questioned (34). However, unlike isopentyl acetate, 2-heptanone has been demonstrated to possess a diversity of important social functions, and its role as an alarm pheromone at the hive entrance (13) probably constitutes nothing more than another example of the parsimonious utilization of a pheromone. Nuñez (35) has observed that foraging bees mark dissipated food sources with 2-heptanone and thus effectively label these sources as "empty" in the event that they are visited in the near future by other bees. Also, Morse (36) has demonstrated that swarming worker bees appear to mark a foreign queen with alarm pheromones and subsequently do not display the normal scenting behavior that is released when they have located their own queen (37). Both 2-heptanone and isopentyl acetate suppress scenting behavior, and it is likely that these compounds may function normally in the context of a swarm encountering a foreign queen. These results further indicate that honeybees use 2-heptanone for a diversity of functions.

## 1. Citral: Varied Pheromonal Functions among Hymenopterous Species

The independent development of chemosociality in different hymenopterous taxa (species) is vividly illustrated by the varied functions of citral among both widely separated and closely related phylogenetic lines. The distribution and roles of the isomers of citral, neral and geranial, exemplify the remarkable evolutionary adaptations that have characterized the utilization of the same exocrine products by different phyletic lines of the Hymenoptera.

Ghent (10) reported that the mandibular-gland products of the formicine ant, *Acanthomyops claviger*—citral and citronellal—functioned both as alarm pheromones and defensive substances. Butenandt et al. (38) identified citral as a mandibular constituent of the myrmicine, *Atta sexdens,* and it was subsequently demonstrated that while citral is used as a defensive product by this species, an alarm signal is generated by aliphatic ketones (39). Thus, citral functions as both an allomone and pheromone for workers of *A. claviger,* whereas it appears to be utilized solely as an allomone by workers of *A. sexdens.*

However, it is chiefly among the bees that citral has been exploited as a multifunctional pheromone and allomone. In the honeybee, citral is synthesized in the Nassanoff gland, an exocrine organ located on the last visible abdominal tergite (40) (41). Frisch (42) reported that workers foraging at a rich food source exposed their Nassanoff glands and scent, thereby attracting other bees to the exploitable food. However, Free and Williams (43) observed that although there is a greater tendency for workers trained to a food source to scent after returning home and dancing, these two activities are not closely correlated. The Nassanoff secretion is utilized also to orient bees in a diversity of other situations. Inexperienced bees are guided to the hive entrance by scenting workers, and lost bees will frequently scent after finding their hive (44). If the location of a hive is changed, workers will scent vigorously after relocating it. Citral is thus employed as an orienting (attractant) pheromone in a wide variety of contexts.

Worker bees that are selecting potential nesting sites during the act of swarming also liberate the Nassanoff pheromones (45). Morse and Boch (37) have demonstrated that worker bees signal the location of a lost queen by vigorously scenting with their Nassanoff volatiles, thus attracting other bees to the interim queen-containing swarm. However, some workers will also return to the queenless cluster and liberate the Nassanoff pheromones, an act contributing to the breaking of this cluster (46). Furthermore, the high concentration of Nassanoff scent in the vicinity of the queenright swarm acts as a beacon for the airborne bees, and thus functions to reunite the swarming bees into a compact mass. The *de novo* evolution of the Nassanoff gland in the honeybee to serve as the producer of key information-bearing agents has played a major role in expanding the social dimensions of this species.

Citral, a mandibular product of the meliponine bee, *Trigona subterranea,* is a powerful attractant for workers of this species and is used to lay chemical trails to food sources (47). Citral is deposited as a series of droplets around new food finds, and successful foragers return to the nest while depositing citral at specific intervals. Citral-treated lures are very attractive to foragers, which grasp them gently in their mandibles. Bees that are flying along chemical trails can be induced to drop to the ground in order to investigate blocks impregnated with citral. On the other hand, high concentrations of citral in the vicinity of the nest are either repellent or release alarm-and-attack behavior. Citral-treated blocks that are placed at the nest entrance trigger immediate attack behavior, which is characterized by the workers vigorously grasping the blocks in their mandibles and shaking them. Low concentrations of citral are thus used as a trail-laying pheromone, whereas high concentrations, produced by numerous workers at the nest entrance, release alarm behavior (47).

Another stingless bee, *Lestrimelitta limao,* has converted citral into a diabolical disarming weapon when unleashed against certain other species of stingless bees. Workers of *L. limao* lack corbiculae on their hind legs and thus cannot collect pollen. These bees obtain their protein-rich food by robbing the nests of selected species of *Trigona* and *Melipona* (48). Citral, which is produced in the well-developed mandibular glands of *L. limao* workers (49), is released by scouts that have penetrated the nests of raid-susceptible species. The terpenes act as both attractants for additional *L. limao* workers as well as cuticular irritants for the workers of the host meliponine species. In addition, citral causes a complete breakdown in organized social behavior of the raided species, and effective resistance to the depredations of the *L. limao* invaders ceases abruptly (47). For *L. limao,* citral is the primary agent for chemical subjugation of foreign bee species.

Introduction of citral-treated blocks into the nests of *Trigona subnuda, T. droryana, T. testaceicornis,* and *Melipona quadrifasciata* resulted in almost identical reactions in all species. Organized activity ceased almost immediately, and many workers flew from the nests and did not return until the treated blocks had been removed. Workers in the nests moved frenetically in all directions but generally avoided contact with the blocks. In general, no manifestations of social cohesiveness were evident as long as the treated blocks were present in the nests. In contrast to this behavior, the social activities of bee species not normally raided by *L. limao* are not appreciably affected by citral (47). Workers of *Apis mellifera,* several *Trigona* species, and *Bombus atratus* either covered the treated blocks with resin, attacked them, or removed them from their nests. In no instance was the social cohesiveness of these species appreciably disrupted.

The host of intra- and interspecific roles played by citral in different hymenopterous species is probably typical of the versatility with which many compounds are utilized by these social insects. Just as slave-raiding in ants is also predicated on the use of disarming chemical stimuli (50), it is almost certain that a variety of exocrine compounds will be demonstrated to subserve both similar and different functions among the various phyletic lines of eusocial Hymenoptera. This may be particularly apparent when the role of the same compound is compared in primitive species and more specialized taxa. For example, alkylpyrazines are produced by both primitive (*Hypoponera*) (51) and advanced (*Odontomachus*) (25) ponerine ant species, but the species in these two taxa react very differently to these compounds. Workers of *Hypoponera opacior* retreat in the presence of the pyrazine, whereas *Odontomachus* workers are attracted to the pheromonal source, which they attack vigorously. Major behavioral differences may characterize the responses of species to the same pheromone as a reflection of the level of chemisociality which has been evolved (51).

## IV. THE CHEMICAL BASES OF RECRUITMENT BEHAVIOR IN HYMENOPTERA

Many ants and bees lay chemical trails to food sources or when they are emigrating to new nest sites. In common with other pheromones, trail pheromones are utilized parsimoniously, and their roles as intraspecific chemical stimuli have shown to be surprisingly varied.

Trail pheromones frequently are used in conjunction with the secretions of other exocrine glands in order to produce highly specific signals. Wilson (52) reported that workers of the fire ant, *Solenopsis invicata* (*saevissima*) discharge a cephalic pheromone from the mandibular glands that releases nonoriented alarm. However, the simultaneous discharge of the trail pheromone from the Dufour's gland results in an oriented response of workers to the emission source of the alarm pheromone. Similarly, workers of *Camponotus pennsylvanicus* secrete an orientation pheromone from their Dufour's gland in admixture with a poison gland product—formic acid—an alarm pheromone that produces a nonoriented signal (15). Hölldobler (53) has reported that workers of *Camponotus socius* lay a recruitment trail with the hind gut secretion, whereas the alarm pheromone—formic acid—is added to the trail in order to keep the recruited workers in a high state of excitement. Similarly, Ayre (54) demonstrated that alarm pheromones were utilized by three species of ants as recruitment stimuli when used in conjunction with trail pheromones. Cammaerts-Tricot (55) has also reported that *Myrmica rubra* workers use both Dufour's and poison-gland secretions in recruiting nestmates.

The harvester ant, *Pogonomyrmex badius*, homes by utilizing visual and chemical cues. The Dufour's gland secretion is used for homing by individual ants, but does not serve as a recruitment stimulus (56). Hölldobler and Wilson (57) had previously reported that *P. badius* workers recruit with a trail pheromone derived from the poison-gland secretion. The possibility that the trail pheromone may be reinforced with the orientation pheromone from Dufour's gland in order to generate a more permanent trail cannot be excluded.

Recently, Maschwitz et al. (58) studied recruitment in the ponerine, *Bothroponera tesserinoda*. This species recruits nestmates by tandem running after pulling the recruit with its mandibles. The recruited ant maintains close contact with the recruiter by frequent antennal probing. The hind legs and gaster of the leader must be frequently stimulated by the recruit in order for tandem running to be maintained. Two types of stimuli—a surface pheromone and mechanical contact—are required to sustain tandem running. Möglich et al. (59) have also observed that recruitment by tandem running in the myrmicine, *Leptothorax acervorum*, required chemical and mechanical stimuli. However, the chemical stimulus is provided by a pheromone secreted from the everted sting. The term *tandem calling* has been proposed to describe the behavior of the recruiter, as a consequence of its assumption of a stereotyped calling posture if contact is not maintained with the recruit. Möglich et al. (59) suggest that tandem calling is the evolutionary precursor of both odor-trail-laying and sex attraction in certain phyletic lines of the Myrmicinae. Buschinger (60) and Hölldobler and Wüst (61) had previously demonstrated that sex pheromones are produced in the poison glands of species in the myrmicine genera *Harpagoxenus* and *Monomorium,* respectively.

Hangartner (62) has established that workers of the formicine, *Lasius fuliginosus*, orient by osmotropotaxis during trail-following. Significantly, it was noted that during trail-following each antenna is alternately inserted in and out of the odor field. It has been suggested that this type of behavior may reflect the utilization of the antennae as nulling devices (1), so that the withdrawal of an antenna from the odor field nulls the signal to the other antenna. The threshold of the out antenna should be decreased in the absence of any strong stimulus, and residual olfactory effects should be avoided. Since only one antenna is inserted in the odor plume at any one instant, an orientation axis to the signal path will be maintained, which would be impossible to achieve otherwise if both antennae were continually bathed with pheromonal molecules (1). Indeed, antennal nulling may constitute a widespread device in insects for maximizing the efficiency of antennal chemoreceptors.

## V. RECOGNITION OF THE EXPANDING SCOPE OF CHEMICAL COMMUNICATION

Pheromones have evolved to act as a major driving force for many of the manifestations of eusociality that are currently comprehended. Recognition of the pheromonal legacies of the hymenopterans has provided biologists with a powerful insight into many of the expressions of social behavior that had been previously illuminated but never clearly understood. The evolution of pheromonal parsimony has enabled many hymenopterans to achieve a level of sociality that probably would not have been possible otherwise. The protean behavior of a species in response to the same chemical signal has now been demonstrated to reflect the preprogrammed ability of hymenopterans to express different reactions as a function of either pheromonal concentration or the context in which the releaser is encountered. Pheromonal synergy, involving the simultaneous secretion of two or more exocrine glands, has also been illuminated as another viable device for modulating the informational content encoded in the signal discharged from a single gland. The evolutionary implications of hymenopterans in different phyletic lines exhibiting a variety of responses to the same pheromone are great, and point to the potential value of these information-bearing agents in phylogenetic studies. It is exciting to contemplate the utilization of exocrine compounds as character states in probing the genesis of eusociality.

The social behavior of hymenopterans, previously described in anecdotal terms for the most part, is now often being analyzed at a dizzying rate as a chemisocial phenomenon. For example, Heinrich (63) has recently described a pheromone that induces brooding behavior in *Bombus* queens incubating larvae in pollen clumps. Ishay (64) has demonstrated that a thermoregulatory pheromone is produced by vespine wasps that evokes abdominal pumping (i.e., warming) behavior in attending adults. It has also been possible to demonstrate that the queen pheromone of *Vespa orientalis* regulates the type of nest-building expressed by workers (65). Exposure of workers to the queen pheromone results in the construction of vertically attached combs, whereas in the absence of this compound horizontally attached combs are built. Detailed investigations on the biology and behavior of hymenopterous species, in combination with chemical studies, promise to elucidate the chemical bases for many facets of social behavior that are already recognized. Ultimately, the eusocial code of the Hymenoptera may be broken by deciphering the information encoded in the chemical signals generated from their primed exocrine glands.

## REFERENCES

(1)    M.S. Blum. 1974. Deciphering the communicative Rosetta Stone. *Bull. Entomol. Soc. Am.* **20**:30-35.

(2)    H.H. Shorey. 1970. Sex pheromones of Lepidoptera. In *Control of Insect Behavior by Natural Products.* D.L. Wood, R.M. Silverstein, and M. Nakajima, Eds. Academic, New York, pp. 249-284.

(3)    J.S. Kennedy. 1972. The emergence of behaviour. *J. Aust. Entomol. Soc.* **11**:19-27.

(4)    W.L. Brown, T. Eisner, and R.H. Whittaker, 1970. Allomones and kairomones: Transspecific chemical messengers. *Bioscience* **20**:21-22.

(5)    S.A. Corbet. 1971. Mandibular gland secretion of larvae of the flour moth, *Anagasta kuhniella*, contains an epideictic pheromone and elicits oviposition movements in a hymenopteran parasite. *Nature* **232**:481-484.

(6)    M. Sternlicht. 1973. Parasitic wasps attracted by the sex pheromone of their coccid host. *Entomophaga* **18**:339-342.

(7)  L.B. Hendry, P.D. Greany, and R.J. Gill. 1973. Kairomone mediated host-finding behavior in the parasitic wasp *Orgilus lepidus*. *Entomol. Exp. Appl.* 16:471-477.

(8)  S.B. Vinson, R.L. Jones, P. Sonnet, V.A. Bierl, and M. Beroza. 1975. Isolation, identification and synthesis of host seeking stimulants for *Cardiochiles nigriceps*, a parasitoid of the tobacco budworm. *Entomol. Exp. Appl.* 18:443-450.

(9)  S.B. Vinson, 1968. Source of a substance in *Heliothis virescens* that elicits a searching response in its habitual parasite, *Cardiochiles nigriceps*. *Ann. Entomol. Soc. Am.* 61:8-10.

(10)  R.L. Ghent. 1961. Adaptive refinements in the chemical defensive mechanisms of certain Formicinae. Ph.D. Thesis, Cornell University, Ithaca, N.Y.

(11)  M.S. Blum and S.L. Warter. 1966. Chemical releasers of social behavior. VII. The isolation of 2-heptanone from *Conomyrma pyramicus* and its modus operandi as a releaser of alarm and digging behavior. *Ann. Entomol. Soc. Am.* 59:774-779.

(12)  M.S. Blum, S.L. Warter, R.S. Monroe, and J.C. Chidester. 1963. Chemical releasers of social behavior. I. Methyl-*n*-amyl ketone in *Iridomyrmex pruinosus* (Roger). *J. Insect Physiol.* 9:881-885.

(13)  U. Maschwitz. 1964. Gefahrenalarmstoffe und Gefahrenalarmierung bei sozialen Hymenoptera. *Z. Vgl. Physiol.* 47:596-655.

(14)  C.P. Haskins, R.E. Hewitt, and E.F. Haskins. 1973. Release of agressive and capture behaviour in the ant *Myrmecia gulosa* F. by exocrine products of the ant *Camponotus*. *J. Entomol.* (A)47:125-139.

(15)  G.L. Ayre and M.S. Blum. 1971. Attraction and alarm of ants *(Camponotus* spp.–Hymenoptera: Formicidae) by pheromones. *Physiol. Zool.* 44:77-83.

(16)  M.S. Blum. 1970. The chemical basis of insect sociality. In *Chemicals Controlling Insect Behavior.* M. Beroza, Ed. Academic, New York, pp. 61-94.

(17)  M.S. Chadha, T. Eisner, A. Monro, and J. Meinwald. 1962. Defense mechanisms of arthropods. VII. Citronellal and citral in the mandibular gland secretions of the ant *Acanthomyops claviger* (Roger). *J. Insect Physiol.* 8:175-179.

(18)  E.O. Wilson and M. Pavan. 1959. Source and specificity of chemical releasers of social behavior in the dolichoderine ants. *Psyche* 66:70-76.

(19)  E.O. Wilson. 1958. A chemical releaser of alarm and digging behavior in the ant *Pogonomyrmex badius* (Latreille). *Psyche* 65:41-51.

(20)  D.J. McGurk, J. Frost, E.J. Eisenbraun, K. Vick, W.A. Drew, and J. Young. 1966. Volatile compounds in ants: Identification of 4-methyl-3-heptanone from *Pogonomyrmex* ants. *J. Insect Physiol.* 12:1433-1441.

(21)  R.M. Crewe and D.J.C. Fletcher. 1974. Ponerine ant secretions: The mandibular gland secretion of *Paltothyreus tarsatus* Fabr. *J. Entomol. Soc. S. Africa* 37:291-298.

(22)  R.H. Leuthold and U. Schlunegger. 1973. The alarm behaviour from the mandibular gland secretion in the ant *Crematogaster scutellaris*. *Insectes Soc.* 20:205-214.

(23)  J. Moser, R.G. Brownlee, and R.M. Silverstein. 1968. The alarm pheromones of *Atta texana*. *J. Insect Physiol.* 14:529-535.

(24)  M.S. Blum, R.E. Doolittle, and M. Beroza. 1971. Alarm pheromones: Utilization in evaluation of olfactory theories. *J. Insect Physiol.* 17:2351-2361.

(25)  J.W. Wheeler and M.S. Blum. 1973. Alkylpyrazine alarm pheromones in ponerine ants. *Science* (Wash.) 182:501-503.

(26)  R. Boch, D.A. Shearer, and B.C. Stone. 1962. Identification of iso-amyl acetate as an active component in the sting pheromone of the honey bee. *Nature* 195:1018-1020.

(27)  D.A. Shearer and R. Boch. 1965. 2-Heptanone in the mandibular gland secretion of the honey-bee. *Nature* 206:530.

(28)  R. Boch, D.A. Shearer and A. Petrasovits. 1970. Efficacies of two alarm substances of the honeybee. *J. Insect Physiol.* 16:17-24.

(29)  J.B. Free. 1961. The stimuli releasing the stinging responses of honeybees. *Anim. Behav.* 9:193-196.

(30)  R.L. Ghent and N.E. Gary. 1962. A chemical alarm releaser in honeybee stings (*Apis mellifera* L.) *Psyche* 69:1-6.

(31)  J.B. Free and J. Simpson. 1968. The alerting pheromones of the honeybee. *Z. Vgl. Physiol.* 61:361-365.

(32)  C.G. Butler. 1966. Mandibular gland pheromones of worker honeybees. *Nature* 212:5061.

(33)  J. Simpson. 1966. Repellency of the mandibular gland scent of worker honeybees. *Nature* 209:531-532.

(34)  C.G. Butler. 1967. Insect pheromones. *Biol. Rev.* 42:42-87.

(35)  J.A. Nuñez . 1967. Sammelbienen markieren versiegte Futterquellen durch Duft. *Naturwissenschaften* 54:322-323.

(36)  R.A. Morse. 1972. Honeybee alarm pheromone: Another function. *Ann. Entomol. Soc. Am.* 65:1430.

(37)  R.A. Morse and R. Boch. 1971. Pheromone concert in swarming honeybees. *Ann. Entomol. Soc. Am.* 64:1414-1417.

(38)  A. Butenandt, B. Linzen, and M. Lindauer. 1959. Über einen Duftstoff aus der Mandibeldrüse der Blattschneiderameise *Atta sexdens rubropilosa* Forel. *Arch. Anat. Microscop. Morphol. Exp.* 48:13-19.

(39)  M.S. Blum, F. Padovani, and E. Amante. 1968. Alkanones and terpenes in the mandibular glands of *Atta* species. *Comp. Biochem. Physiol.* 26:291-299.

(40)  D.A. Shearer and R. Boch. 1966. Citral in the Nassanoff pheromone of the honeybee. *J. Insect Physiol.* 12:513-521.

(41)  C.G. Butler and D.H. Calam. 1969. Pheromones of the honeybee—The secretion of the Nassanoff gland of the worker. *J. Insect Physiol.* 15:237-244.

(42)  K. von Frisch. 1923. Über die "Sprache" der Bienen. *Zool. Jb., Allg. Zool. Physiol.* 40:1-186.

(43)  J.B. Free and I.H. Williams. 1972. The role of the Nasanov gland pheromone in crop communication by honeybees (*Apis mellifera* L.). *Anim. Behav.* 20: 314-318.

(44)  M. Renner. 1960. Das Duftorgan der Honigbiene und die physiologische Bedeutung ihre Lockstoffes. *Z. Vgl. Physiol.* 43:411-468.

(45)  M. Lindauer. 1951. Bienentänze in der Schwarmtraube. *Naturwissenschaften* 38:509-513.

(46)  D.D. Mautz, R. Boch, and R.A. Morse. 1972. Queen finding by swarming honeybees. *Ann. Entomol. Soc. Am.* 65:440-443.

(47)  M.S. Blum, R.M. Crewe, W.E. Kerr, L.H. Keith, A.W. Garrison, and M.M. Walker. 1970. Citral in stingless bees: Isolation and functions in trail-laying and robbing. *J. Insect Physiol.* 16:1637-1648.

(48)  J.S. Moure, P. Nogueira-Neto, and W.E. Kerr. 1956. Evolutionary problems among the Meliponinae. *Proc. 10th Int. Congr. Entomol.* 2:481-493.

(49)  M.S. Blum. 1966. Chemical releasers of social behavior. VIII. Citral in the mandibular gland secretion of *Lestrimelitta limao. Ann. Entomol. Soc. Am.* 59:962-964.

(50)  F.E. Regnier and E.O. Wilson. 1971. Chemical communication and "propaganda" in slave-maker ants. *Science* (Wash.) 172:267-269.

(51)  R.M. Duffield, M.S. Blum, and J.W. Wheeler. 1976. Alkylpyrazine alarm pheromones in primitive ants with small colonial units. *Comp. Biochem. Physiol.* **54B**:439-440.

(52)  E.O. Wilson. 1962. Chemical communication among workers of the fire ant *Solenopsis saevissima* (Fr. Smith). 3. The experimental induction of social responses. *Anim. Behav.* **10**:159-164.

(53)  B. Hölldobler. 1971. Recruitment behavior in *Camponotus socius* (Hym. Formicidae). *Z. Vgl. Physiol.* **75**:123-142.

(54)  G.L. Ayre. 1968. Comparative studies on the behavior of three species of ants (Hymenoptera: Formicidae). I. Prey finding, capture, and transport. *Can. Entomol.* **100**:165-172.

(55)  M.C. Cammaerts-Tricot. 1974. Piste et phéromone attractive chez la fourmi *Myrmica rubra. J. Comp. Physiol.* **88**:373-382.

(56)  B. Hölldobler. 1971. Homing in the harvester ant *Pogonomyrmex badius. Science* (Wash.) **171**:1149-1151.

(57)  B. Hölldobler, and E.O. Wilson. 1970. Recruitment trails in the harvester ant *Pogonomyrmex badius. Psyche* **77**:385-399.

(58)  U. Maschwitz, B. Hölldobler, and M. Möglich. 1974. Tandemlaufen als Rekrutierungsverhalten bei *Bothroponera tesserinoda* Forel (Formicidae: Ponerinae). *Z. Tierpsychol.* **35**:113-123.

(59)  M. Möglich, U. Maschwitz, and B. Hölldobler. 1974. Tandem calling: A new kind of signal in ant communication. *Science* (Wash.) **186**:1046-1047.

(60)  A. Buschinger. 1972. Giftdrüsensekret als Sexualpheromon bei der Ameise *Harpagoxenus sublaevis. Naturwissenschaften* **59**:313-314.

(61)  B. Hölldobler and M. Wüst. 1973. Ein Sexualpheromon bei der Pharaoameise *Monomorium pharaonis* (L.). *Z. Tierpsychol.* **32**:1-9.

(62)  W. Hangartner. 1967. Spezifität und Inaktivierung des Spurpheromons von *Lasius fuliginosus* Latr. und Orientierung der Arbeiterinnen im Duftfeld. *Z. Vgl. Physiol.* **57**:103-136.

(63)  B. Heinrich. 1974. Pheromone induced brooding behavior in *Bombus vosnesenskii* and *B. edwardsii* (Hymenoptera: Bombidae). *J. Kansas Entomol. Soc.* **47**:396-404.

(64)  J. Ishay. 1972. Thermoregulatory pheromones in wasps. *Experientia* **28**:1185-1187.

(65)  J. Ishay. 1973. The influence of cooling and queen pheromone on cell building and nest architecture of *Vespa orientalis* (Vespinae: Hymenoptera). *Insectes Soc.* **20**:243-252.

# CHAPTER 11

# Behavioral Responses of Coleoptera to Pheromones, Allomones, and Kairomones

J. H. Borden
*Department of Biological Sciences*
*Simon Fraser University*
*Burnaby, British Columbia, Canada*

We are beginning to understand how insects behave. What is understood can be manipulated and managed. This chapter will explore our understanding of the biology and practical implications of behavioral responses of Coleoptera to pheromones, allomones, and kairomones. Rather than to attempt total, comprehensive coverage as did Jacobson (1), I intend to use pertinent, illustrative examples. It is not the intention of this chapter to cover the biogenesis, chemistry, or perception of chemical messengers, or to examine in detail pest-management schemes employing such chemicals.

In a behavioral sense, as well as in its implications for pest management, the most important and significant pheromone-induced behavioral response in the Coleoptera is orientation. Although it is popular to think of orientation as behavior that results in movement toward a stimulus, I have chosen to follow the more precise behavioral definition, taxes toward *or* away from a stimulus—that is, *positive* and *negative* orientation. In most cases, these responses, which are considered in Sections I.A and I.B, are respectively equivalent to Shorey's categories of aggregation and dispersion behavior (2). The majority of positive orientation responses lead to some form of reproductive behavior (e.g., courtship, mating, and/or oviposition), considered in Section I.C.

Behavioral responses to chemical messengers produced by other species are considered in Section II. The chemical messengers include allomones, kairomones, and a third category which I have termed allomone-kairomones.

1. An allomone is "a chemical substance, produced or acquired by an organism, which, when it contacts an individual of another species in the natural context, evokes in the receiver a behavioral or physiological reaction adaptively favorable to the emitter" (3). The majority of allomones identified in the Coleoptera are defensive secretions that convey an important but general chemical message. Because of this generality in effect, behavioral response to defensive allomones is deemphasized in this chapter.

2. A kairomone is "a transspecific chemical messenger the adaptive benefit of which falls on the recipient rather than the emitter" (4). Kairomones are quite significant releasers of behavior in several beetle species, and are extensively covered in this chapter.

3. An allomone-kairomone is a chemical messenger which serves as an allomone to the emitter and as a kairomone to the perceiver. Such dual or multiple functions were recognized by Brown et al. (4), and with more research, may increasingly be considered important in insect communication.

An additional section of this chapter will examine the strategies (not tactics) by which man can or could exploit pheromones, allomones, and kairomones in coleopteran pest management.

## I. BEHAVIOR RELEASED BY PHEROMONES

### A.  Positive Orientation

Many species of beetles employ pheromones that induce positive orientation of other members of their own species to a pheromone source (1). On the basis of the behavior of the responding insects, such chemicals are termed *aggregation pheromones* (2). Aggregation in the Coleoptera may function to bring prospective mates in close proximity, in which case the pheromones may be termed *sex pheromones* (1). In numerous instances, however, aggregations of many beetles of both sexes occur, regardless of whether the pheromone is produced by one or both sexes. Shorey (2) speculates that aggregation pheromones in the Coleoptera arose as mechanisms to cause aggregation at a suitable food source, apparently implying that a sex pheromone function arose secondarily. Indeed, such would appear to be the case in many Scolytidae, in which large aggregations enable a population to overcome a potentially resistant host tree (5) and/or to utilize fully a temporary habitat (6). Similarly, the response of both sexes of the boll weevil, *Anthonomus grandis* Boheman, serves to aggregate populations on cotton (7) (8). This response may be particularly important in establishing early-season populations that must orient from an overwintering site to locate new host plants. A population-aggregation pheromone occurs also in the khapra beetle, *Trogoderma granarium* (Everts) (9), but curiously, only sex pheromones appear to occur in other *Trogoderma* spp. (1) (10), which might also benefit from aggregation in a temporary habitat. Another function of aggregation occurs in *Lycus loripes,* in which both sexes apparently aggregate for mutual defense in response to a male-produced pheromone. The beetles taste bad, and the development of a negative feeding image by predators may protect the entire aggregation from attack once one individual is sampled (11).

The following two sections on response sequences and on maturation and physiological readiness for response contain many aspects that apply to numerous types of pheromone-released behavior. They are considered under "positive orientation" for convenience, and because there is less research on either phenomenon in relation to other types of behavior.

## 1. Response Sequences

It is generally accepted that behavioral responses of insects to chemical messengers take place in a sequence of genetically controlled behavioral events. The responding insect will encounter a series of "take-it-or-leave-it situations" (12). It will proceed from one type of behavior to another (a take-it situation) only when its internal physiological condition and the external environmental factors are favorable. Behavioral orientation sequences almost invariably include initiation of locomotory activity, random and/or directed movement, and cessation of movement (i.e., arrestment). Proposed sequences have been advanced for orientation behavior in host selection and mass attack by bark and timber beetles (Scolytidae) in general (13), and for several species in the genus *Dendroctonus* (14) (15).

## 2. *Maturation of Response Ability, and Physiological Readiness for Response*

### a. Reproductive Maturation

Behavioral response sequences are predicated on completion of maturation processes and on a state of physiological readiness. Immediately following eclosion, newly emerged insects are often unable to respond to pheromones. The inability is of adaptive significance, since at this time they are seldom reproductively mature. For example, *Tenebrio molitor* L. males were unresponsive to female pheromone in the first day of adulthood, but positive orientation increased to a maximum level during the next 4 days (16). A similar maturation period is required for pheromone emission rate to reach a maximum in females (17) and for successful mating to occur (16). Synchronized maturation of both sexes also occurs in other species. *A. grandis* females became responsive to males and males became attractive to females when 2 days old (18). Peak responsiveness occurred in 5 days, corresponding to observations of age-dependent mating frequency (19). Male *T. granarium* reached peak responsiveness when 6 to 7 days old, coinciding with maturation of females and their emergence from puparia (20).

Control mechanisms for the development of response ability are apparently unknown. The increase in pheromone production by female *T. molitor* appears to be in part intrinsic (17), but also appears to be promoted by exposure to primer pheromone(s) produced by either sex (21).

Reproductive maturation may not necessarily coincide with the ability to respond to pheromones. Male *T. granarium* were somewhat responsive to females, even when less than 1 day old (20). Both male and female callow adult *Ips paraconfusus* Lanier excised from pine bark were able to orient to the male-produced pheromone as soon as they were sufficiently sclerotized to walk (22). In nature, however, such beetles would not have been exposed to pheromone until completion of a maturation feeding period and emergence from their host trees as mature beetles (23). Moreover, callow males did not produce attractive frass until 7 days after eclosion (22).

### b. Seasonal Variation in Responsiveness

Seasonal variations in responsiveness may also occur. Female *Dendrocto-nus brevicomis* Le Conte were unresponsive to the pheromone, *exo-brevicomin* during the autumn (24), and female *I. paraconfusus* showed a pronounced decline in response to male-produced frass and frass extract from September to April (22) (25). The seasonal decline in response by female *Ips pini* (Say) stems from two causes (26). Female response to male frass extract made in July was least during the winter, indicating a change in responsiveness, while response to extract made in November was consistently lower than that to the July extract, indicating that overwintering males produce less pheromone. Seasonal variation in pheromone production or sexual responsiveness may be due in part to an overwintering reproductive diapause. Overwintering brood adult *Dendroctonus pseudotsugae* Hopkins in diapause (27) are unresponsive to attractive female frass when excised prematurely from host bark (28). Diapausing male *A. grandis* are definitely pheromone-deficient (29). Female cereal leaf beetles, *Oulema melanopus* (L.) (30) and male *Trypodendron lineatum* (Olivier) (31) are sexually unresponsive during diapause.

## c. Diel Periodicity in Responsiveness

Positive orientation responses to pheromones may follow distinct, species-specific diel periodicity. For example, bark beetles respond to field traps baited with pheromone-producing insects with definite diel periodicity (32)-(34). Rather than representing diel changes in response ability, these patterns may simply reflect periodicity in emergence and subsequent flight from host trees or overwintering sites (35)-(37). In support of this hypothesis, Borden (22) was unable to show definite diel periodicity in laboratory response to male frass extract by *I. paraconfusus* held before testing in natural or constant conditions. In *Trogoderma inclusum* Le Conte and *T. glabrum* (Herbst), however, there is definite, species-specific diel periodicity. Response levels to standard extracts were 50 to 100 times greater at the third and sixth hours of the 14-hr photophase for *T. inclusum* and *T. glabrum,* respectively, than at its beginning or end (38).

Almost no research has been done to determine whether diel periodicity in pheromone-related behavior in the Coleoptera represents true circadian rhythms. However, pheromone-release periodicity in *T. glabrum* persists under constant light, temperature and humidity, can be entrained by imposed photoperiod, and thus appears to be under the control of a truly endogenous circadian rhythm (39).

## d. Dispersal-Dependent, Pheromone-Positive Behavior

In laboratory experiments, many Coleoptera initiate positive orientation on exposure to pheromones. In nature, however, initiation of locomotion may not be in response to a pheromone stimulus. In many groups—for example, the Scolytidae (13)—dispersal flights by physiologically "ready" insects are initiated in appropriate environmental conditions, particularly temperature, relative humidity, and light, as, for example, Henson (40) described for the white pine cone beetle, *Conophthorus coniperda* (Schwarz). Already on the move, the dispersing insect encounters pheromone odor. A behavioral transition must then occur from dispersal to pheromone and/or host orientation. Such transitions were first recognized and investigated in aphids (41) (42), and can be explained by shifting stimulus threshold levels (43). The shift in response from dispersal-inducing to other stimuli is flight-dependent. Only after a requisite flight period will the insect be able to respond to a pheromone stimulus.

Flight-dependent behavioral reversals in the Coleoptera have been investigated only in the Scolytidae. After their discovery in *T. lineatum,* Graham (44) (45) hypothesized that postdiapause beetles in the spring are dominated by photopositive responses (to dispersal-inducing stimuli), and that only after flight experience are they released from photopositive domination and able to respond to host stimuli. Although some unsuccessful attempts have been made to verify the hypothesis in other insects—for example, *I. paraconfusus* (22) (34)—more recent work has disclosed that flight-dependent behavioral reversals do in fact occur in *T. lineatum* and also in *D. pseudotsugae.* Atkins (46) found that female *D. pseudotsugae* became host-positive only after varying amounts of flight exercise. On a laboratory flight mill, *T. lineatum* and *D. pseudotsugae* ceased flight on exposure to the odor of pheromone-laden frass only after elapsed flight time of 30 and 90 min, respectively (47).

Two hypotheses have been advanced to explain the physiological basis for such behavioral reversals. Both depend on internal feedback mechanisms. The first hypothesis (48) proposed that air-swallowing in flight by *T. lineatum* created a large ventricular air bubble that provided an internal feedback indicator (via stretch receptors) of flight duration. *T. lineatum* may, in fact, swallow air during flight. However, large ventricular air bubbles occurred in *T. lineatum* that were denied flight (49), and a photographic record has been made of air-swallowing and the development of a large ventricular air bubble in stationary *D. pseudotsugae* (47). While the ventricular air bubble may have some function as an indicator of the physiological condition of a beetle, it more probably functions as an indicator of barometric pressure changes (47). The alternative hypothesis is that the volume of lipid reserves and their consumption in flight exercise provide biochemical feedback indicative of the physiological condition of the beetle (50). This hypothesis has been investigated in *D. pseudotsugae.* Beetles with more than 30% lipid (per dry weight) were flight-positive and host-negative, whereas those with less than 20% lipid were capable of flight but were host-positive if given the appropriate stimuli (46). In flight, male *D. pseudotsugae* selectively oxidize the monounsaturated fatty acids C16:1 and C18:1 (51), suggesting that specific metabolites of lipid oxidation may be biochemical feedback indicators of physiological condition. The adaptive advantage of control of behavior by indicators of lipid metabolism is obvious. Beetles with great energy reserves could afford the luxury of extended dispersal before responding to pheromone and/or host odors. Those with lesser lipid reserves would conserve them for reproductive metabolism by responding to the first pheromone and/or host stimulus encountered.

### e. Effect of Previous Reproductive Experience

Previous sexual or reproductive experience may also influence the physiological readiness to respond to pheromones. Mated male *T. granarium* were significantly less responsive than unmated males to female sex pheromone extract (20). Mated female *A. grandis* were less responsive than virgin females to males in laboratory bioassays (52) and field tests (53). Borden (22) found that reproducing female *I. paraconfusus* excised from host logs were significantly less responsive than emerged brood females to male frass extract. The control of such response inhibition is unknown, but it is unlikely to be due to either sensory adaptation or habituation since the inhibition is persistent. Rather, there is more likely to be a hormonally controlled feedback inhibition of responsiveness following reproductive activity.

### 3. Orientation Mechanisms

### a. Characterization of Taxes

There is very little definitive research on the precise type of behavioral response by which beetles orient positively to pheromones. Most of the evidence consists of observations made in the course of other studies. For example, male *Diabrotica balteata* (Le Conte) respond to a pheromone lure by rising from the plants in which they were resting and approaching upwind in a characteristic hovering flight (54). The upwind orientation flight of *T. lineatum* has been observed to be steady, with frequent turns not exceeding 2 yd to either side of the straight line of approach (55). Wood and Bushing (25) describe the orientation of walking *I. paraconfusus* as being characterized by alternating movements to the left and right.

There is apparent, still-unresolved confusion as to whether olfactory orientation responses by insects are tropotactic or klinotactic (2). Fraenkel and Gunn (56) speculate that most olfactory orientations are klinotactic. An animal, usually with a single-intensity receptor, would compare stimulus intensities by a succession of lateral deviations. Alternatively, in tropotactic orientation, responding insects with paired sensory organs (antennae) make successive simultaneous comparisons of stimulus intensity on either side. The two taxes are not separated on the basis of behavioral observations. Therefore, arguments for one mechanism or the other tend to be more speculative than definitive. For example, Kennedy (57) finds it difficult to accept that odor gradients would be steep or steady enough over the minute distance between a beetle's antennae for it to make the definitive and reliable simultaneous comparisons of odor concentration that are a prerequisite for tropotactic orientation. He cites Wright's hypothesis (58) that, due to small-scale turbulence, insects are exposed to pulses of smell, and suggests that detection of increasing frequency of odor pulses as the odor source is approached is sufficient to induce klinotactic orientation. However, the antennal morphology of many insects is such that, regardless of the distance between antennae, one lateral receptor area could be shielded from olfactory stimulation almost as effectively as a lateral photoreceptor placed in the shadow of a body on its lateral deviation from the stimulus source. In the genus *Ips,* only one side of the antennal club bears olfactory sensilla (59). Thus, the beetle can hold its antennae such that the sensilla on only one side are directed toward the odor source following lateral deviation of the beetle to either side. In the genus *Dendroctonus,* however, the olfactory sensilla extend in bands completely around the antennal club (60). Even this morphology does not preclude differential intensity comparison, since the sensilla on either side of both clubs could be differentially stimulated.

Experimental testing of Fraenkel and Gunn's criteria (56) has yielded helpful, but not conclusive, evidence in favor of klinotactic orientation. Unilateral removal of receptors in tropotactically orienting insects should lead to circus movements in a uniform stimulus field, and often should produce the same result in a directed stimulus or stimulus gradient (56). In fact, *I. paraconfusus* orients quite successfully with only one antenna, both in the field (61) and the laboratory (59). These results led Borden and Wood (59) to designate the orientation response by *I. paraconfusus* as a chemoklinotaxis.

Almost invariably, laboratory tests of behavioral response to pheromones have involved deploying attractive pheromones in some sort of airflow apparatus. It has been generally assumed that an airflow was necessary, leading to the inclusion of wind in response terminology. For example, Borden (13) described the characteristic response of scolytids as an "anemo-chemo-klino-taxis." However, beetles have never been tested to determine if wind is necessary for successful positive orientation to occur. Experimental examination of this question, as in the pink bollworm moth, *Pectinophora gossypiella* Saunders (62), may have a considerable influence on the design and operation of pest-management systems employing pheromones.

## b. Role of Vision in Olfactory Orientation

Visual cues may also act in the complex of stimuli governing positive orientation responses of beetles to pheromones. *C. coniperda* oriented to silhouettes following an initially photopositively oriented flight (40). Lilly and Shorthouse (63) concluded that in the 10-lined June beetle, *Polyphylla decemlineata* (Say), response to the visual stimulus of silhouetted trees supplements the response of males to female-emitted pheromone. The mountain pine beetle, *Dendroctonus ponderosae* Hopkins, oriented more readily to horizontal, rather than vertical, "log" configuration olfactometers (64). Color has been implicated as an important factor only in *A. grandis*. Green and yellow pheromone traps are superior to those of other colors (65)-(67). Both colors presumably correspond with the spectral reflectance of the cotton plants on which pheromone-emitting males are found.

### 4. Nature and Function of the Chemical Message

The chemical message that induces positive orientation ranges from a very simple, one-compound message to a chemical complex involving more than on῾ pheromone compound, and even host-plant compounds. Examples of insects in which there is a one-compound pheromone message are: the black carpet beetle, *Attagenus megatoma* (F.) (68); the sugar beet wireworm, *Limonius californicus* (Mannerheim) (69); the grass grub beetle, *Costelytra zealandica* (White) (70); and the furniture carpet beetle, *Anthrenus flavipes* (Le Conte) (71). Many beetle species utilize complex pheromone messages. These include: several scolytids, as reviewed by Borden, VanderSar, and Stokkink (72) and Borden (13); *A. grandis* (73); *T. inclusum* (74); and *T. glabrum* (75).

Host-plant compounds have been implicated in the attractive chemical message in several species. Apparently they are utilized as kairomones by the beetles, promoting the orientation to host plants, as one component of the host plant-pheromone orientation process. The addition of sugar cane pieces to chambers with pheromone-producing male New Guinea sugar cane weevils, *Rhabdoscelus obscurus* (Boisduval), increased the female response (76). Similarly, more *A. grandis* of both sexes responded to the pheromone complex, grandlure, when it was deployed with an extract of cotton square (77). Although host compounds and beetle pheromones may act in concert, it is not necessarily a synergistic mechanism. For example, the increased response of *R. obscurus* when sugar cane odor is added to that of male beetles can be accounted for by the response to sugar cane alone (76).

Specific host compounds that contribute to a pheromone-based, orientation-inducing chemical message have been identified only in the Scolytidae. These are myrcene (78) (79) and 3-carene (80) in *D. brevicomis*; α-pinene in the southern pine beetle *Dendroctonus frontalis* Zimmermann (81), *D. ponderosae* (80), and *D. pseudotsugae* (82); camphene in *D. pseudotsugae* (83); (-)α-cubebene in *Scolytus multistriatus* (Marsham) (84); and ethanol in *D. pseudotsugae* (85), and *Gnathotrichus sulcatus* (Le Conte) (86).

The exact behavioral roles in orientation played by individual host or beetle-produced compounds acting alone or in combination with other compounds is unknown. A single compound could have multiple behavioral roles. For example, it might function to induce locomotory activity, to act with wind in guiding positive orientation, and, at higher concentrations, to induce arrestment or to stimulate other responses in addition to orientation—for example, mating or oviposition. A mixture of compounds could act as a single chemical. Alternatively, individual components of the mixture might control or share in the control of various components of the behavioral response sequence. The role of individual compounds could be elucidated in part by definitive behavioral experiments. In addition, electrophysiological studies might precisely define the behavioral role of individual compounds by monitoring not only sensory perception—for example, in the antennae (87) (88)—but also motor neuron and muscle activity—for example, in flight muscles, legs or the genitalia.

### d. Arrestment

Regardless of the mechanisms, successful, directed, positive orientation does occur. At some point, taxes must cease—that is, the beetle must be arrested. In some cases, it is possible to bypass the normal behavioral phases of dispersal and directed orientation in the laboratory by the use of an arrestment olfactometer (89)-(91). The insects are usually induced to orient to some other stimulus—for example light—and are arrested on suddenly encountering an olfactory stimulus. While such responses can be termed *true arrestment,* and indicate a physiological readiness to respond, they are probably a laboratory artifact, since dispersal and/or positive orientation to pheromones would rarely be bypassed in nature.

Arrestment behavior in nature is usually a consequence of prior orientation. A physiologically ready beetle—that is, after sufficient flight experience and long-range orientation to an odor source—must encounter a chemical and/or visual message that indicates that directed orientation must cease, and some other type of behavior commence.

Form and/or color of a substrate could be important in arrestment behavior, in addition to their role in long-range orientation. However, there is no evidence to support this hypothesis in the Coleoptera, and it would be difficult indeed to devise a situation in which the role of vision in directed orientation and arrestment behavior was experimentally separated.

If chemicals were to act as stimuli for arrestment, they could act in at least three ways:

1. A relatively nonvolatile host chemical or pheromone could signal immediate proximity to the source of a more volatile, attractive pheromone to which a beetle has oriented. There is no conclusive evidence that such a phenomenon occurs in any coleopteran species.

decline in attraction if pheromone-emitting beetles simply ceased to produce and/or release pheromones that induce positive orientation.

There are many reports in the literature of decline in positive attraction after mating. It is possible that many of the observations could be accounted for by the production of repellent or antiattractive pheromones. Despite the obvious possible uses of such pheromones in pest management, almost no research on the phenomenon (outside of the Scolytidae) has been initiated. Only when both sexes, or their pheromone-producing organs are tested for pheromone activity alone, and in all possible combinations as for *T. molitor* (99) or *T. lineatum* (100), will conclusive behavioral evidence for repellent or antiattractive pheromones be collected. However, some evidence does exist. Mated or unreceptive females of the ground beetle *Pterostichus lucublandus* Say release a pheromone that somehow deters male response (101). The effect is to protect the eggs from predation by males. In *T. molitor* a similar phenomenon occurs (99). Males produce an "antiaphrodisiac" that deters positive orientation by males to females paired with males or to which the pheromone has been transferred during mating. Happ (99) suggests that the antiaphrodisiac pheromone acts to ensure that newly mated females utilize freshly transferred spermatozoa before remating.

Pheromones that induce negative orientation or inhibit positive orientation have proven most often to be associated with beetle habitat. Most research has been on the Scolytidae in which the need for mechanisms that govern gallery spacing and terminate aggregation, so as to prevent overpopulation of a limited host resource, has long been recognized.

The galleries of *D. ponderosae* (103) and *Scolytus ventralis* Le Conte (103) are evenly distributed over available bark surface. The discovery that female *Dendroctonus* stridulate when initiating an attack (104) suggests that at least part of this spacing could be controlled by audio communication. However, spacing could also be regulated by relatively nonvolatile, repellent pheromones or by close-range repellency of otherwise attractive pheromones. Such a communication system appears to exist in the stored-product pest, *Tribolium confusum* Duval. Female *T. confusum* apparently release a pheromone that repels other females, thus causing them to be dispersed in a nonrandom way throughout a medium (105).

Control of the termination of mass attack in the Scolytidae has been investigated most extensively in *D. pseudotsugae*. The research followed two separate routes.

Behavioral research established that a *D. pseudotsugae* female can chemically "mask" her aggregation pheromone in response to a male stridulating at the gallery entrance, or to a nonstridulating male placed in her gallery (106)-(109). The mask is lost within 9 to 14 min after males are removed from the galleries (106) or after stridulation ceases (109). The mask

2. Arrestment may simply be a consequence of encountering a suffic
   high concentration of an otherwise attractive pheromone. Su
   phenomenon would appear to be the case in *C. zealandica* in which
   numbers of beetles were arrested within 1 to 2 ft of a concen
   attractive odor source (92). In laboratory olfactometers, *I. pa*
   *fusus* (93) and *G sulcatus* (93a) are arrested by very concen
   pheromone, and fail to orient upwind. Such a phenomenon could se
   arrest scolytids near the entrances of galleries containing potential ma

3. An arrestant chemical present in extremely small amounts m:
   perceived in sufficient concentration only in close proximity to a sou
   an attractive pheromone. This hypothesis could account for dat
   indicate that verbenone, 3-methyl-2-cyclohexen-1-one (MCH) and *exc*
   comin are multifunctional pheromones, respectively, for *D. frontal*
   *D. pseudotsugae* (95), and *D. ponderosae* (96). However, these dat
   not be conclusive. In all three cases the compounds were diluted in e
   which is now known to function as an active chemical in conjunction
   *pseudotsugae* pheromones (85).

### e. Inhibition of Dispersal

Following arrestment, other types of essential behavior—for
reproductive activity—must occur. Therefore, the beetles must be de1
least temporarily from response to dispersal stimuli. Reports of
sensory adaptation in *I. paraconfusus* (22) and habituation in *T. inclu:*
on exposure to pheromone, and consequently reduced or absent or
responses, suggest that these beetles might then regain responsiv
dispersal stimuli. However, even though sensory adaptation results in (
orientation to a pheromone source in *I. paraconfusus*, exposure to ph
odor also inhibits flight initiation (22). Thus, in *I. paraconfusus*
adaptation to pheromones and inhibition of response to dispersal st
not mutually exclusive. In the Scolytidae, which remain in their hos
logs for some time following orientation to them, reversible degene
the flight muscles at least temporarily precludes further dispersal (13)

## B. Negative Orientation and/or Inhibition of Positive Orientat

Many beetle species employ pheromones that cause behavior in
to positive orientation. The effect may be to repel others of the sar
or to act as "antiattractants" (98) that inhibit perception of or r
attractive pheromones. Such pheromones may signal that a prospect
no longer a virgin, or that a given habitat or region thereof is full t
They provide for a more rapid and efficient effect than the m(

functions within a limited area around the producing female's gallery entrance (108). Thus, it may function at least partially as a spacing control, since females orienting to a source of secondary attraction (aggregation pheromone plus host odor) could still respond to host volatiles and initiate attack on uninfested tissue. Moreover, males would not be inhibited from orienting to unmated females. When the host tree or log is "full," the mask would be emitted from all available infestation sites, and attack would be terminated.

Coincident with behavioral studies, the isolation of volatile chemicals from the hindguts of female *D. pseudotsugae* was under way. One of these isolated volatiles was methylcyclohexenone (MCH), which was shown to have an arrestant effect on males in the laboratory (110). Field tests with MCH and other attractive pheromones disclosed a startling result. It was not attractive when tested alone, and inhibited response to pheromone-baited host trees and to other pheromones and host-tree volatiles when added to a bait mixture (111) (112).

The exact mechanism by which MCH is employed as an antiaggregative pheromone is still unclear. Rudinsky et al. (113) present data indicating that MCH is, in fact, the pheromone mask discovered in earlier behavioral experiments, and that it is released by females in response to male stridulation. However, Pitman and Vité (114) discovered that emergent male *D. pseudotsugae* contain large quantities of MCH, whereas females do not. Thus, the males could play a significant role in inhibiting the response of other males to a female that has already been joined by a male.

Pheromones that are repellent or antiattractive apparently occur in several scolytid species. Verbenone inhibits positive orientation by *D. frontalis* (81) and *D. brevicomis* (14) (15). MCH appears to have the same effect on *Dendroctonus rufipennis* (Kirby) as on *D. pseudotsugae* (115); *exo*-brevicomin at high concentrations inhibits attraction of *D. ponderosae* to secondary attraction (96); and *endo*-brevicomin apparently functions as an antiaggregation pheromone in *D. frontalis* (96) (116). Field and laboratory studies indicate that female *T. lineatum* produce a pheromone mask (100) (117) (118), although no chemicals have been isolated or identified.

In addition to their role in attack termination, repellent or antiattractant pheromones may be responsible for other phenomena. For example, verbenone might control the "switching phenomenon" by which the center of *D. frontalis* aggregation is shifted from saturated to unattacked trees (119). MCH and verbenone have been implicated in complex chemical and sonic communication systems in *D. pseudotsugae* and *D. frontalis,* respectively (94) (95), and reportedly act in a concentration-dependent manner to induce attraction, arrestment, and stridulation, as well as attack termination. *Exo*-brevicomin and *endo*-brevicomin have been implicated also in rivalry behavior of *D. ponderosae* and *D. frontalis,* respectively (96) (116).

The question as to whether pheromones such as MCH are repellent or antiattractant has not been investigated. Yet it is potentially of considerable significance in pest management. A truly repellent pheromone could be used as an olfactory "barrier" to protect uninfested host material. Alternatively, an antiattractant pheromone could function as a "blanket." Beetles would fly through an area in which such a pheromone was dispersed, but be inhibited from perceiving and/or responding to an attractant pheromone. The effect would be the same, but the method of deployment would differ. Field tests with MCH deployed more or less as a "blanket" have been at least partially successful in deterring attack by *D. pseudotsugae* (120) (121) and *D. rufipennis* (121). However, the mode of action is not evident from the results. Electrophysiological recording of both sensory and motor activity, coupled with precise behavioral investigation, might elucidate the mode of action of these pheromones.

## C. Reproductive Behavior

Pheromonal control of reproductive behavior—that is, courtship, copulation, and oviposition—has not been extensively investigated in the Coleoptera. In many cases, research has been concentrated on positive orientation responses. Even when reproductive behavior has been observed in apparent response to pheromones, it has almost invariably been ignored as a separate phenomenon. Almost never have attempts been made to isolate the role of pheromones from other stimuli in controlling reproductive behavior—for example, by inducing courtship or copulation attempts with pheromone-treated inanimate objects or members of the same sex.

Several observations have been made of the apparent induction of courtship and copulatory behavior by pheromones. For examples, pheromone odor induces stridulation in *I. paraconfusus* females (122) and *D. pseudotsugae* males (109), the first step in close-range communication and behavior leading to recognition and admission of the opposite sex to the nuptial gallery, and ultimately, copulation. Sexual recognition of female lucanids by males is apparently by close-range chemotactile means (123). In courtship, some male meloids draw the female antennae through their frontal grooves in which there are many pores through which an aphrodisiac pheromone could be released (124). Malachiid beetles exhibit an elaborate courtship controlled by dermally secreted female pheromones (125). *T. molitor* males will attempt to copulate with a glass rod treated with female pheromone extract (126) (127). Copulatory movements and homosexual activity in response to pheromone extracts have been observed in the elaterid, *Limonius californicus* (Mannerheim) (128) and the cigarette beetle, *Lasioderma serricorne* (F.) (129).

For *T. confusum,* three copulation-inducing pheromones have been identified as *1*-pentadecene, *n*-hexadecene and *1*-heptadecene (130). All three compounds are found in both sexes, but appear to excite only males.

Oviposition may also be pheromone-induced. In laboratory bioassays, female *T. molitor* often were observed to extrude their ovipositors in response to male odor (99). A *Tribolium castaneum* (Herbst) female left with a choice of two elderberry pith disks, respectively impregnated with the extract of *T. castaneum* and *T. confusum* pupae, was found 50 days later with 11 larvae in the disk impregnated with *T. castaneum* pupal extract (131). However, the response could have been accounted for by repellancy of the disk impregnated by *T. confusum* pupae, and a relative neutrality of the *T. castaneum* disk.

Aggression associated with reproductive activity also may be controlled by pheromones. Lucanid males, alerted by visual detection of a conspecific beetle, apparently recognize its male odor and commence aggressive activity (123). The scolytid pheromones, *exo*-brevicomin and *endo*-brevicomin at appropriate concentrations, evoke rivalry behavior, including stridulation, in male *D. ponderosae*; and verbenone and *endo*-brevicomin induce a similar rivalry in male *D. frontalis* (96). In the presence of females, aggression by *I. paraconfusus* males is increased and directed preferentially against other males (132). As in vertebrates, such aggressive activity could be of selective advantage in that it would result in maintenance of pair bonds, and the most vigorous males mating preferentially with receptive females.

Control of pheromone-associated reproductive behavior may be at several levels. First, it may require a maturation period, and often completion of diapause. Reference was made previously to a requisite maturation period in several species prior to emission of or response to attractive pheromones that signal female receptivity. In *O. melanopus* (30) and *T. lineatum* (31) mating will occur only after a requisite diapause by females and males, respectively. In both species, however, no direct evidence is reported for pheromonal control of reproductive behavior. Diel periodicity, possibly reflecting circadian rhythms, may also occur. For example, mating frequency by both *T. inclusum* and *T. glabrum* followed distinct diel curves throughout the photophase, very similar to the corresponding frequency curve for male response to pheromone, but with peak frequencies slightly later for each species (38).

Once active, a pheromone-positive beetle may go through a sequence of behavioral events leading to copulation. Apparently, these events are dependent on concentration of the same pheromone. Thus, *L. californicus* males orient excitedly toward an attractive wind-borne extract of female beetles, but at a certain point (presumably reflecting a threshold pheromone concentration) they cease orientation and commence milling about, moving their antennae, extruding the genitalia, and mounting each other (128). Similarly, *L. serricorne* males exhibit a series of discrete activities, each dependent on increased exposure to female extract, as follows: antennal elevation, leg extension, locomotion, and copulatory movements (129).

## II. BEHAVIOR RELEASED BY ALLOMONES AND KAIROMONES

### A. Allomones

Many beetles produce defensive secretions that obviously function as allomones. In general, they lack species specificity in their targets or type of behavior elicited. There are few data on the behavioral response of beetles to defensive allomones, and such responses will not be considered in this chapter.

### B. Kairomones

In the Coleoptera, most of the behavioral responses to kairomones enable certain species to find a habitat infested by its commensal or host species. Thus, the myrmecophilic staphylinid, *Atemeles pubicollis* Bris., uses the odor of its host ant colonies in host-finding (133), while *Amphotis marginata* F. follows its host's trail pheromones (134). Similarly, response by some stored-products insects to *T. granarium* pheromones (135) may be a kairomone response by which suitable habitats may be located, although there is no evidence that such a response occurs in nature. Predaceous clerids and trogositids utilize pheromones of their host bark beetles as kairomones (13). I was able to find no evidence that other coleopteran predators—for example coccinellids—orient to their hosts by employing kairomones. Should such phenomena occur, they could have great utility in pest management, particularly in attracting general predators to specific crops and/or hosts, or preventing their departure.

Curiously, there is much more known about specific kairomone chemicals to which scolytid predators respond than the behavioral-response mechanisms involved. This is because the phenomenon was discovered as a "spin-off" from field research on synthetic bark beetle pheromones. The first kairomone response by a scolytid predator was discovered by Wood et al. (136), who trapped numerous *Enoclerus lecontei* (Wolcott) on field traps baited by three synthetic *I. paraconfusus* pheromones. *E. lecontei* also responds to the odor produced by *I. pini* (137). Three other predators have been captured by the thousands in response to bark beetle pheromones. The clerid, *Thanasimus dubius* (F.) responds to the *D. frontalis* pheromone, frontalin, alone or in combination with oleoresin or *trans*-verbenol, and exhibits the same diel periodicity as its prey (138). Addition of verbenone, which is hypothesized to be an attack termination factor in *D. frontalis* (81), caused an increase in the ratio of male response, but lowered the overall response level (138). Another clerid, *Thanasimus undatulus* (Wolcott), has been attracted in very large numbers to frontalin used as an attractant for *D. pseudotsugae* (139), and *D. rufipennis* (115) (140). The trogositid, *Temnochila chlorodia* (Mannerheim), responds readily to the *D. brevicomis* pheromone, *exo*-brevicomin (79) (141). The response can be truly amazing. In a 65-km² population suppression test for *D. brevicomis*, the attractant mixture, *exo*-brevicomin, frontalin, and myrcene in a 1:1:1 ratio attracted 86,000 *T. chlorodia* in a 2-month period (142). This response represented one *T. chlorodia* for every 6.9 *D. brevicomis* captured.

A commentary on the lack of behavioral data on the kairomone response by scolytid predators is that only very rarely have the captured beetles even been sexed. Although there has been a small amount of research on flight periodicity (143) (144), there has been very little research done specifically to investigate the kairomone response by any predator. Moreover, although some studies have approached the reproductive biology of predators—for example, *E. lecontei* (145) (146)—no research has attempted to investigate intraspecific pheromone communication by kairomone-responding predators.

## C. Allomone-Kairomones and the Maintenance of Species Isolation

The problem of species-specific response mechanisms to insect pheromones and the maintenance of species isolation has long intrigued entomologists. There are frequent records of the natural attractants or synthetic pheromones of one species attracting members of another species in the same genus, or in different genera or even families.

If pheromone compounds and/or habitats are shared by more than one species, specificity could be maintained by such mechanisms as temporal and geographic separation, or by the use of multichemical pheromone or host compound-pheromone complexes, which may vary in the presence of particular components, or in the relative concentrations and/or isomerism of the same components.

Multichemical attractant complexes that ensure species specificity have been shown to function in the Coleoptera by either of two mechanisms. In the first mechanism, members of a given species may not respond unless the "right" bouquet is present on the wind. In such a case, there is no transspecific chemical message. An insect would simply be unresponsive unless each component of its entire pheromone complex were present in at least approximately the "correct" proportions. An example of such a case can be found in the relationship between *Ips latidens* (Le Conte) and *I. paraconfusus*. When ipsenol and *cis*-verbenol, two of the three *I. paraconfusus* pheromones (147), were deployed in the forest, *I. latidens* responded (148). However, when the third pheromone component, ipsdienol, was present, only *I. paraconfusus* responded (136) (148).

The alternative mechanism of maintaining species specificity in pheromone response or habitat utilization is to employ an additional compound as an allomone-kairomone. The behavioral effect of such a compound would be to curtail orientation or actively to repel insects in one or more species. It would function as an allomone since it would be of benefit to the emitting species not to attract other species to, for example, a desirable but already-occupied habitat. Interspecific competition for food and shelter would be avoided. For the same reasons the chemical would function as a kairomone to the perceiving species, which would also benefit by avoiding interspecific competition and the energy wasted responding to potential "mates" of the "wrong" species.

Although there is circumstantial evidence for several allomone-kairomone mechanisms [e.g., *endo*-brevicomin in frontalin-responding *Dendroctonus* (149)], only one has been elucidated in detail, that occurring between *I. paraconfusus* and *I. pini* (150). The attractive pheromone components of *I. paraconfusus* are known to be *cis*-verbenol, ipsenol, and ipsdienol (147), whereas *I. pini* is reported to produce *cis*-verbenol and ipsdienol (151). In laboratory and field tests, ipsenol (which is produced by male *I. paraconfusus*) inhibited the response of *I. pini* to *I. pini* pheromone extracts (laboratory tests) or male beetles (field tests). Linalool, which is produced by male *I. pini* (152), at least partially inhibited the field response of *I. paraconfusus* to *I. paraconfsus* males. Thus, ipsenol and linalool function as transspecific chemical messengers that ensure sole occupation of a new habitat by the first-arriving species (150). This effect is of obvious benefit to both species; both compounds can truly be termed allomone-kairomones. It should be noted that linalool, produced by *I. paraconfusus* females (152), the responding sex, probably also functions as a pheromone that signals that a given male is adequately supplied with mates and that a given host is saturated with established galleries containing reproducing insects of both sexes.

## III. STRATEGIES FOR THE USE OF PHEROMONES, ALLOMONES, AND KAIROMONES IN COLEOPTERAN PEST MANAGEMENT

Our knowledge of the behavioral response of beetles to insect-produced chemicals, and of the chemicals themselves, is at best fragmentary. Therefore, it is not surprising that pest-management practices utilizing insect-produced, behavior-modifying chemicals are not widely used and, to date, are relatively unrefined. However, there have been several promising developments, and there is now evidence that there is a real use for such chemicals in pest management.

Table I summarizes the strategies for application of pheromones, allomones, and kairomones in coleopteran pest management. In each case,

TABLE I.

Tested or Hypothetical Strategies for the Use of Pheromones,
Allomones, and Kairomones in Coleopteran Pest Management

| General Strategy | False Biological Message Conveyed or Condition Created | Specific Strategy, with Selected References |
|---|---|---|
| 1. Exploitation of positive orientation and/or arrestment to pheromones | A potent source of natural attraction is present; pheromone-positive behavior will be of advantage | a. Use of pheromones in detection and survey (7) (10) (153) (154) |
| | | b. Use of pheromones to reduce population numbers by mass traping (8) (140) (142) (153) (155)-(157) |
| | | c. Attraction of beetle populations to selected trap crops (158)-(163) |
| 2. Disruption of positive orientation and/or arrestment to pheromones | The level of natural attraction is so great and/ or widespread that pheromone-positive behavior is impossible, or would not be of advantage | a. Saturation of an environment with attractive pheromone such that directional orientation is precluded, and/or sensory adaptation or habituation occurs (164) |
| 3. Exploitation of response to antiattractive or repellent pheromones | All potential mates in the area are mated, or all available habitats are fully utilized by members of the same species | a. Saturation of an environment with an antiattractive pheromone that would preclude positive orientation (112) (115) (121) |
| | | b. Creation of selected sources of repellency to protect potential hosts from attack (112) (115) (121) |
| 4. Exploitation of kairomone response by entomophagous beetles | Potential host insects in considerable numbers are present; kairomone-positive behavior will be of advantage in host selection | a. Use of kairomones to concentrate both host insects and naturally present entomophagous insects in selected habitats |
| | | b. Concentration of entomophagous insects in selected habitats in an inundative release program |
| 5. Exploitation of response to allomone-kairomones | All available habitats are occupied by a competitor species | a. Deployment of allomone-kairomones to protect selected hosts from attack by either or both species |

effective implementation of the strategy depends on "fooling" natural populations by the creation of a false biologic message or condition. The success of each strategy will depend on the extent to which the natural populations are diverted from their normal activities. The first 3 general strategies in Table 1 have been researched to at least some extent, and in some cases demonstrated to have real management potential or efficacy. Many of these demonstrations, identified by selected literature references, are considered in more detail in Section V of this book.

The vulnerability of beetles to behavior-modifying chemicals in pest management depends on their natural response mechanisms. If beetles orient toward a natural pheromone source, they should orient to an artificial facsimile of that source. Thus, they become vulnerable to the use of pheromones in detection and survey, mass trapping, and trap-crop strategies.

It is not known if pheromone concentrations in nature ever build up to the point at which disruption of positive orientation and/or arrestment occurs. However, it would be of adaptive advantage in that intraspecific competition might, thereby, be avoided. Disruption techniques have rarely been considered in coleopteran pest management. For example, Hardee (157) makes particular note of this omission in *A. grandis* pheromone research. Perhaps the omission is because of the demonstration of population aggregation pheromones in many species, suggesting that population suppression by trapping both sexes would be more effective than disrupting orientation of one sex to the other. Yet, if broadcast deployment of pheromones can be used successfully to disrupt positive orientation in the Lepidoptera, it may be equally effective in the Coleoptera. When detailed behavioral research is done, the data suggest that disruption could be used. For example, *I. paraconfusus* orientation to attractive frass extract was significantly lowered after a 15-min exposure to the odor of attractive frass, suggesting that fairly rapid sensory adaptation occurred (22). When *L. californicus* males were tested for response to females with 30 min between tests, their response was normal at each test. However, when the between-test interval was reduced to 10 min, the response ceased after four or five tests, suggesting inhibition of sensory neurons (165). The response of *T. inclusum* to synthetic pheromone was lowered after as little as a 1-min exposure to the pheromone odor; "habituation" (lowered response) persisted for at least 4.5 hr (97).

Antiattractive and/or repellent pheromones have obvious potential applications in which a natural response is artificially induced (Table 1). However, their use would be facilitated by research that identifies whether they are antiattractive, repellent, or both.

The utilization of kairomones to manipulate entomophagous Coleoptera has not been investigated. In fact, the great efficiency of orientation by trogositids and clerids has been considered a possible impediment to the use of pheromones in scolytid pest management because of the large numbers of predators caught and killed (13). However, this orientation efficiency could conceivably be exploited selectively to concentrate predators on their hosts, as has been demonstrated in parasitic hymenoptera (166,167).

Although there have been many defensive allomones identified in the Insecta, none are currently used in pest management (168). The more probable candidates for use in pest management are compounds used as allomone-kairomones in the maintenance of species specificity in habitat utilization. This as yet unresearched strategy might employ single- or multiple-compound applications depending on the number of target species present in the area.

## IV. CONCLUSIONS

It should be clear that we know relatively little about the responses of beetles to behavior-modifying chemicals. What information we do have is primarily from a handful of species: some scolytids of importance in the forest, a few stored-products' insects primarily in the families Dermestidae, Anobiidae, and Tenebrionidae, and in agriculture, mainly the boll weevil, and, to a lesser extent, a few elaterids. The information is fragmentary and often based on circumstantial evidence. However, there are sufficient data to allow some generalizations to be made about response sequences, physiological control of response behavior, environmental factors influencing responses, and the various roles played by different behavior-modifying chemicals. From these data it is also possible to propose certain strategies for the use of pheromones, allomones, and kairomones in coleopteran pest management.

To date there are no species about which one can state that its responses to behavior-modifying chemicals are truly understood. Moreover, there are multitudes of species about which nothing whatsoever is known. The needs are primarily for data in three basic areas: physiology, behavior, and applied biology.

There is a great need for data on the physiological prerequisites for response to behavior-modifying chemicals. These include the physiology of maturation, the physiological basis of seasonal and diel periodicity, and the metabolic effects of dispersal and prior reproductive experience. In particular there is a need to investigate the hormonal control of behavior and the interaction between pheromones and hormones. There is also a need for electrophysiological studies, which will provide specific information on sensory mechanisms, neural coordination, and motor responses of beetles exposed to behavior-modifying chemicals.

To be able to apply pheromones, allomones, and kairomones as pest-management tools we must understand the behavioral mechanisms utilized by responding (and nonresponding) insects. We must understand behavioral-response sequences and the mechanisms by which one type of behavior is replaced by another. There must be comprehensive data on the effects of environmental factors on behavioral responses, the interaction between chemical, visual, and other stimuli, and on the variation in response to chemicals at different concentrations and as individual components of complex chemical messages. The response behavior of many more species must be studied, including the kairomone response by predators to pheromones produced by their prey. Lest we be fooled, we must also know when and how insects communicate by other than chemical means—for example, in dense natural populations in which naturally produced pheromones build up to potentially disruptive concentrations.

Finally, all possible strategies for applying pheromones, allomones, and kairomones in coleopteran pest management must be explored, developed, and continually refined. An array of tested behavioral techniques based on behavior-modifying chemicals must be available in the modern pest-management arsenal.

Without sufficient behavioral data and techniques, our predictive and management capacities will be poor. We will not really know when beetles in a given population are able to respond to behavior-modifying chemicals. We will not understand their sensory and motor capabilities, or their response mechanisms and how they are controlled or modified. There will be many successful applications about which we will wonder, and there will be many failures that we will not be able to explain.

## REFERENCES

(1)  M. Jacobson. 1972. Insect Sex Pheromones. Academic, New York.
(2)  H.H. Shorey. 1973. Behavioral responses to insect pheromones. *Ann. Rev. Entomol.* 18:349-380.
(3)  W.L. Brown, Jr. 1968. An hypothesis concerning the function of the metaplural glands in ants. *Am. Natur.* 102:188-191.
(4)  W.L. Brown, Jr., T.E. Eisner, and R.H. Whittaker. 1970. Allomones and kairomones: transpecific chemical messengers. *Bio Sci.* 20:21-22.
(5)  R.F. Anderson. 1948. Host selection by the pine engraver. *J. Econ. Entomol.* 41:596-602.
(6)  M.D. Atkins. 1966. Behavioral variation among scolytids in relation to their habitat. *Can. Entomol.* 98:285-288.
(7)  D.D. Hardee. 1972. A review of literature on the pheromone of the boll weevil, *Anthonomus grandis* Boheman (Coleoptera: Curculionidae). *USDA, Coop. Econ. Insect Rep.* 22:200-207.
(8)  W.H. Cross. 1973. Biology, control and erradication of the boll weevil. *Ann. Rev. Entomol.* 18:17-46.

(9)   H.Z. Levinson and A.R. Bar Ilan. 1967. Function and properties of an assembling scent in the khapra beetle *Trogoderma granarium*. *Riv. Parassitol.* 28:27-42.

(10)  W.E. Burkholder. 1970. Pheromone research with stored-product Coleoptera. In *Control of Insect Behavior by Natural Products*. D.L. Wood, R.M. Silverstein, and M. Nakajima, Eds. Academic, New York, pp 1-20.

(11)  T.E. Eisner and F.C. Kafatos. 1962. Defense mechanisms of arthropods. X. A pheromone promoting aggregation in an aposematic distasteful insect. *Psyche* 69:53-61.

(12)  A.J. Thorsteinson. 1960. Host selection in phytophagous insects. *Ann Rev. Entomol.* 5:193-218.

(13)  J.H. Borden. 1974. Aggregation pheromones in the Scolytidae. In *Pheromones*. M.C. Birch, Ed. North-Holland, Amsterdam, pp. 135-160.

(14)  J.A.A. Renwick and J.P. Vité. 1970. Systems of chemical communication in *Dendroctonus*. *Contrib. Boyce Thompson Inst.* 24:283-292.

(15)  D.L. Wood. 1972. Selection and colonization of ponderosa pine by bark beetles. In *Insect Plant Relationships*. H.F. van Enden, Ed. Symp. Roy. Entomol. Soc., London, pp. 101-117.

(16)  G.M. Happ. 1970. Maturation of the response of male *Tenebrio molitor* to the female sex pheromone. *Ann. Entomol. Soc. Am.* 63:1782.

(17)  G.M. Happ and J. Wheeler. 1969. Bioassay, preliminary purification, and effect of age, crowding, and mating on the release of sex pheromone by female *Tenebrio molitor*. *Ann. Entomol. Soc. Am.* 62:846-851.

(18)  D.D. Hardee, E.B. Mitchell, and P.M. Huddleston. 1967. Laboratory studies on sex attraction in the boll weevil. *J. Econ. Entomol.* 60:1221-1224.

(19)  M.S. Mayer and J.R. Brazzel. 1963. The mating behavior of the boll weevil, *Anthonomus grandis*. *J. Econ. Entomol.* 56:605-609.

(20)  C. Adeesan, G.W. Rahalkar, and A.J. Tamhankar. 1969. Effect of age and previous mating on the response of khapra beetle males to female sex pheromone. *Entomol. Exp. & Appl.* 12:229-234.

(21)  G.M. Happ, M.E. Schroeder, and J.C.H. Wang. 1970. Effects of male and female scent on reproductive maturation in young female *Tenebrio molitor*. *J. Insect Physiol.* 16:1543-1548.

(22)  J.H. Borden. 1967. Factors influencing the response of *Ips confusus* (Coleoptera: Scolytidae) to male attractant. *Can. Entomol.* 99:1164-1193.

(23)  G.R. Struble. 1966. California five-spined ips. U.S.D.A., Forest Serv., *Forest Pest Leaf.* No. 102.

(24)  R.M. Silverstein, R.G. Brownlee, T.E. Bellas, D.L. Wood and L.E. Browne. 1968. Brevicomin: Principal sex attractant in the frass of the female western pine bettle. *Science* 159:889-890.

(25)  D.L. Wood and R.W. Bushing. 1963. The olfactory response of *Ips confusus* (Le Conte) (Coleoptera: Scolytidae) to the secondary attraction in the laboratory. *Can. Entomol.* 95:1066-1078.

(26)  M.C. Birch. 1974. Seasonal variation in pheromone-associated behavior and physiology of *Ips pini*. *Ann. Entomol. Soc. Am.* 67:58-60.

(27)  R.B. Ryan. 1959. Termination of diapause in the Douglas-fir beetle, *Dendroctonus pseudotsugae* Hopkins (Coleoptera: Scolytidae), as an aid to continuous laboratory rearing. *Can. Entomol.* 91:520-525.

(28)  J.H. Borden. Unpublished results.

(29)  E.J. Villavaso and N.W. Earle. 1974. Attraction of female boll weevils to diapausing and reproducing males. *J. Econ. Entomol.* 67:171-172.

(30) R.V. Connin and R.A. Hoopingarner. 1971. Sexual behavior and diapause or the cereal leaf beetle, *Oulema melanopus. Ann. Entomol. Soc. Am.* 64:655-660.

(31) C.E. Fockler and J.H. Borden. 1972. Sexual behavior and seasonal mating activity of *Trypodendron lineatum* (Coleoptera: Scolytidae) *Can. Entomol.* 104:1841-1853.

(32) R.I. Gara and J.P. Vité. 1962. Studies on the flight patterns of bark beetles (Coleoptera: Scolytidae) in second growth ponderosa pine forests. *Contrib. Boyce Thompson Inst.* 21:275-289.

(33) R.I. Gara. 1963. Studies on the flight behavior of *Ips confusus* (Le C.) (Coleoptera: Scolytidae) in response to attractive material. *Contrib. Boyce Thompson Inst.* 22:51-66.

(34) J.P. Vité, R.I. Gara and H.D. von Scheller. 1964. Field observations on the response to attractants of bark beetles infesting southern pines. *Contrib. Boyce Thompson Inst.* 22:461-470.

(35) G.E. Daterman, J.A. Rudinsky and W.P. Nagel. 1965. Flight patterns of bark and timber beetles associated with coniferous forests of western Oregon. *Ore. State Univ. Agric, Exp. Sta. Tech. Bull.* 87.

(36) E.A. Cameron and J.H. Borden. 1967. Emergence patterns of *Ips confusus* (Coleoptera: Scolytidae) from ponderosa pine. *Can. Entomol.* 99:236-244.

(37) J.H. Borden and E. Stokkink. 1973. Laboratory investigation of secondary attraction in *Gnathotrichus sulcatus* (Coleoptera: Scolytidae). *Can. J. Zool.* 51:469-473.

(38) K.W. Vick, P.C. Drummond and J.A. Coffelt. 1973. *Trogoderma inclusum* and *T. glabrum:* Effects of time of day on production of female pheromone, male responsiveness, and mating. *Ann. Entomol. Soc. Am.* 66:1001-1004.

(39) L. Hamack and W.E. Burkholder. Personal communication.

(40) W.R. Henson. 1962. Laboratory studies on the adult behavior of *Conophthorus coniperda* (Coleoptera: Scolytidae). III. Flight. *Ann. Entomol. Soc. Am.* 55:524-530.

(41) J.S. Kennedy and C.O. Booth. 1963. Free flight of aphids in the laboratory. *J. Exp. Biol.* 40:67-85.

(42) J.S. Kennedy and C.O. Booth. 1963. Co-ordination of successive activities in an aphid. The effect of flight on the settling responses. *J. Exp. Biol.* 40:351-369.

(43) C.G. Johnson. 1969. *Migration and Dispersal of Insects by Flight.* Methuen, London.

(44) K. Graham. 1959. Release by flight exercise of a chemotropic response from photopositive domination in a scolytid beetle. *Nature* 184:283-284.

(45) K. Graham. 1962. Photic behavior in the ecology of the ambrosia beetle *Trypodendron lineatum. Proc. 11th. Int. Congr. Entomol.* (Vienna, 1960) 2:226.

(46) M.D. Atkins. 1966. Laboratory studies on the behavior of the Douglas-fir beetle, *Dendroctonus pseudotsugae* Hopkins. *Can. Entomol.* 98:953-991.

(47) R.B. Bennett and J.H. Borden. 1971. Flight arrestment of tethered *Dendroctonus pseudotsugae* and *Trypodendron lineatum* (Coleoptera: Scolytidae) in response to olfactory stimuli. *Ann. Entomol. Soc. Am.* 64:1273-1286.

(48) K. Graham. 1961. Air swallowing: A mechanism in photic reversal of the beetle *Trypodendron. Nature* 191:519-520.

(49) C.E. Fockler and J.H. Borden. Unpublished results.

(50) M.D. Atkins. 1969. Lipid loss with flight in the Douglas-fir beetle. *Can. Entomol.* 101:164-165.

(51) S.N. Thompson and R.B. Bennett. 1971. Oxidation of fat during flight of male Douglas-fir beetles. *Dendroctonus pseudotsugae. J. Insect Physiol.* 17:1555-1563.

(52)   D.D. Hardee, E.B. Mitchell and P.M. Huddleston. 1967. Procedure for bioassaying the sex attractant of the boll weevil. *J. Econ. Entomol.* 60:169-171.

(53)   D.D. Hardee, W.H. Cross, E.B. Mitchell, P.M. Huddleston, H.C. Mitchell, M.E. Merkl and T.B. Davick. 1969. Biological factors influencing responses of the female boll weevil to the male sex pheromone in field and large-cage tests. *J. Econ. Entomol.* 62:161-165.

(54)   F.P. Cuthbert, Jr. and W.J. Reid, Jr. 1964. Studies of sex attractant of banded cucumber beetle. *J. Econ. Entomol.* 57:247-250.

(55)   J.A. Rudinsky and G.E. Daterman. 1964. Field studies on flight patterns and olfactory responses of ambrosia beetles in Douglas-fir forests of western Oregon. *Can. Entomol.* 96:1339-1352.

(56)   G.S. Fraenkel and D.L. Gunn. 1961. *The Orientation of Animals.* Dover, New York.

(57)   J.S. Kennedy. 1966. Some outstanding questions in insect behaviour. In *Insect Behaviour.* P.T. Haskell, Ed. Roy. Entomol. Soc. Lond., pp. 97-112.

(58)   R.H. Wright. 1964. *The Science of Smell.* Allen and Unwin, London.

(59)   J.H. Borden and D.L. Wood. 1966. The antennal receptors and olfactory response of *Ips confusus* (Coleoptera: Scolytidae) to male sex attractant in the laboratory. *Ann. Entomol. Soc. Am.* 59:253-261.

(60)   T.L. Payne, H.A. Moeck, C.D. Willson, R.N. Coulson and W.J. Humphreys. 1973. Bark beetle olfaction–II. Antennal morphology of sixteen species of Scolytidae (Coleoptera). *Int. J. Insect Morphol. & Embryol.* 2:177-192.

(61)   W.D. Bedard. 1964. Variation in capacity of *Ips confusus* to reach attractive hosts. In *Breeding Pest-Resistant Trees.* H.D. Gerhold, E.J. Schreiner, R.E. McDermott and J.A. Winieski, Eds. Pergammon Press, Oxford, pp. 137-142.

(62)   S.R. Farkas and H.H. Shorey. 1972. Chemical trail-following by flying insects: A mechanism for orientation to a distant odor source. *Science* 178:67-68.

(63)   C.E. Lilly and J.D. Shorthouse. 1971. Responses of males of the 10-lined June beetle, *Polyphylla decemlineata* (Coleoptera: Scarabaeidae), to female sex pheromone. *Can. Entomol.* 103:1757-1761.

(64)   G.B. Pitman and J.P. Vité. 1969. Aggregation behavior of *Dendroctonus ponderosae* (Coleoptera: Scolytidae) in response to chemical messengers. *Can. Entomol.* 101:143-149.

(65)   W.H. Cross, D.D. Hardee, F. Nichols, H.C. Mitchell, E.B. Mitchell, P.M. Huddelston and J.H. Tumlinson. 1969. Attraction of female boll weevils to traps baited with males or extracts of males. *J. Econ. Entomol.* 62:154-161.

(66)   W.H. Cross, J.E. Leggett and D.D. Hardee. 1971. Improved traps for capturing boll weevils. *USDA, Coop. Econ. Insect Rep.* 21:367-368.

(67)   S.H. Roach, H.R. Agee and L. Ray. 1972. Influence of position and color of male-baited traps on captures of boll weevils. *Environ. Entomol.* 4:530-532.

(68)   R.M. Silverstein, J.O. Rodin, W.E. Burkholder and J.E. Gorman. 1967. Sex attractant of the black carpet beetle. *Science* 157:85-87.

(69)   M. Jacobson, C.E. Lilly and C. Harding. 1968. Sex attractant of sugar beet wireworm: Identification and biological activity. *Science* 159:208-210.

(70)   R.F. Henzell and M.D. Lowe. 1970. Sex attractant of the grass grub beetle. *Science* 168:1005-1006.

(71)   H. Fukui, F. Matsumura, M.C. Ma and W.E. Burkholder. 1974. Identification of the sex pheromone of the furniture carpet beetle, *Anthrenus flavipes* Le Conte. *Tetrahedron Letts.* 40:3563-3566.

(72)   J.H. Borden, T.J.D. VanderSar and E. Stokkink. 1975. Secondary attraction in the Scolytidae: An annotated bibliography. *Simon Fraser Univ. Pest Mgt. Pap.* 4.

(73) J.H. Tumlinson, D.D. Hardee, R.C. Gueldner, A.C. Thompson, P.A. Hedin and J.P. Minyard. 1969. Sex pheromones produced by male boll weevil: Isolation, identification and synthesis. *Science* 166:1010-1012.

(74) J.O. Rodin, R.M. Silverstein, W.E. Burkholder and J.E. Gorman. 1969. Sex attractant of female dermestid beetle, *Trogoderma inclusum* Le Conte. *Science* 165:904-906.

(75) R.G. Yarger, R.M. Silverstein and W.E. Burkholder. 1976. Sex pheromone of the female dermestid beetle *Trogoderma glabrum* (Herbst). *J. Chem. Ecol.* (in Press).

(76) V.C.S. Chang and G.A. Curtis. 1972. Pheromone production by the New Guinea sugarcane weevil. *Environ. Entomol.* 1:478-481.

(77) D.D. Hardee, N.M. Wilson, E.B. Mitchell and P.M. Huddleston. 1971. Factors affecting activity of grandlure, the pheromone of the boll weevil, in laboratory bioassays. *J. Econ. Entomol.* 64:1454-1456.

(78) R.M. Silverstein. 1970. Methodology for isolation and identification of insect pheromones—examples from Coleoptera. In *Control of Insect Behavior by Natural Products.* D.L. Wood, R.M. Silverstein, and M. Nakajima, Eds. Academic, New York, pp. 285-299.

(79) W.D. Bedard, P.E. Tilden, D.L. Wood, R.M. Silverstein, R.G. Brownlee and J.O. Rodin. 1969. Western pine beetle: Field response to its sex pheromone and a synergistic host terpene, myrcene. *Science* 164:1284-1285.

(80) G.B. Pitman. 1969. Pheromone response in pine bark beetles: Influence of host volatiles. *Science* 166:905-906.

(81) J.A.A. Renwick and J.P. Vité. 1969. Bark beetle attractants: Mechanism of colonization by *Dendroctonus frontalis. Nature* 224:1222-1223.

(82) M.M. Furniss and R.F. Schmitz. 1971. Comparative attraction of Douglas-fir beetles to frontalin and tree volatiles. U.S.D.A., Forest Ser. Res. Pap. INT-96.

(83) G.B. Pitman and J.P. Vité. 1970. Field response of *Dendroctonus pseudotsugae* (Coleoptera: Scolytidae) to synthetic frontalin. *Ann. Entomol. Soc. Am.* 63:661-664.

(84) G.T. Pearce, W.E. Gore, R.M. Silverstein, J.W. Peacock, R. Cuthbert, G.N. Lanier and J.B. Simeone. 1975. Chemical attractants for the smaller European elm bark beetle, *Scolytus multistriatus* (Coleoptera: Scolytidae). *J. Chem. Ecol.* 1:115-124.

(85) G.B. Pitman, R.L. Hedden and R.I. Gara. 1975. Synergistic effects of ethyl alcohol on the aggregation of *Dendroctonus pseudotsugae* (Col., Scolytidae) in response to pheromones. *Z. angew. Entomol.* 78:203-208.

(86) J.A. McLean and J.H. Borden. 1976. Attack by *Gnathotrichus sulcatus* on stumps and logs baited with sulcatol and ethanol. *Can. Entomol.* (in press).

(87) T.L. Payne. 1970. Electrophysiological investigations on response to pheromones in bark beetles. *Contrib. Boyce Thompson Inst.* 24:275-282.

(88) T.L. Payne. 1971. Bark beetle olfaction. I. Electroantennogram responses of the southern pine beetle (Coleoptera: Scolytidae) to its aggregation pheromone frontalin. *Ann. Entomol. Soc. Am.* 64:266-268.

(89) O.K. Jantz and J.A. Rudinsky. 1965. Laboratory and field methods for assaying olfactory responses of the Douglas-fir beetle, *Dendroctonus pseudotsugae* Hopkins. *Can. Entomol.* 97:935-941.

(90) J.H. Borden, R.G. Brownlee and R.M. Silverstein. 1968. Sex pheromone of *Trypodendron lineatum* (Coleoptera: Scolytidae): Production, bio-assay, and partial isolation. *Can. Entomol.* 100:629-636.

(91) J.W. Peacock, R.M. Silverstein, A.C. Lincoln and J.B. Simeone. 1973. Laboratory investigations of the frass of *Scolytus multistriatus* (Coleoptera: Scolytidae) as a source of pheromone. *Environ. Entomol.* 2:355-359.

(92)     G.O. Osborne and C.P. Hoyt. 1970. Phenolic resins as chemical attractants for males of the grass grub beetle, *Costelytra zealandica. Ann. Entomol. Soc. Am.* 63:1145-1147.

(93)     D.L. Wood, L.E. Browne, R.M. Silverstein and J.O. Rodin. 1966. Sex pheromones of bark beetles–I. Mass production, bio-assay, source, and isolation of the sex pheromone of *Ips confusus* (Le C.). *J. Insect Physiol.* 12:523-536.

(93a)    J.A. McLean and J.H. Borden. Unpublished results.

(94)     J.A. Rudinsky. 1973. Multiple functions of the southern pine beetle pheromone, verbenone. *Environ. Entomol.* 2:511-514.

(95)     J.A. Rudinsky. 1973. Multiple functions of the Douglas fir beetle pheromone, 3-methyl-2-cyclohexen-1-one. *Environ. Entomol.* 2:579-585.

(96)     J.A. Rudinsky, M.E. Morgan, L.M. Libbey and T.B. Putnam. 1974. Antiaggregative-rivalry pheromone of the mountain pine beetle, and a new arrestant of the southern pine beetle. *Environ. Entomol.* 3:90-98.

(97)     K.W. Vick, D.C. Drummond, L.L. Sower and J.A. Coffelt. 1973. Sex pheromone habituation: the effects of habituation on the pheromone response level of *Trogoderma inclusum* (Coleoptera: Dermestidae). *Ann. Entomol. Soc. Am.* 66:667-670.

(98)     R.H. Wright, D.L. Chambers and I. Keiser. 1971. Insect attractants, antiattractants, and repellents. *Can. Entomol.* 103:627-630.

(99)     G.M. Happ. 1969. Multiple sex pheromones of the mealworm beetle, *Tenebrio molitor. Nature* 222:180-181.

(100)    J.H. Borden. 1974. Pheromone mask produced by male *Trypodendron lineatum* (Coleoptera: Scolytidae). *Can. J. Zool.* 52:533-536.

(101)    V.M. Kirk and B.J. Dupraz. 1972. Discharge by a female ground beetle, *Pterostichus lucublandus*, used as a defense against males. *Ann. Entomol. Soc. Am.* 65:513.

(102)    R.F. Shepherd. 1965. Distribution of attacks by *Dendroctonus ponderosae* Hopk. on *Pinus contorta* Dougl. var. *latifolia* Engelm. *Can. Entomol.* 97:207-215.

(103)    A.A. Berryman. 1968. Distributions of *Scolytus ventralis* attacks, emergence, and parasites in grand fir. *Can. Entomol.* 100:57-68.

(104)    J.A. Rudinsky and R.R. Michael. 1973. Sound production in Scolytidae: Stridulation by female *Dendroctonus* beetles. *J. Insect Physiol.* 19:689-705.

(105)    A.F. Naylor. 1959. An experimental analysis of dispersal in the flour beetle, *Tribolium confusum. Ecology* 40:453-465.

(106)    J.A. Rudinsky. 1968. Pheromone-mask by the female *Dendroctonus pseudotsugae* Hopk., an attraction regulator. *Pan-Pac. Entomol.* 44:248-250.

(107)    J.A. Rudinsky. 1969. Masking of the aggregating pheromone in *Dendroctonus pseudotsugae* Hopk. *Science* 166:884-885.

(108)    J.A. Rudinsky. 1970. Sequence of Douglas-fir beetle attraction and its ecological significance. *Contrib. Boyce Thompson Inst.* 24:311-314.

(109)    J.A. Rudinsky and R.R. Michael. 1972. Sound production in the Scolytidae: Chemostimulus of sonic signal by the Douglas-fir beetle. *Science* 175:1386-1390.

(110)    G.W. Kinzer, A.F. Fentiman, Jr., R.L. Foltz and J.A. Rudinsky. 1971. Bark beetle attractants: 3-Methyl-2-cyclohexen-1-one isolated from *Dendroctonus pseudotsugae. J. Econ. Entomol.* 64:970-971.

(111)    J.A. Rudinsky, M.M. Furniss, L.N. Kline and R.F. Schmitz. 1972. Attraction and repression of *Dendroctonus pseudotsugae* (Coleoptera: Scolytidae) by three synthetic pheromones in traps in Oregon and Idaho. *Can. Entomol.* 104:815-822.

(112)    M.M. Furniss, L.N. Kline, R.F. Schmitz and J.A. Rudinsky. 1972. Tests of three pheromones to induce or disrupt aggregation of Douglas-fir beetles (Coleoptera:

Scolytidae) on live trees. *Ann. Entomol. Soc. Am.* 65:1227-1232.

(113)   J.A. Rudinsky, M.E. Morgan. L.M. Libbey and R.R. Michael. 1973. Sound production in Scolytidae: 3-Methyl-2-cyclohexen-1-one released by the female Douglas-fir beetle in response to male sonic signal. *Environ. Entomol.* 2:505-509.

(114)   G.B. Pitman and J.P. Vité. 1974. Biosynthesis of methylcyclohexenone by male Douglas-fir beetle. *Environ. Entomol.* 3:886-887.

(115)   L.N. Kline, R.F. Schmitz, J.A. Rudinsky and M.M. Furniss. 1974. Repression of spruce beetle (Coleoptera) attraction by methylcyclohexeone in Idaho. *Can. Entomol.* 106:485-491.

(116)   J.A. Rudinsky and R.A. Michael. 1974. Sound production in Scolytidae: "Rivalry" behaviour of male *Dendroctonus* beetles. *J. Insect Physiol.* 20:1219-1230.

(117)   W.W. Nijholt. 1970. The effect of mating and the presence of the male ambrosia beetle, *Trypodendron lineatum* on "secondary" attraction. *Can. Entomol.* 102:894-897.

(118)   W.W. Nijholt. 1973. The effect of male *Trypodendron lineatum* (Coleoptera: Scolytidae) on the response of field populations to secondary attraction. *Can. Entomol.* 105:583-590.

(119)   R.I. Gara and J.E. Coster. 1968. Studies on the attack behavior of the southern pine beetle. III. Sequence of tree infestation within stands. *Contrib. Boyce Thompson Inst.* 24:77-85.

(120)   M.M. Furniss, G.E. Daterman, L.N. Kline, M.D. McGregor, G.C. Trostle, L.F. Pettinger, and J.A. Rudinsky. 1974. Effectiveness of the Douglas-fir beetle antiaggregative pheromone methylcyclohexenone at three concentrations and spacings around felled host trees. *Can. Entomol.* 106:381-392.

(121)   J.A. Rudinsky, C. Sartwell, Jr., T.M. Graves and M.E. Morgan, 1974. Granular formulation of methylcyclohexenone: An antiaggregative pheromone of the Douglas fir and spruce bark beetles. *Z. angew. Entomol.* 75:254-263.

(122)   B.A. Barr. 1969. Sound production in the Scolytidae (Coleoptera) with emphasis on the genus *Ips. Can. Entomol.* 101:636-672.

(123)   J.M. Mathieu. 1969. Mating behavior of five species of Lucanidae (Coleoptera:Insecta). *Can. Entomol.* 101:1054-1062.

(124)   J.D. Pinto. 1975. Intra- and interspecific courtship behavior in blister beetles of the genus *Tegrodera* (Meloidae). *Ann. Entomol. Soc. Am.* 68:275-285.

(125)   D. Matthes. 1962. Excitatoren und Paarungsverhalten mitteleuropäisher Malachiiden (Coleopt., Malacodermata). *Z. Morphol. Okol. Tiere.* 51:375-546.

(126)   W.R. Tschinkel, C.D. Willson, and H.A. Bern. 1967. Sex pheromone of the mealworm beetle, *Tenebrio molitor. J. Exp. Zool.* 164:81-85.

(127)   W.R. Tschinkel. 1970. Chemical studies on the sex pheromone of *Tenebrio molitor* (Coleoptera: Tenebrionidae). *Ann. Entomol. Soc. Am.* 63:626-627.

(128)   C.E. Lilly and A.J. McGinnis. 1965. Reactions of male click beetles in the laboratory to olfactory pheromones. *Can. Entomol.* 97:317-321.

(129)   J.A. Coffelt and W.E. Burkholder. 1972. Reproductive biology of the cigarette beetle, *Lasioderma serricorne.* I. Quantitative laboratory bioassay of the female sex pheromone from females of different ages. *Ann. Entomol. Soc. Am.* 65:447-450.

(130)   R. Keville and P.B. Kannowski. 1975. Sexual excitation by pheromones of the confused flour beetle. *J. Insect Physiol.* 21:81-84.

(131)   M.F. Ryan, T. Park, and D.B. Mertz, 1970. Flour beetles: Responses to extracts of their own pupae. *Science* 170:178-180.

(132)  J.H. Borden and E. Stokkink. Unpublished results.

(133)  B. Hölldobler. 1969. Host finding by odor in the myrmecophilic beetle *Atemeles pubicollis*. *Science* 166:757-758.

(134)  B. Holldobler. 1968. Der Glanzkafer als "Wegelagerer" an Ameisenstrassen. *Naturwissenshaften* 55:397.

(135)  U. Yinon and A. Shulov. 1969. Response of some stored-product insects to *Trogoderma granarium* pheromones. *Ann. Entomol. Soc. Am.* 62:172-175.

(136)  D.L. Wood, L.E. Browne, W.D. Bedard, P.E. Tilden, R.M. Silverstein, and J.O. Rodin. 1968. Responses of *Ips confusus* to synthetic sex pheromones in nature. *Science* 159:1373-1374.

(137)  G.N. Lanier, M.C. Birch, R.F. Schmitz, and M.M. Furniss. 1972. Pheromones of *Ips pini* (Coleoptera: Scolytidae): Variation in response among three populations. *Can. Entomol.* 104:1917-1923.

(138)  J.P. Vité and D.L. Williamson. 1970. *Thanasimus dubius:* Prey perception. *J. Insect Physiol.* 16:233-239.

(139)  R.F. Schmitz. Personal communication.

(140)  E.D.A. Dyer. 1973. Spruce beetle aggregated by the synthetic pheromone frontalin. *Can. J. Forest Res.* 3:486-494.

(141)  G.B. Pitman and J.P. Vité. 1971. Predator-prey response to western pine beetle attractants. *J. Econ. Entomol.* 64:402-404.

(142)  W.D. Bedard and D.L. Wood. 1974. Programs utilizing pheromones in survey or control. Bark beetles—the western pine beetle. In *Pheromones*. M.C. Birch, Ed. North-Holland, Amsterdam, pp. 441-449.

(143)  W.G. Harwood and J.A. Rudinsky. 1966. The flight and olfactory behavior of checkered beetles (Coleoptera:Cleridae) predatory on the Douglas-fir beetle. *Oregon St. Univ. Agric. Expt. Sta. Tech. Bull.* 95.

(144)  R.E. Rice. 1971. Flight characteristics of *Enoclerus lecontei, Temnochila virescens,* and *Tomicobia tibialis* in central California. *Pan-Pac. Entomol.* 47:1-8.

(145)  A.A. Berryman. 1966. Studies on the behavior and development of *Enoclerus lecontei* (Wolcott), a predator of the western pine beetle. *Can. Entomol.* 98:519-526.

(146)  A.A. Berryman. 1966. Factors influencing oviposition and the effect of temperature on development and survival of *Enoclerus lecontei* (Wolcott) eggs. *Can. Entomol.* 98:579-585.

(147)  R.M. Silverstein, J.O. Rodin, and D.L. Wood. 1966. Sex attractants in frass produced by male *Ips confusus* in ponderosa pine. *Science* 154:509-510.

(148)  D.L. Wood, R.W. Stark, R.M. Silverstein, and J.O. Rodin. 1967. Unique synergistic effects produced by the principal sex attractant compounds of *Ips confusus* (Le Conte) (Coleoptera: Scolytidae). *Nature* 215:206.

(149)  J.P. Vité and J.A.A. Renwick. 1971. Inhibition of *Dendroctonus frontalis* response to frontalin by isomers of brevicomin. *Naturwissenschaften* 58:418.

(150)  M.C. Birch and D.L. Wood. 1975. Mutual inhibition of the attractant pheromone response by two species of *Ips* (Coleoptera: Scolytidae). *J. Chem. Ecol.* 1:101-113.

(151)  J.P. Vité, A. Bakke, and J.A.A. Renwick. 1972. Pheromones in *Ips* (Coleoptera: Scolytidae): Occurrence and production. *Can. Entomol.* 104:1967-1975.

(152)  J.C. Young, R.G. Brownlee, J.O. Rodin, D.N. Hildebrand, R.M. Silverstein, D.L. Wood, M.C. Birch, and L.E. Browne. 1973. Identification of linalool produced by two species of bark beetles of the genus *Ips*. *J. Insect Physiol.* 19:1615-1622.

(153)  W.E. Burkholder. 1974. Programs utilizing pheromones in survey or control.

Stored product pests. In *Pheromones.* M.C. Birch, Ed. North-Holland, Amsterdam, pp. 449-451.

(154) J.A. McLean and J.H. Borden. 1976. Survey for *Gnathotrichus sulcatus* (Coleoptera: Scolytidae) in a commercial sawmill with the pheromone, sulcatol. *Can. J. Forest Res.* 5:586-591.

(155) R.G. Cox. 1972. Cooperative research on control of the mountain pine beetle in western white pine. Montana-Northern Idaho Forest Pest Action Council Prog. Rept. 8.

(156) E.P. Lloyd, M.E. Merkl, F.C. Tingle, W.P. Scott, D.D. Hardee, and T.B. Davich. 1972. Evaluation of male-baited traps for control of boll weevils following a reproduction-diapause program in Monroe County, Mississippi. *J. Econ. Entomol.* 65:552-555.

(157) D.D. Hardee. 1974. Programs utilizing pheromones in survey or control. Cotton—the boll weevil. In *Pheromones.* M.C. Birch, Ed. North-Holland, Amsterdam, pp. 427-431.

(158) J.P. Vité. 1970. Pest management systems using synthetic pheromones. *Contrib. Boyce Thompson Inst.* 24:343-350.

(159) J.A.E. Knopf and G.B. Pitman. 1971. Aggregation pheromone for manipulation of the Douglas-fir beetle. *J. Econ. Entomol.* 65:723-726.

(160) L.A. Rasmussen. 1972. Attraction of mountain pine beetles to small-diameter lodgepole pines baited with *trans*-verbenol and *alpha*-pinene. *J. Econ. Entomol.* 65:1396-1399.

(161) G.B. Pitman. 1973. Further observations on Douglure in a *Dendroctonus pseudotsugae* management system. *Environ. Entomol.* 2:109-112.

(162) B.H. Baker and G.C. Trostle. 1973. Douglas fir beetle attraction and tree-group response. *J. Econ. Entomol.* 66:1002-1005.

(163) W.P. Scott, E.P. Lloyd, J.O. Bryson and T.B. Davich. 1974. Trap plots for suppression of low density overwintered populations of boll weevils. *J. Econ. Entomol.* 67:281-283.

(164) W.E. Burkholder. 1973. Black carpet beetle: Reduction of mating by megatomic acid, the sex pheromone. *J. Econ. Entomol.* 66:1327.

(165) C.E. Lilly and A.J. McGinnis. 1968. Quantitative responses of males of *Limonius californicus* (Coleoptera:Elateridae) to female sex pheromone. *Can. Entomol.* 100:1071-1078.

(166) R.L. Jones, W.J. Lewis, M. Beroza, B.A. Bierl, and A.N. Sparks. 1973. Host-seeking stimulants (kairomones) for the egg parasite, *Trichogramma evanescens. Environ. Entomol.* 2:593-596.

(167) D.A. Nordlund, W.J. Lewis, H.R. Gross, Jr., and E.A. Harrell. 1974. Description and evaluation of a method for field application of *Heliothis zea* eggs and kairomones for *Trichogramma. Environ. Entomol.* 3:981-984.

(168) M. Beroza. 1972. Attractants and repellents for insect pest control. In *Pest Control Strategies for the Future.* Natl. Acad. Sci., Washington, D.C., pp. 226-253.

## CHAPTER 12

# Behavioral Responses of Lepidoptera to Pheromones

R. J. Bartell

*Division of Entomology*
*CSIRO*
*Canberra, A.C.T., Australia*

Lepidoptera display a wide and varied range of responses to chemical substances produced by other insects in their environment. Most research thus far has been concerned with elucidating the chemical nature of these materials and investigating their possible use in field programs designed for the control of economic pest species. To a certain extent these practical aims have dominated the field and thus have tended to hamper the analyses of behaviors mediated by the various classes of biologically active compounds. The mainstream of research in this field lies with the class of compounds known as pheromones, especially sex pheromones.

## I. SEX-PHEROMONE-INDUCED SEQUENCE OF BEHAVIOR

The release of sex pheromone by the females of many species elicits in the conspecific male a hierarchic sequence of behavioral steps. It is a matter of convenience to the observer which of these behavioral steps are selected for record, frequently they include: activation of a resting male, the orientation of that male toward the source of the pheromone, and an often complicated series of behavioral responses at short range in the presence of the female, which usually lead to copulation.

Most experimental work that has been designed to evaluate levels of responsiveness of males has depended heavily upon the use of laboratory techniques that assay a particular type of response (usually activation), or it has depended upon field-trapping programs that attempt to measure the orientation response. Both approaches use some form of behavioral end-point, which an observer can readily quantify. In the laboratory bioassay the measure of response is often the proportion of individual males that move in a small population exposed to the active material, and in the field it is simply a count of the numbers of insects that become stuck to the sticky coating of the trap. In either case the conditions are usually sufficiently limited as to render a complete analysis of the behavior of the insects impossible. However, each approach makes some manipulations of conditions possible, which enable the observer to assess the effects of various physiologic and environmental limiting factors on some defined behavioral response.

Two basic methods seem to emerge in the study of whole-organism behavior of insects responding to pheromones: first, the empirical study of behavior patterns and of specific responses under comparative environmental conditions; and second, the investigation of the physiologic mechanisms underlying the behavior. The second method seems to be the more difficult, as sense organs are integrated into a system of regulatory components whose synthesis of the total sensory input can result in complex patterns of behavior that vary when external circumstances are changed by only a very small degree. Behavior as a whole is a very complex phenomenon, and the inference of physiologic mechanisms from behavioral studies or even from basic stimulus-response relationships can lead, and has led, to severe problems of interpretation.

A large body of evidence has accrued which shows that such environmental variables as the physical parameters of light intensity, temperature. and wind speed influence the responses of male Lepidoptera to the female sex pheromones and and have optima for the performance of each step in the male behavioral sequence (1)-(8).

Similarly a number of physiologic variables exert a controlling influence on the expression of the behaviors. The circadian rhythm of the insect determines the time of day at which certain responses will occur (5) (6) (8)-(21). Likewise the responsiveness of insects to pheromones varies in relation to the age of the individual, presumably as a reflection of the timing of sexual maturation (5). The occurrence of previous bouts of mating activity of an individual also influences its responsiveness to pheromone, both within the same diel cycle during which mating occurred and over successive days. Previous exposure of male insects to pheromone results in decreased responsiveness to the pheromone on subsequent occasions (5) (22)-(24).

In 1955, Schwinck (25) proposed that the initial step in the behavioral sequence has the lowest threshold of pheromone concentration and that each successive step has a threshold higher than the preceding one. In 1969, Shorey and I (26) demonstrated that successive steps in the response sequence of males of the light-brown apple moth (LBAM), *Epiphyas postvittana*, did indeed fit the scheme proposed by Schwinck. Recently, Cardé et al. (27) have shown that the response hierarchy may also be mediated by different chemical components in the female secretion. Working with the oriental fruit moth, *Grapholitha molesta,* they were able to show that while a mixture of a specific ratio of two isomers of the pheromone resulted in a maximum number of approaches by males to the source, the addition of a third, related compound elicited close-range, precopulatory behavior in the male.

## II. ACTIVATION

The first recognizable step in the sequence of response is the activation of a resting male. Such activity may be represented by elevation of the antennae, grooming of the antennae, "restlessness," and wing vibration. The latter is most marked in the larger species. Presumably it serves the function of raising the temperature in the flight muscles preparatory to take off. These behavior patterns have been characterized for a large number of species, for the most part under laboratory conditions. It is doubtful whether they may be an important component of pheromone communication in the field because at the usual period for release of pheromone by females most males in the local population are probably already in flight. This probability would be extremely difficult to test experimentally, for it does not seem feasible to perform exhaustive studies on male flight behavior under natural conditions in an atmosphere free of pheromone, even if it could be established that apheromonal conditions prevailed. Furthermore, it appears that there would be selective advantage in utilizing a mobile population of males whose preorientative flight would enable the sampling of a greater airspace and thus facilitate the location of an aerial odor-trail.

## III. ORIENTATION

Orientative behavior results in the male locating the pheromone source: the conspecific female. Of course, the analysis of insect orientative behavior is not restricted to orientations to pheromones, nor even to other chemical cues. A vast library of chemical knowledge is now available to us. It should serve as a versatile tool for the study and the manipulation of orientative behavior.

Farkas and Shorey (28) have reported that males of the pink bollworm, *Pectinophora gossypiella,* in still air remained capable of orienting along an odor plume that had hitherto been established in a moving airstream. They proposed that the insects were capable of gaining chemotactic orientation cues from the stationary plume along which the moths progressed in the former upwind direction. Kennedy and Marsh (29), however, challenged this view in the light of their own experiments with the moth, *Plodia interpunctella,* and of previous experiments with other insects, which indicated that the insects maintained upwind orientation by means of anemotaxis regulated by optomotor responses to the visual surroundings and modulated by the pheromone flowing downwind. They concluded "that, although the mechanism of odour regulated optomotor anemotaxis remains largely unanalysed, it is still the most plausible guidance mechanism for the male moths studied so far."

Farkas et al. (30) looked more closely at environmental factors affecting the aerial odor-trail following of males of *P. gossypiella.* They found that the behavior of the males was regulated by the concentration (at source) of the pheromone, the spatial disposition and intensity of an illuminating light source, and visual cues from nearby objects. Increasing pheromone concentration tended to reduce the forward air speed of the responding male, both along the zigzag flight path and along the axis of the trail. The reduction in speed occurs, in part, as a result of reduction in wing-beat frequency and represents a negative orthokinesis. Visual cues and the concentration of the pheromone appear to work jointly in stimulating the slowing of flight and landing.

## IV. MALE-FEMALE CLOSE-RANGE BEHAVIOR

Precopulatory behavior is manifested by a variety of specific behaviors. Here again pheromones play an important role in a two-way exchange between the sexes, stimulating close-range behaviors either by increasing concentrations of the "attractant" or by additional components in the secretion (27). Brady et al. (31) isolated a compound from *Cadra cautella* females that, at certain concentrations, stimulated males but did not elicit orientative behavior, and that, at higher concentrations, inhibited male activity.

Secretions produced by male insects after they locate the female have an aphrodisiac effect on the females, either by inhibition of flight or by promotion of cessation of flight. In the day-flying butterflies of the subfamily Danaiinae, males are initially attracted by visual orientation to the flying female. They approach and overtake the female and, flying slightly above and ahead of her, distribute scent particles in an attempt to induce her to settle (32). In the moths the male pheromone appears to prevent the female from taking flight and/or to induce her to take up an "acceptance" posture.

Noctuid males have scent brushes situated at various locations on the abdomen or legs. They may be on the ventral surface of the eighth abdominal segment, as in the cabbage looper, *Trichoplusia ni,* or on the posterior of the abdomen, as in *Phlogophora meticulosa.* Normally these brushes are concealed in abdominal pockets, and a glandular secretion is deposited on them (33). In the presence of calling females, male moths are stimulated to perform a sequence of responses (34) (35) in which the abdominal brushes are everted from their pouches. In many species there appears to be no overt female response to the contact with these brushes, which the male applies to the region of her genitalia momentarily before attempting copulation.

In other families (Pyralidae, Phycitidae, Tortricidae, etc.) the males possess invaginations of the wing surfaces, forming pouches and folds that probably produce pheromones. In many species within these families the excited males may be observed conducting elaborate and frenzied mating "dances" in the presence of the calling female. In *E. postvittana* the males make repeated excursions to the anterior end of the female and, facing her, maintain a rapid fanning of the wings in a manner likely to cause an airstream to be directed over the female's antennae. The female responds by raising her wings slightly and extending her abdomen, thus enabling the male to take up position beneath the posterior edge of the female's wings and make genital contact.

Recent work on *Plodia interpunctella* (36) lends support to the notion that the wing pouches produce pheromones that have an aphrodisiac effect on the female: excision of the wing pouches of the male or removal of the female's antennae results in elimination of receptivity by the female.

Having outlined briefly some of the processes involved in pheromone-mediated sexual communication, I wish now to turn to what I consider to be the current position of research in this field, and deal the three key areas—namely, newly identified chemicals termed *synergists* and *inhibitors;* multiple chemical systems; and sensory adaptation and habituation.

## V. SYNERGISTS AND INHIBITORS

In the concluding plenary session of the 14th International Congress of Entomology in Canberra, 1972, Kennedy (37) discussed in broad terms the emergence of behavior as a science in its own right within entomology. In this address, Kennedy took issue with anthropomorphism, reductionism, and teleology as having been instrumental in hindering the analysis of insect behavior. He applauded the dawning recognition that insects have central nervous systems that synthesize and integrate a whole array of sensory inputs, which result in complex patterns of motor output. I agree completely with this viewpoint, and I wish to elaborate further upon teleology.

In recent years a steadily growing number of chemical identifications have begun to provide some insights into several aspects of the pheromone field. Not the least interesting of these newly identified chemicals are those termed *synergists* and *inhibitors*, chemical compounds that modify (under certain conditions) the responsiveness of male moths to what we might call *primary pheromone components* (i.e., those compounds which singly or in obligative combination will elicit the complete hierarchical sequence of sexual behavior in the males). Although the term *synergist* is used with the same rationale as *inhibitor,* and is therefore equally suspect, I wish to refer to inhibitors, particularly since certain difficulties have arisen because of this unfortunate choice of terminology.

The term *inhibitor* was adopted for those chemicals that, when liberated with a normally attractive single compound or mixture of compounds, somehow prevented males from becoming trapped in the sticky coating of the baited traps. The choice of this term might have been justifiable had it been restricted to use when describing what happened to trap catches, but the logical (or illogical?) next step was to extrapolate this knowledge to ascribe some sensory function to the chemical and to thereby generate models for chemoreception (38).

As Kennedy points out (37): "It is not so much the behavior of insects as the behavior of substances that is being studied here." It also seemed perfectly logical to look to *inhibitors* for possible use in field control programs. Rothschild (39) conducted such tests with the oriental fruit moth, *G. molesta,* for which there was already a known trap inhibitor, dodecyl acetate (40). As field trials were already under way for the investigation of communication-disruption techniques involving the dispersion of the pheromone itself (*cis*-8-dodecenyl acetate), and, as dodecyl acetate is simpler and cheaper to synthesize, it was thought desirable to set up a trial in which the inhibitor was broadcast from dispensers. The results were unexpected, to say the least: when incorporated in a trap or dispensed within a few centimeters of a trap, dodecyl acetate resulted in no males being trapped; but when present as a background odor, the numbers of males that became trapped increased significantly above the level with pheromone alone. Observations on traps containing dodecyl acetate revealed that males approached the traps but veered away at about 15 cm from them. Rothschild concluded: "It does not seem profitable to speculate at length on the possible types of behavioral mechanism that might explain the results obtained. To do so it would be necessary to have more detailed information on the actual behavior of the moth in the field as well as behavioral, electrophysiological and chemical data provided by laboratory experiments." Nevertheless, Rothschild has pointed to enormously profitable fields of study, which should help elucidate behavior and furnish further clues for the analysis of orientation mechanisms.

Although there is no evidence to suggest that dodecyl acetate is part of the natural pheromone system of *G. molesta,* one possibility that does arise is that the so-called "inhibitor" may turn out, in fact, to be one of several components in the female secretion, as was found to be the case for the red-banded leaf roller, *Argyrotaenia velutinana* (41). This naturally leads to the second area currently being explored, multiple chemical systems.

## VI. MULTIPLE CHEMICAL SYSTEMS

It is now becoming increasingly apparent that many moth species use multiple chemical systems in their communicatory repertoire (41)-(49). Quite apart from the growing mass of information regarding two or more component systems in which one compound alone will not achieve attraction at any concentration, the recent work of Cardé et al. (27) shows that different compounds in the secretion may mediate different behaviors, and nicely demonstrates the need for care in devising laboratory and field experiments.

As laboratory investigations usually bioassay only a single step in the response hierarchy (generally not orientation), it may be quite possible to miss significant components in the pheromone system when attempts are made to elucidate the chemical nature of the secretion. Similarly the absence of one or more components from a field-sampling trap may severely impair its efficiency. If the pattern of sensory input is important in multicomponent systems, then the interpretation of electrophysiologic data may be complicated when stimuli are presented as single compounds over ranges of concentration. But more seriously yet, the dispensation of chemically incomplete pheromones in control measures aimed at disrupting male orientation may apply sufficient selective pressure to establish quickly new strains of insect that operate on former secondary components of the secretion.

Recent work on LBAM (50) indicates that there are at least four biologically active components in the female secretion. The actual behavioral roles of each of these have not yet been characterized; suffice it to say that these components act in two couplets, each eliciting sexually specific responses in the male moth. Preliminary indications are that two genetic strains (of several) in LBAM, each characterized by a color phenotype in the male, have different spectra of electroantennogram (EAG) response to standard, synthetic homologues and analogues of the "primary components." If this indication is borne out by subsequent investigation, a very real possibility may exist for the selection of divergent strains, which are separated by a premating isolating mechanism, based upon variations of the chemical composition of the pheromone.


## VII. SENSORY ADAPTATION AND HABITUATION

Finally, I wish to consider the nervous-system phenomena of sensory adaptation and habituation. I earlier alluded to the reduction in responsiveness of males to the sex pheromone of the conspecific female, which was brought about by previous exposure to the pheromone. The results of a number of experiments (22)-(24) suggest that the lowered responsiveness may be due to an interaction of these phenomena.

First, sensory adaptation has the more transitory effect; Payne et al. (51) showed that the olfactory neurons in the antennae of the male cabbage looper moth recover their threshold for the transmission of action potentials within 60 sec following the cessation of stimulation. Preliminary work with LBAM based upon EAG studies (50) has revealed a similar time course for the recovery of the D.C. resting potential even after prolonged stimulation of over several minutes at moderately high concentrations.

The second cause of reduction in responsiveness (habituation) has a considerably more prolonged effect upon the subsequent pheromonal elicitation of behavioral responses. Elevated thresholds for male response following prior exposure to female pheromones have been widely demonstrated for a number of lepidopterous species (2) (5) (23) (24) (52), but recent work with LBAM (50) indicates a broad spectrum of possibilities by which insects may become habituated to the sex pheromone. Single, brief pre-exposures which were made up to 5 hr before subsequent stimulation resulted in significant reduction of male responsiveness (23). The longer the duration of the pre-exposure, the lower the subsequent response: thus it was concluded that the degree of reduction is dependent upon both the duration of pre-exposure and the timing of this in relation to the subsequent bioassay. To quote from the published work: "The observed effects of pre-exposure could result from influence on one or more levels in the nervous system of the insect. These may be in the afferent peripheral system, at some integrative centre within the central nervous system or in the motor centres." Furthermore, according to the results of Payne et al (51), "provided there is no lasting effect which prevents disadaptation following prolonged pre-exposure to stimulus. . .it seems that the effects of pre-exposure are probably unrelated to sensory adaptation. As the effects are long term and are not significantly dependent upon the performance of some behavioral response at the time of the pre-exposure, we attribute them to inhibition in some integrative centre in the CNS." The work on LBAM was done using crude extracts of the pheromone glands and, as it is now known that a multiple chemical system is involved, not only the time course of sensory recovery but also the pattern of input and of recovery may be important components in characterizing the effects of pre-exposure.

A comparative assay of pre-exposures to either one of what may be considered to be the primary components of LBAM sex pheromone does not result in any of the long-term effects (50), which further points to the implication of the Central Nervous System (CNS). It is thus apparent that the input signals elicited by each chemical alone must be integrated into a total recognizable signal before central inhibitory effects are established. If these conclusions are valid, then it is to be expected that sensory adaptation will also occur at the time of pre-exposure and thus, logically there should be an optimum duration of pre-exposure to produce the greatest central effect. If, however, the pre-exposure is pulsed in such a way that disadaptation might occur in the interval between each stimulus pulse, and if the summation of pulses produces the same measurable dose as that for a single, prolonged stimulus, then one may predict that there will be a greater central effect. This is, in fact, borne out by experimentation (50). It appears that a 3-min-duration pre-exposure is about the maximum time over which the insects will continue to perceive (or at least respond to) the pheromone; the decay of behavioral response thereafter is not significantly different from that under sustained exposure. The recovery of responsiveness after 2 min from the cessation of a 3-min pre-exposure is about 40% when compared with controls. If the pre-exposure is divided into three 1-min pulses, with 2-min intervals between pulses, the recovery of responsiveness is negligible. Moreover, preliminary studies indicate that the shorter the duration of each pulse the greater the overall effect.

It is not possible at this stage to speculate usefully upon the possible role of concentration of pheromone, except to say that it does play a part in promoting habituation, and that reduction in responsiveness equivalent to a tenfold reduction in concentration for an unadapted and unhabituated insect may be restored by stimulation with a tenfold higher concentration than that of the pre-exposure. Within certain limits the product of concentration and time is equivalent to a given dose.

The implications of these phenomena may place in entirely new perspective the study of insect-orientation mechanisms. For, if we are to accept the models for the structure of the aerial odor trail proposed by Wright (53) and Farkas and Shorey (28), within which the orientating insect would be exposed to a pulsed sequence of stimuli, then this appears to be the most efficient way to promote habituation in the flying insect. Naturally the closer the individual approaches the pheromone source, the higher the concentration and the greater the frequency of odor pulses. Increased concentrations may then compensate for raised thresholds of responsiveness due to progressive habituation.

## VIII. CONCLUSION

Investigations into pheromone-mediated behavior are still fragmentary in character. If we acknowledge the fact that the insect itself integrates and synthesizes a whole array of sensory inputs, then research workers must, on their part, use an integrated approach to the problem; systems thinking must predominate. And Farkas and Shorey (54) point out:

"Knowledge of how insects use olfactory orientation mechanisms is still rudimentary. . . . The way in which the mechanisms integrate into the complex scheme of orientation for any one species is strictly hypothetical at this time. Many more questions need to be answered before we can accumulate enough information to understand just one integrated system for guidance of an insect to its pheromone-releasing mate."

I find I can add nothing further to this observation; the future of research in this field will depend not only upon analysis of individual components of the behavioral hierarchy and the factors that modify them, but upon the piecing together of these components with all their interactions into the integrated whole—a not altogether undaunting prospect, but one I believe to be scientifically, if not practically, rewarding.

## REFERENCES

(1)  C.W. Collins and S.F. Potts. 1932. Attractants for the flying gipsy moths as an aid to locating new infestations. *U.S. D.A. Tech. Bull.* 336:1-43.

(2)  H.H. Shorey and L.K. Gaston. 1964. Sex pheromones of noctuid moths. III. Inhibition of male responses to the sex pheromone in *Trichoplusia ni* (Lepidoptera: Noctuidae). *Ann. Entomol. Soc. Am.* 57:775-779.

(3)  H.H. Shorey. 1966. The biology of *Trichoplusia ni* (Lepidoptera: Noctuidae). IV. Environmental control of mating. *Ann. Entomol. Soc. Am.* 59:502-506.

(4)  J.A. Klun. 1968. Isolation of a sex pheromone of the European corn borer. *J. Econ. Entomol.* 61:484-487.

(5)  R.J. Bartell and H.H. Shorey. 1969. A quantitative bioassay for the sex pheromone of *Epiphyas postvittana* (Lepidoptera) and factors limiting male responsiveness. *J. Insect Physiol.* 15:33-40.

(6)  W.C. Batiste. 1970. A timing sex-pheromone trap with special reference to codling moth collections. *J. Econ. Entomol.* 63:915-918.

(7)  R.S. Kaae and H.H. Shorey. 1973. Sex pheromones of Lepidoptera. XLIV. Influence of environmental conditions on the location of pheromone communication and mating in *Pectinophora gossypiella*. *Environ. Entomol.* 2:1081-1084.

(8)  W.C. Batiste, W.H. Olson, and A. Berlowitz. 1973. Codling moth: Diel periodicity of catch in synthetic sex attractants vs. female-baited traps. *Environ. Entomol.* 2:673-676.

(9)  P. Rau and N. Rau. 1929. The sex attraction and rhythmic periodicity in giant saturniid moths. *Trans. Acad. Sci. St. Louis* 26:83-221.

(10)  H.B.D. Kettlewell. 1942. The assembling scent of *Arctia villica* and *Parasemia plantaginis*. *Entomol. Rec.* 54:62-63.

(11)  H.B.D. Kettlewell. 1946. Female assembling scents with reference to an important paper on the subject. *Entomologist* 79:8-14.

(12)  B. Götz. 1951. Die Sexualduftstoffe an Lepidopteren. *Experientia* 7:406-418.

(13)  H.H. Shorey and L.K. Gaston. 1965. Sex pheromones of noctuid moths. V. Circadian rhythm of pheromone-responsiveness in males of *Autographa californica, Heliothis virescens, Spodoptera exigua,* and *Trichoplusia ni* (Lepidoptera: Noctuidae). *Ann. Entomol. Soc. Am.* 58:597-600.

(14)    J.A. George. 1965. Sex pheromone of the oriental fruit moth *Grapholitha molesta*-Busck (Lepidoptera, Tortricidae). *Can. Entomol.* 97:1002-1007.

(15)    J.D. Solomon and R.C. Morris. 1966. Sex attraction of the carpenter-worm moth. *J. Econ. Entomol.* 59:1534-1535.

(16)    C.A. Saario, H.H. Shorey, and L.K. Gaston. 1970. Sex pheromones of noctuid moths. XIX. Effect of environmental and seasonal factors on captures of males of *Trichoplusia ni* in pheromone-baited traps. *Ann. Entomol. Soc. Am.* 63:667-672.

(17)    R.M.M. Traynier. 1970. Sexual behaviour of the Mediterranean flour moth, *Anagasta kuhniella:* Some influence of age, photo-period and light intensity. *Can. Entomol.* 102:534-540.

(18)    R. Lange, D. Hoffman, and M. Weissinger. 1971. Untersuchungen über das flugverhalten männlicher Falter von *Rhyacionia buoliana* (Lep., Tortricidae). *Oecologia* 6:156-163.

(19)    R.K. Sharma, R.E. Rice, H.T. Reynolds, and H.H. Shorey. 1971. Seasonal influence and effect of trap location on catches of pink bollworm males in sticky traps baited with hexalure. *Ann. Entomol. Soc. Am.* 64:102-105.

(20)    T.T.Y. Wong, M.L. Cleveland, D.F. Ralston, and D.G. Davis. 1971. Time of sexual activity of codling moths in the field. *J. Econ. Entomol.* 64:553-554.

(21)    R.T. Cardé, C.C. Doane, and W.L. Roelofs. 1974. Diel periodicity of male sex pheromone response and female attractiveness in the gipsy moth. *Can. Entomol.* 106:479-484.

(22)    R.M.M. Traynier. 1970. Habituation of the response to sex pheromone in two species of Lepidoptera, with reference to a method of control. *Entomol. exp. & appl.* 13:179-187.

(23)    R.J. Bartell and L.A. Lawrence. 1973. Reduction in responsiveness of males of *Epiphyas postvittana* (Lepidoptera) to sex pheromone following previous brief pheromonal exposure. *J. Insect Physiol.* 19:845-855.

(24)    R.J. Bartell and W.L. Roelofs. 1973. Inhibition of sexual response in males of the moth *Argyrotaenia velutinana* by brief exposures to synthetic pheromone or its geometrical isomer. *J. Insect Physiol.* 19:655-661.

(25)    I. Schwinck. 1955. Wietere Untersuchungen zur frage Geruchsorientierung der Nachtschmetterlinge: Partielle Fühleramputation bei Spinnermännchen, insbesondere am Siedenspinner *Bombyx mori L. Z. vgl. Physiol.* 37:439-458.

(26)    R.J. Bartell, and H.H. Shorey. 1969. Pheromone concentrations required to elicit successive steps in the mating sequence of males of the light-brown apple moth, *Epiphyas postvittana. Ann. Entomol. Soc. Am.* 62:1206-1207.

(27)    R.T. Cardé, T.C. Baker, and W.L. Roelofs. 1975. Behavioural role of individual components of a multichemical attractant system in the oriental fruit moth. *Nature* 253:348-349.

(28)    S.R. Farkas, and H.H. Shorey. 1972. Chemical trail-following by flying insects: A mechanism for orientation to a distant odor source. *Science* 178:67-68.

(29)    J.S. Kennedy and D. Marsh. 1974. Pheromone-regulated anemotaxis in flying moths. *Science* 184:999-1001.

(30)    S.R. Farkas, H.H. Shorey, and L.K. Gaston. 1974. Sex pheromones of Lepidoptera. XLV. Influence of pheromone concentration, and visual cues on aerial odor trail following by males of *Pectinophora gossypiella. Ann. Entomol. Soc. Am.* 67:633-638.

(31)    U.E. Brady, J.H. Tumlinson, R.G. Brownlee, and R.M. Silverstein. 1971. Sex stimulant and attractant in the Indian meal moth and in the almond moth. *Science* 171:802-804.

(32)   L.P. Brower, J.v.Z. Brower, and F.P. Cranston. 1965. Courtship behaviour of the queen butterfly, *Danaus gilippus berenice* (Cramer). *Zoologica* 50:1-39.

(33)   R.H. Stobbe. 1912. Die abdominalen Duftorgane der männlichen Sphingiden und Noctuiden. *Zool. Jb.* 32:493-632.

(34)   M.C. Birch. 1970. Structure and function of the pheromone-producing brush-organs in males of *Phlogophora meticulosa* (L.) (Lepidoptera: Noctuidae). *Trans. Roy. Entomol. Soc.* (Lond.) 122:277-292.

(35)   G.G. Grant. 1970. Evidence for a male sex pheromone in the noctuid *Trichoplusia ni. Nature* 227:1345-1346.

(36)   G.G. Grant. 1974. Male sex pheromone from the wing gland of the Indian meal moth, *Plodia interpunctella* (Hbn) (Lepidoptera: Phycitidae). *Experientia* 30:917-918.

(37)   J.S. Kennedy. 1972. The emergence of behaviour. (Address to closing plenary session of 14 Int. Congr. Entomol., 1972.) *J. Aust. Entomol. Soc.* 11:19-27.

(38)   W.L. Roelofs and A. Comeau. 1971. Sex pheromone perception: Synergists and inhibitors for the redbanded leafroller attractant. *J. Insect Physiol.* 17:435-449.

(39)   G.H.L. Rothschild. 1974. Problems in defining synergists and inhibitors of the oriental fruit moth pheromone by field experimentation. *Entomol. exp. & appl.* 17:294-302.

(40)   W.L. Roelofs, R.T. Cardé, and J.P. Tette. 1973. Oriental fruit moth attractant synergists. *Environ. Entomol.* 2:252-254.

(41)   W.L. Roelofs, A. Hill, and R.T. Cardé. 1975. Sex pheromone components of the redbanded leafroller, *Argyrotaenia velutinana* (Lepidoptera: Tortricidae). *J. Chem. Ecol.* 1:83-89.

(42)   M. Jacobson, R.E. Redfern, W.A. Jones, and M.W. Aldridge. 1970. Sex pheromones of the southern armyworm moth: Isolation, identification and synthesis. *Science* 170:542-544.

(43)   Y. Tamaki, H. Noguchi, T. Yushima, C. Hirano, K. Honma, and H. Sugawara. 1971. Sex pheromone of the summerfruit tortrix: Isolation and identification. *Konchu* 39:338-340.

(44)   Y. Tamaki, H. Noguchi, T. Yushima, and C. Hirano. 1971. Two sex pheromones of the smaller tea tortrix: Isolation, identification and synthesis. *Appl. Entomol. Zool.* 6:139-141.

(45)   G.M. Meijer, F.J. Ritter, C.J. Persoons, A.K. Minks, and S. Voerman. 1972. Sex pheromones of summer fruit tortrix moth *Adoxophyes orana:* Two synergistic isomers. *Science* 175:1469-1470.

(46)   A.K. Minks, W.L. Roelofs, F.J. Ritter, and C.J. Persoons. 1973. Reproductive isolation of two tortricid moth species by different ratios of a two-component sex attractant. *Science* 180:1073-1074.

(47)   U.E. Brady. 1973. Isolation, identification and stimulatory activity of a second component of the sex pheromone system (complex) of the female almond moth, *Cadra cautella* (Walker). *Life Sci.* 13:227-235.

(48)   B.F. Nesbitt, P.S. Beevor, R.A. Cole, R. Lester, and R.G. Poppi. 1973. Sex pheromones of two noctuid moths. *Nature* 244:208-209.

(49)   A. Hill, R.T. Cardé, A. Comeau, W. Bode, and W.L. Roelofs. 1974. Sex pheromones of the tufted apple bud moth *(Platynota idaeusalis). Environ. Entomol.* 3:249-252.

(50)   R.J. Bartell. Unpublished information.

(51)   T.L. Payne, H.H. Shorey, and L.K. Gaston. 1970. Sex pheromones of noctuid moths: Factors influencing antennal responsiveness in males of *Trichoplusia ni. J. Insect Physiol.* 16:1043-1055.

(52)   C.M. Ignoffo, R.S. Berger, H.M. Graham, and D.F. Martin. 1963. Sex attractant of cabbage looper, *Trichoplusia ni* (Hubner). *Science* **141**:902-903.

(53)   R.H. Wright. 1958. The olfactory guidance of flying insects. *Can. Entomol.* **90**:51-89.

(54)   S.R. Farkas and H.H. Shorey. 1974. In *Pheromones.* M.C. Birch, Ed. North-Holland, Amsterdam.

## CHAPTER 13

# Behaviorally Discriminating Assays of Attractants and Repellents

J. S. Kennedy*

*Agricultural Research Council Insect Physiology Group*
*Department of Zoology and Applied Entomology*
*Imperial College, London, England*

The desirability of assay methods for behavior-modifying chemicals more analytical than those now in use is a recurrent theme in this book. This chapter considers what behavior is measured in existing laboratory assays for attractants and repellents only, and offers some very general comments on the methodology of more discriminating assays.

There have been many previous calls for a move in this direction, with R.H. Wright (1) as the movement's chief prophet. But the call is acquiring some practical urgency as the existing field and laboratory assay methods present more and more complexities, problems, and paradoxes—multicomponent chemical signals; compound and contradictory behavioral effects even of single chemicals; "masks" and "inhibitors"; adaptation and habituation; and discrepancies between field and laboratory results. There is a practical interest in knowing not just how many insects arrive at an odor source, for which purpose any convenient experimental arrangement will do, but also by what maneuvers they get there. There is thus a need to segregate, characterize, and measure the behavioral components of the system no less than the chemical ones, and to know which chemical affects which response in an excitatory or

---

*Address: Imperial College Field Station, Silwood Park, Ascot, Berks SL5 7PY, England.

inhibitory fashion. One can no longer assume that a positive flutter response to a candidate attractant allows one to infer all the other ways the insect will respond to it (1) (Chapter 14), let alone which ones can best be manipulated chemically.

If lack of time or money permit only one or two responses to be used for assaying candidate attractant or repellent chemicals, then there is an advantage in using an assay method that displays each one unconfused by others. Only thus can one know what is being measured so that quantitative comparisons are meaningful. But it will often be advantageous to assay for more of the responses that make up the natural sequence—for instance, when multicomponent pheromones are studied, or when trap action needs to be understood to improve the design, or when repellents/inhibitors are being sought and applied. Here, too, unambiguous assays are needed so that the component responses can be measured. Olfactometers generally have provided highly ambiguous assays, and more discriminating ones are possible.

## I. ATTRACTION AS A UNITARY PROCESS

Two recent findings with the Oriental fruit moth referred to by Bartell (Chapter 12) will suffice to epitomize the present changing requirements. Given the predominantly chemical approach to the subject hitherto, it was natural that attention was focused upon the interface between chemical and insect, the chemoreceptive surface of a "black box." Thus it was supposed that the "synergistic" and "primary" components of attractant pheromones identified from field-trap assays acted on the same process in the insect, probably chemoreception. Cardé et al. (2) have shown, however, that the "synergistic" component of the female pheromone complex in the Oriental fruit moth has its own positive effect on male behavior—that of stimulating close-range courtship activities. Without the addition of these responses, arriving males may not enter a trap. The "primary" components, on the other hand, stimulate the males' upwind approach from a distance; without this the "synergist" cannot trap them. The effects are in reality additive: "synergism" was a meaningful description only in terms of trap catch, and quite misleading in terms of behavior or sensory physiology.

"Inhibitor" substances, which reduce pheromone trap catches, were likewise taken to act on a unitary process, presumptively chemoreception (3) (4). So it seemed paradoxical, as Bartell relates, when such a substance was found to "inhibit" trap catches of males only when placed in or close to the pheromone trap, whereas when generally disseminated in the vicinity it actually increased the trap catch (5). This kind of result could have been predicted many years ago from the observation that "repellent" chemicals which turn mosquitoes aside at close range also act like "attractants" such as $CO_2$ in stimulating them to take off and fly upwind (6)-(8). In fact, wind-tunnel experiments (9) have shown that a similar two-stage mechanism explains the action of a trap-catch "inhibitor" for the cabbage looper moth. Again, this substance inhibits only close-range approach and does not prevent upwind approach from a distance.

Thus the working idea of attraction/repulsion as a unitary process has become counterproductive, and recent work on "synergists" and "inhibitors" has brought out a primary distinction between the distant and close-range components of the overall behavioral mechanism.

## II. COMPONENTS OF ATTRACTION/REPULSION

The terms *attraction* and *repulsion* are used here only because conventionally they are still the ones used to embrace all the variety of locomotory responses that, when elicited or conditioned by odorous chemicals, bring an insect to the source of the odor, or, alternatively, somehow prevent it from finding that source. Saxena (10) dissected these portmanteau concepts of *attraction* and *repulsion* and followed tradition (11) (12) in combining all the components under the single term *orientation,* but justly noted that this included arriving and staying without benefit of any orientation! One useful outcome of present concern with these processes would be the adoption of some less ambiguous terms. The simple term *finding*—of host, mate, and so on—perhaps best covers what is meant without begging any questions of mechanism.

The known kinds of locomotory response involved in finding an odor source were categorized in Chapter 5. If one takes that as read, assay methods can be surveyed in terms of the stimulus situations they present to the test insects and thus of the classes of locomotory response they may measure.

The first major division among such responses is between those operating at a distance and those operating at close range in the field, already mentioned. The second major division is between directed responses (taxes) and undirected ones (kineses). This second division cuts across the first inasmuch as both distant and close-range responses include both kineses and taxes. Nevertheless the taxis mechanisms involved in distant and close-range responses seem to be distinct (Chapter 5, Sections IV and V). Close-range responses commonly include chemotactic ones to directional cues provided by the steep local gradient of odor, whereas distant responses can be defined operationally as those occurring where the odor gradients are too disrupted and shallow to permit a chemotactic approach, and any taxes depend on directional cues provided by other features of the environment such as wind or a visual target (Chapter 5, Sections II, VI, and VII). Ortho- and klinokinetic responses (Chapter 5, Section III) may occur both at a distance and at close range, but as distance increases they become less and less efficient as target-finding mechanisms.

The responses involved in "repellency" remain poorly understood although they are of great interest for the control of insects with behavior-modifying chemicals. Despite the name they are not necessarily directed, but may be either kineses or taxes (13). In either case they may occur at close range, as in negative chemotaxis; or they may occur at some distance, as in negative anemotaxis and direct klinokinesis. The direct klinokinetic response to a repellent consists in turning at random on entering the odorous zone, whereas the inverse klinokinetic response to an attractant consists in turning at random on leaving the odorous zone (7).

## III. UNDISCRIMINATING ASSAYS

Most olfactometers designed to measure behavioral components of "attraction" beyond the basic, direct orthokinesis (activation, flutter, etc.) permit several components to occur together. What is measured is the net result of their several contributions in unknown, uncontrolled proportions. This is mainly because there are steep odor gradients to which the insects may respond orthokinetically, klinokinetically, and chemotactically; if there is air flow in the plane of maneuver the insects may respond anemotactically as well. These olfactometers do not therefore discriminate between what could be distant and what could be close-range responses when transposed to field conditions. For example, in assays based on counting those insects that arrive at air entry-ports in the side of a cage [e.g., (14)-(16)], a higher count or trap catch at one port could be due to the insects slowing down or stopping there (inverse chemoorthokinesis), or to increased or decreased random turning (direct or inverse chemoklinokinesis), or to directed turning (chemotaxis), or to odor-conditioned anemotaxis, or to all of these. Thus no single one of them is knowingly measured in such an assay.

### A. Horizontal Airflow Systems

In assaying specifically for distant responses (Chapter 5, Sections II, VI, and VII) the evident advantages of allowing the insects to maneuver in a horizontal airstream carrying the odor are: that it effectively flattens the odor gradient along the line toward the source, thus creating a stimulus situation more like that at a distance from a field source; and that it allows the insect to use positive anemotaxis for guidance toward the source in the way that the same insect could, in principle, use it from a distance in the field. Nevertheless this crucial mechanism of attraction is not unequivocally assayed, even in walking insects, in the classical Y-tube olfactometer or its various descendants also embodying enclosed, convergent airstreams [e.g., (17)-(20)] because of the steep odor gradients at the interface between the airstreams. Such a steep gradient would permit chemotactic and chemoklinokinetic responses to help the insects stay in one stream, and this is unlikely at a distance in the field.

Nor is potentially distant anemotactic guidance displayed unambiguously in another well-known type of olfactometer, with two or more airstreams entering a wider arena by separate ports and then converging upon the insect's starting point [e.g., [(19) (21)-(23)]. The detailed pattern of air currents in such systems must be quite complex, but there will again be steep odor gradients encountered by the insect as it crosses in and out of an odor stream, to which again it may react chemoklinokinetically and chemotactically, as was noted by Wood and Bushing (21). In addition, the insect presumably encounters steep gradients of air velocity, shear, across each airstream, and these unnaturally steep gradients of mechanical stimulation could themselves, conceivably, enable a walker or tethered flier to keep in line with the stream.

Such ambiguities are well illustrated in the classical example of odor-conditioned anemotaxis in walking *Drosophila* (24), where the insect crosses a narrow odorless airstream but turns and walks up the same airstream when it is suitably scented. It seems a reasonable inference that the upwind turn at the moment when the insect first enters the scented airstream from one side is truly anemotactic and not chemotactic, because at that moment the odor concentrations on its two sides will be virtually the same. But from that moment on, during the insect's progression along the narrow column of scent and air, we cannot exclude some contribution from chemotactic responses to the steep odor gradients running across the airstream, or from what might be called "mechanotactic" responses to the steep air velocity gradients across the stream. To an unknown extent the insect could be "trail-following" in both these respects.

In the wide, shallow, wind-tunnel olfactometer developed by Moeck (25) for walking bark beetles the arena is swept by uniform airflow across its full width. This is an advance on systems offering a choice of discrete airstreams in that it rules out mechanical trail-following in response to wind shear, so that any orientation to the wind must be a true anemotaxis cued solely by the direction of flow. On the other hand, the possibility of chemotactic trail-following was not ruled out in Moeck's system because a point source of odor was used so the odor plume was narrow. This narrow plume gave clear-cut results which must have been due partly to anemotaxis, but perhaps modified by an additional chemotactic component.

In Schwink's (26) and Francia and Graham's (27) systems of fully wind-swept, open arenas, the odor permeated the air but the airflow was presumably rough and thus less well-defined than in Moeck's system. In Daterman's (28) small-cylinder wind tunnels the airflow was probably smoother than that over the open arenas but the odor distribution in the airflow was not even and the insects' mixed walking and flying movements in the small space were difficult to disentangle, although an upwind bias was clearly measurable.

For flying orientation assays, wind tunnels using free fliers are cumbersome, and one alternative that has been tried is to tether a wing-flapping insect, free to turn horizontally, at the meeting point of four horizontal airstreams converging at right angles (29). This unfortunately carries serious ambiguities. The tethered insect's orientation reactions to the mechanical stimuli it receives from the air movement will not be those that a free-flying insect uses for anemotactic orientation (these are visual; Chapter 5, Section II.B), but rather those used for flight-stabilizing corrections of yaw (30). Secondly, the steep odor gradients lying transversely to the airstreams will permit chemotactic orientation along them; and lastly, the air- and odor-flow situation at the point where the four streams meet is indefinable.

## B. Alternative Chambers

If there is no airflow in the plane of maneuver then anemotaxis is of course ruled out, and to that extent the arrangement favors what in the field would be close-range responses rather than distant ones. In many such olfactometers there is no air movement whatever (unless it be unintended convective circulation); these are the numerous "alternative chambers," "choice chambers," "split arenas," and other arrangements of odorous and nonodorous regions in a still-air arena (10) (27) (31)-(39). In other alternative chambers there is air movement, but in the form of two contiguous airstreams rising vertically through the perforated floor of a horizontal arena so that horizontal anemotaxis is again ruled out (40)-(43). This permits better control of the odor conditions and ensures a well-defined odor boundary between the two regions of the arena.

Both forms of alternative chamber permit several types of response to occur in uncontrolled proportions so that assays of different chemicals are not strictly comparable. "The response" or "intensity of reaction" is commonly measured by the difference between the numbers of insects found in the two halves of the arena after a standard time, which leaves unmeasured the relative contributions of chemokineses, which can be both distant and close-range reactions depending on odor concentration and gradient steepness (Chapter 5, Section III), and chemotaxes that are close-range reactions (Section IV). Moreover the size of this difference depends critically on how active the insects happen to be at the time. If they move little, then few of them will have sampled both sides of the arena in the time allowed and a "weak response" will be recorded even if any insects that did cross the boundary responded strongly. If they are very active there will be little orthokinetic arrestment on the "preferred" side, and again a "weak response" may be recorded. Thus there will be an optimum level of general activity to produce a maximum difference in numbers on the two sides, and any departure from this optimum will produce a spuriously weaker "response."

Tracking and timing the movements of the insects in an alternative chamber reveals more about the reactions at work. But since it is left to the free insects to determine the frequency, speed, and direction of their boundary-crossings, the stimulation they receive is not controlled by or known to the observer. Thus in one of a series of papers on assaying pine-bark fractions for their effect on bark beetle "orientation" (10), Perttunen et al. (44) tracked and timed the beetles in their standard still-air alternative chamber to determine what response mechanisms might underlie the simple distribution counts used for routine assays. This proved rather difficult. They found, first, a weak orthokinetic component in the form of slower movement on the "preferred" side, and secondly that turnings-back on crossing the boundary line were much more frequent when the insects were going from that side to the other, than vice versa. From the tracks they concluded that these turnings were in part due to directed reactions interpreted as klinotactic (they could also have been tropotactic), but were in part perhaps due to sharp klinokinetic reactions. Wigglesworth (31) and Daykin et al. (45) inferred definite klinokinetic reactions from the tracks of their insects walking and flying, respectively, but it is not easy to distinguish these from chemotactic turns when the insects are moving freely about in an alternative chamber.

The odor stimulus has been presented in a much more standardized and repeatable manner by using an independent orienting stimulus—a horizontal light beam—to set the insects on a definite path across an arena with odorous and nonodorous floor areas (36) (46). Orthokinetic arrest in response to the odor then becomes clear; but chemotactic and chemoklinokinetic reactions at the boundary are still confused. If the boundary between the nonodorous and odorous regions runs at right angles to the light beam, and if the insect meets that boundary at right angles as intended, then there will be no difference of odor concentration between its two sides, so it cannot respond chemotactically. If it meets the boundary more or less obliquely, there will be some bilateral difference of odor concentration which may be sufficient for it to respond chemotactically, but the difference will be an uncontrolled variable. At the same time the insect may respond by klinokinetic turning when it crosses the boundary at any angle.

## IV.  DISCRIMINATING ASSAYS

Undiscriminating assays that compound a number of different responses among many free-moving insects, as above, will presumably continue in use because they can often be done with quite simple apparatus and provide a quick means of accumulating data for statistical comparisons. Despite the fact that the results are to some extent misleading for the reasons outlined above, they are usually helpful when it comes to field applications. Nevertheless, surer and perhaps even quicker field application would come from devising discriminating assays, not least from the spin-off in better understanding of the behavior to be controlled (Chapter 14). Contrary to what might be expected from what has been said above about the multiplicity of behavioral mechanisms involved, discriminating laboratory assays do not call for a similar multiplicity of olfactometer types or much more time to accumulate results once the system is working.

## A. Principles

A few basic principles of design for discriminating assays emerge from the foregoing survey of undiscriminating ones:

1. The odor stimulus should be presented either in a form that will elicit only one kind of response, or, if more than one kind is elicited, then in a form that allows separate measurement of each. This is desirable because different attractants, repellents, and "antiattractants" (47) may act through different components of the target-finding process.
2. The form and frequency of the stimulus presentation needs to be standardized under the observer's control and not left to depend on the uncontrolled activity of the insects themselves.
3. Provision should be made for testing responses not only to the onset and strengthening of the odor stimulus but also to its weakening and cessation. "Off"-responses are not necessarily the opposite of "on"-responses, a reversion to the previous behavior. Distinct "off"-responses and after-effects play an essential part both in the inverse chemoklinokinesis mechanism and in the odor-conditioned anemotaxis mechanism of finding an odor source (Chapter 5, Sections III and VII).
4. The pretreatment of the test insects needs attention to ensure that they are as highly and uniformly responsive to the odor as possible and not, for instance, in varying degree still in a migratory state (Chapters V and XI). The test insects must also have adequate space to execute the expected maneuver. The chemoklinokinetic behavior of free-flying insects, for instance, may need considerable space.

Controlling the presentation of the stimulus has two operational aspects: first, control of the composition and concentration of the odor and changes in them, and, second, control of the insects' exposure to these. Flowing air has the advantage over still air on both counts. The maintenance of a constant odor composition and concentration, and the controlled variation of them, are easier with flowing air, as many workers have found [e.g., (42) (48)]. More important, flowing air makes it possible to simulate the spatial variations of odor encountered in nature, by introducing variations over time in the composition of uniformly odorous air. Then the insect's own movements cannot alter the odor stimulation it receives, which is thus brought under the observer's complete control. Such a system of temporal rather than spatial variation of the odor stimulus is suitable for assays of all the types of response mentioned in Section II above and in Chapter 5, excepting chemotaxis. These types of response may now be considered seriatim.

## B. Chemokineses

An olfactometer something like that which Daykin and Kellogg (42) devised with two contiguous, equal-velocity, airstreams passing vertically up through the whole area of a horizontal arena of adequate size, excludes anemotaxis and allows separate measurement of several other types of response. In measuring such responses, aside from chemotaxis, it is best to arrange for identical odor compositions in the two airstreams. This provides one large arena for insect maneuver in air that is always uniformly odorous (or nonodorous) at any one time, but varies in odor content with time according to a definite program. Using this system, orthokinetic and klinokinetic responses can be measured simultaneously but separately by timed tracking of walkers or fliers. With due regard to the free insects' walking or flying speeds, their recorded responses to odor changes presented over time can be translated roughly into spatial maneuvers in natural, nonuniform odor fields. Orthokinetic responses alone can be measured with a suitable actograph such as a flight mill (48).

If desired, the twin-stream vertical-flow apparatus can be used to record spatial maneuvers directly by arranging for different odor conditions in the two contiguous airstreams. This has often been done, but the results depend very much on the particular dimensions and other characteristics of the given arena. The results are not therefore of such general application as results obtained by using purely temporal changes in a uniform odor field. So the latter seems preferable for routine assay, and it is a more flexible system.

In any case, in order to record klinokinetic responses unambiguously it is necessary to expose the insects to purely temporal changes in a uniform odor field. Otherwise, when the insects move through odor gradients in space, their random klinokinetic turning responses may be mixed with and obscured by chemotactic turns. Moreover, when the insects are moving about in a nonuniform odor field it is not possible to measure an important variable in klinokinesis—namely, the rate of adaptation of the random turning response to an odor change. It is also difficult to measure the separate effects of increases and decreases of odor on turning, an important point because one way that repellent substances act is simply to invert these effects (7) (45).

## C. Odor-Conditioned Responses to Visual Targets

The same type of olfactometer, again using purely temporal changes of odor concentration, is also suitable for the measurement of odor-conditioned visual responses to target objects (Chapter 5, Section VI), with the advantage that the olfactory effect of distance from the target can be simulated simply by varying the odor concentration.

## D. Chemotaxes

In principle, the assays methods so far considered (Sections IV.A-IV.C) do not necessitate any constraint on the movements of the insects other than by the walls of the arena. In assays for chemotactic responses, on the other hand, the arrangement for assay must provide a difference of odor concentration between the two sides of the insect—that is, an odor gradient in space. This would be available in precisely controlled form along the boundary between the two vertical airstreams in the same vertical-flow olfactometer. In order to present the given odor gradient to the insect in a controlled, repeatable manner which maximizes the likelihood that it will show the looked-for response, and which also makes measurements comparable, the insect must be oriented along the boundary—that is, transversely to the odor gradient. At the same time this constraint must not be rigid; the insect should remain free to turn to one side or the other in response to the odor difference it may detect between its two sides.

This situation can be achieved by means of another, opposing stimulus situation, such as a light beam, or a raised "cat-walk" for a walker, which orients the insect's path along the odor boundary. This provides a standard-strength orienting stimulus against which one can "titrate," as it were, the strength of the odor gradient as a stimulus for turning to one side. This "titration" method, deliberately providing a standard-stimulus situation for a response that is antagonistic to the one being measured, is a useful one that has been applied in other contexts [e.g., (25) (28)].

If the twin airstream olfactometer is not available, or is in some way inconvenient to use, then the required transverse odor gradient can be set up by diffusion in still air, for walking insects only, by arranging an odor source along one side, and an odor "sink" (activated charcoal) along the other side, of the insect's path, which is restricted to a straight line as before. The source and sink can form the walls of a corridor or (less neatly) the two halves of the floor of an alternative chamber. In either case the path between them should be fairly long to give the insect some time to react to the gradient by turning, thus increasing the sensitivity of the assay. The still-air system has the disadvantage that it is less flexible in use than the twin airstream system because the odor gradient is less easily changed.

An untried but neater method of measuring chemotactic responses would be to keep an insect walking in one place on a feedback-rotated sphere (see below) or a "Y-maze globe" (49) with twin nonmixing odor-conditioned airstreams impinging vertically on the two sides of the insect. As far as flying insects are concerned, it hardly seems worth attempting the difficult task of measuring their chemotactic responses until there is evidence that these play a significant part in nature.

## E. Odor-Conditioned Anemotaxis

Since odor-conditioned anemotaxis can operate both at a distance and at close range in both walkers and fliers, it is probably the most important mechanism to use in assaying attractants and repellents. The most elegant technique so far devised for measuring these responses during walking uses a captive insect in a horizontal airstream on top of a feedback-rotated sphere, where its every movement is automatically compensated and recorded. The path these movements would follow in space, if the insect were free, is traced out by an X-Y recorder (50) (51). The development of a similar device for flying insects would present formidable technical problems because of the complex mechanical and more especially visual feedbacks required, in order to simulate in a tethered flier the sensory input patterns of a free flier in wind (Chapter 5, Sections II.B and VII).

Otherwise, assays using walkers and fliers must continue to use free insects in some kind of wind tunnel. For the reasons already discussed (Section III.A) unambiguous display of odor-conditioned anemotactic behavior calls for uniform airflow across the tunnel and uniform permeation of the air with the odor under test (examples were given in Chapter 5, Section II.A). The purely anemotactic effects of entering and emerging from a natural wind-borne odor plume from a point source could then be simulated, once more, by temporal instead of spatial variations in the odor content of the airstream. This would obviate any possible admixture of chemotactic responses; and, for the purposes of laboratory identification and assay of odor-conditioned anemotaxis, there is probably no need to reproduce the zigzagging behavior of many insects flying up a wind-borne plume of "attractant." The reversing anemomenotactic responses that apparently cause the zigzagging are more conspicuously and reproducibly displayed when the odor is simply removed from the air (Chapter 5, Section VII).

The method of alternating between uniformly odorous and uniformly clean air would be simpler in every way—physically, procedurally, and in recording and repeatability of results—once the facility has been created. This method is therefore much to be preferred when enough of the substance to be assayed is available for uniform permeation of the wind tunnel. When enough is not available, it will be necessary to fall back on a point source and thus an odor plume. But this is to be avoided if at all possible even with fliers, where the risk of chemotactic admixture is probably negligible (Chapter 5, Sections IV.B and V) because it means much more variable results: the flier's encounters with the odor then depend too much on the flier, and the pattern of odor stimulation it receives in the filamentous plume is variable and unknown.

It is undoubtedly easier to build and to operate a wind tunnel for walking insects than for flying ones. Substantial progress has already been made in the field application of chemicals controlling the behavior of flying insects, notably bark beetles (Chapter 22), on the basis of field-trapping backed by laboratory assays using walking insects only. Assays for routine chemical screening must be simple and of course cannot include separate measurements of every possible kind of response. Nevertheless, routine assay using walking insects can itself be made simpler, less ambiguous, truly comparative, and more informative, as indicated above, by the use of a wind tunnel with full-width uniform flow of uniformly odorous or nonodorous air.

The response that is chosen for routine screening of chemicals will always be in large part a matter of convenience; the point is for the investigator to know exactly what response he has chosen, by using a behaviorally unambiguous assay. This will of course leave other important responses to be tested separately, but this can be done selectively. The discriminating principles outlined apply equally to in-depth analyses of target-finding mechanisms in a few model species: it is on such studies that any further improvement of routine assays will depend.

## V. SUMMARY

Recent work on chemical attractants and repellents has underlined the need for bioassay methods that distinguish behavioral as well as chemical components of insect behavior. Most existing types of olfactometers do not do this because the odor stimulus is presented in such a way that several different types of response, having different roles in the field situation, may contribute simultaneously and in unknown proportions to what is measured. It is suggested that all of the known types of response, except chemotaxis, could be assayed unambiguously by presenting the insects with temporal instead of spatial changes of odor stimulation in a uniform arena.

## REFERENCES

(1)    R.H. Wright. 1964. After pesticides—what? *Nature* (Lond.) **204**:121-125.
(2)    R.T. Cardé, T.C. Baker, and W.L. Roelofs. 1975. Behavioural role of individual components of a multichemical attractant system in the Oriental fruit moth. *Nature* (Lond.) **253**:348-349.

(3)   M. Jacobson. 1972. *Insect Sex Pheromones.* Academic, New York.

(4)   R.T. Cardé, W.L. Roelofs, and C.C. Doane. 1973 Natural inhibitor of the gypsy moth sex attractant. *Nature* (Lond.) 241:474-475.

(5)   G.H.L. Rothschild. 1974. Problems in defining synergists and inhibitors of the Oriental Fruit Moth pheromone by field experimentation. *Entomol. exp. appl.* 17:294-302.

(6)   J.S. Kennedy. 1940. The visual responses of flying mosquitoes. *Proc. Zool. Soc. Lond., A,* 109:221-242.

(7)   R.H. Wright. 1968. Tunes to which mosquitoes dance. *New Sci.* (Lond.) 28 March: 694-697.

(8)   R.H. Wright. 1975. Why mosquito repellents repel. *Sci. Am.* 233:104-111.

(9)   J.R. McLaughlin, E.R. Mitchell, D.L. Chambers, and J.H. Tumlinson. 1974. Perception of Z-7-dodecen-1-ol and modification of the sex pheromone response of male loopers. *Environ. Entomol.* 3:677-680.

(10)  K.H. Saxena. 1969. Patterns of insect-plant relationships determining susceptibility or resistance of different plants to an insect. *Entomol. exp. appl.* 12:751-766.

(11)  V.G. Dethier. 1947. *Chemical Insect Attractants and Repellents.* Blakiston, Philadelphia.

(12)  G. Fraenkel and D.L. Gunn. 1961. *The Orientation of Animals. Kineses, Taxes and Compass Reactions.* Dover, New York.

(13)  V.G. Dethier, L.B. Browne, and C.N. Smith. 1960. The designation of chemicals in terms of the responses they elicit from insects. *J. Econ. Entomol.* 53:134-136.

(14)  J.O.G. Wieting and W.M. Hoskins. 1939. The olfactory responses of flies in a new type of insect olfactometer. II. Responses of the housefly to ammonia, carbon dioxide and ethyl alcohol. *J. Econ. Entomol.* 32:24-29.

(15)  H.K. Gouck and C.E. Schreck. 1965. An olfactometer for use in the study of mosquito attractants. *J. Econ. Entomol.* 58:589-590.

(16)  H.J. Bos and J.J. Laarman. 1975. Guinea pig, lysine, cadaverine and estradiol as attractants for the malaria mosquito *Anopheles stephensi. Entomol. exp. appl.* 18:161-172.

(17)  G.C. Varley and R.L. Edwards. 1953. An olfactometer for observing the behaviour of small animals. *Nature* (Lond.) 171:789-790.

(18)  A.A. Guerra. 1968. New techniques to bioassay the sex attractant of pink bollworms with olfactometers. *J. Econ. Entomol.* 61:1252-1254.

(19)  W.E. Burkholder. 1970. Pheromone research with stored-product Coleoptera. In *Control of Insect Behavior by Natural Products.* D.L. Wood, R.M. Silverstein, and M. Nakajima, Eds. Academic, New York, pp. 1-20.

(20)  R.S. Rejesus and H.T. Reynolds. 1970. Demonstration of the presence of a female sex pheromone in the cotton leaf perforator. *J. Econ. Entomol.* 63:961-964.

(21)  D.L. Wood and R.W. Bushing. 1963. The olfactory response of *Ips confusus* (LeConte) (Coleoptera: Scolytidae) to the secondary attraction in the laboratory. *Can. Entomol.* 95:1066-1078.

(22)  H.J. Meyer and D.M. Norris. 1967. Behavioral responses by *Scolytus multistriatus* (Coleoptera: Scolytidae) to host- (*Ulmus*) and beetle-associated chemotactic stimuli. *Ann. Entomol. Soc. Am.* 60:642-646.

(23)  D.D. Hardee, E.B. Mitchell, and P.M. Huddleston. 1967. Procedure for bioassaying the sex attractant of the boll weevil. *J. Econ. Entomol.* 60:167-171.

(24)  C. Flügge. 1934. Geruchliche Raumorientierung von *Drosophila melanogaster. Z. vgl. Physiol.* 20:463-500.

(25)   H.A. Moeck. 1970. An olfactometer for the bioassay of attractants for Scolytids. *Can. Entomol.* **102**:792-796.

(26)   I. Schwinck. 1954. Experimentelle Untersuchungen über Geruchssinn und Strömungswahrnehmung in der Orientierung bei Nachtschmetterlingen. *Z. vgl. Physiol.* **37**:19-56.

(27)   C. Francia and K. Graham. 1967. Aspects of orientation behavior in the ambrosia beetle *Trypodendron lineatum* (Olivier). *Can. J. Zool.* **45**:985-1002.

(28)   G.E. Daterman. 1972. Laboratory bioassay for sex pheromone of the European pine shoot moth, *Rhyacionia buoliana. Ann. Entomol. Soc. Am.* **65**:119-123.

(29)   P.R. Hughes and G.B. Pitman. 1970. A method for observing and recording the flight behavior of tethered bark beetles in response to chemical messengers. *Contrib. Boyce Thompson Inst. Plant Res.* **24**:329-336.

(30)   M. Gewecke. 1974. The antennae of insects as air-current sense organs and their relationship to the control of flight. In *Experimental Analysis of Insect Behaviour.* L.B. Browne, Ed. Springer-Verlag, Berlin, pp. 100-113.

(31)   V.B. Wigglesworth. 1941. The sensory physiology of the human louse *Pediculus humanis corporis* de Geer (Anoplura). *Parasitology* **33**:67-109.

(32)   S.R. Loschiavo, S.D. Beck and D.M. Norris. 1963. Behavioural responses of the smaller European elm bark beetle, *Scolytus multistriatus* (Coleoptera: Scolytidae) to extracts of elm bark. *Ann. Entomol. Soc. Am.* **56**:764-768.

(33)   E. Kangas, V. Perttunen, H. Oksanen, and M. Rinne. 1965. Orientation of *Blastophagus piniperda* L. (Coleoptera: Scolytidae) to its breeding material. Attractant effect of *a*-terpineol isolated from pine rind. *Ann. Entomol. Fenn.* **31**:61-73.

(34)   E. Kangas, V. Perttunen, H. Oksanen, and M. Rinne. 1967. Laboratory experiments on the olfactory orientation of *Blastophagus piniperda* L. (Coleoptera: Scolytidae) to substances isolated from pine rind. *Acta Entomol. Fenn.* **22**:7-87.

(35)   O.K. Jantz and J.A. Rudinsky. 1965. Laboratory and field methods for assaying olfactory responses of the Douglas-fir beetle, *Dendroctonus pseudotsugae* Hopkins. *Can. Entomol.* **97**:935-941.

(36)   J.H. Borden, R.G. Brownlee and R.M. Silverstein. 1968. Sex pheromone of *Trypodendron lineatum* (Coleoptera: Scolytidae): Production, bio-assay and partial isolation. *Can. Entomol.* **100**:629-636.

(37)   C.J. August. 1970. The role of male and female pheromones in the mating behaviour of *Tenebrio molitor. J. Insect Physiol.* **17**:739-751.

(38)   D.G.H. Halstead. 1973. Preliminary biological studies on the pheromone produced by male *Acanthoscelides obtectus* (Say) (Coleoptera: Bruchidae). *J. Stored Prod. Res.* **9**:109-117.

(39)   O.R.W. Sutherland, R.F.N. Hutchins, and C.H. Wearing. 1974. The role of the hydrocarbon α-farnesene in the behaviour of codling moth larvae and adults. In *Experimental Analysis of Insect Behaviour.* L.B. Browne, Ed. Spring-Verlag, Berlin, pp. 249-263.

(40)   W.F. Chamberlain. 1956. A simplified quantitative olfactometer for use with agriculturally important insects. *J. Econ. Entomol.* **49**:659-633.

(41)   W.F. Chamberlain. 1959. The behavior of agricultural insects toward olfactory repellents in the olfactometer and in split-arena tests. *J. Econ. Entomol.* **52**:286-289.

(42)   P.N. Daykin and F.E. Kellogg. 1965. A two-air-stream observation chamber for studying responses of flying insects. *Can. Entomol.* **97**:264-268.

(43)  I. Yamamoto and R. Yamamoto. 1970. Host attractants for the rice weevil and the cheese mite. In *Control of Insect Behavior by Natural Products.* D.L. Wood, R.M. Silverstein, and M. Nakajima, Eds. Academic, New York, pp. 331-345.

(44)  V. Perttunen, E. Kangas, and H. Oksanen. 1968. The mechanisms by which *Blastophagus piniperda* L. (Coleoptera: Scolytidae) reacts to the odour of an attractant fraction isolated from pine phloem. *Ann. Entomol. Fenn.* **34**:205-222.

(45)  P.N. Daykin, F.E. Kellogg, and R.H. Wright. 1965. Host-finding and repulsion of *Aëdes aegypti. Can. Entomol.* **97**:239-263.

(46)  D. Marsh. 1975. Responses of male aphids to the female sex pheromone in *Megoura viciae* Buckton. *J. Entomol.* (A) **50**:43-64.

(47)  R.H. Wright, D.L. Chambers, and I. Keiser. 1971. Insect attractants, anti-attractants, and repellents. *Can. Entomol.* **103**:627-630.

(48)  J.H. Borden and R.B. Bennett. 1969. A continuously recording flight mill for investigating the effect of volatile substances on the flight of tethered insects. *J. Econ. Entomol.* **62**:782-785.

(49)  B. Hassenstein. 1961. Wie sehen Insekten Bewegungen? *Naturwissenschaften* **48**:207-214.

(50)  E. Kramer. 1975. Orientation of the male silkmoth to the sex attractant Bombykol. In *Olfaction and Taste* 5. D. Denton and J.D. Coghlan, Eds. Academic, New York, pp. 329-335.

(51)  E. Kramer. 1976. The orientation of walking honeybees in odour fields with small concentration gradients. *Physiol. Entomol.* **1**: 27-37.

## CHAPTER 14

# Complexity, Diversity, and Specificity of Behavior-Modifying Chemicals: Examples Mainly from Coleoptera and Hymenoptera

R. M. Silverstein

*SUNY College of Environmental Science and Forestry*
*Syracuse, N. Y.*

Several conventional reviews—very good current ones—of chemicals that modify the behavior of insects are available; my theme—complexity, diversity, and specificity of these chemicals—provides an umbrella for a special kind of review. It makes no claim to being comprehensive, for it will draw examples mainly from the Coleoptera and Hymenoptera. Moreover, I would find it difficult, after a decade in this highly controversial field, to present an unbiased account, and so shall make no serious attempt to do so. Inordinate emphasis is placed on work done by myself and my collaborators, and we hope that our first-hand experience may compensate for any possible deficiencies in what some might consider too specific a presentation.

## I. EXAMPLES TAKEN MAINLY FROM COLEOPTERA BUT WITH SOME REFERENCES TO LEPIDOPTERA

The approach to attractant pheromones by actually isolating and identifying the chemical compounds involved dates back to the classical study by Butenandt *et al.* (1) of the sexual stimulating pheromone secreted by the female silkworm moth, *Bombyx mori*. These investigators, without benefit of the chemical instrumentation now widely available, isolated a single compound by fractionating the complex mixture from an extract of the secretory gland and monitoring the fractionation with a laboratory bioassay. The compound, *trans*-10, *cis*-12-hexadecadien-1-ol (1) was synthesized and shown to be active in the laboratory bioassay at fantastically low concentrations. And thus the "magic bullet" concept used in pharmaceutical chemistry (2) was firmly implanted in the developing pheromone field: one insect, one specific compound.

1

This concept was reinforced by the report in 1960 by Jacobson et al. (3) [retracted by Jacobson et al. (4)] that the sex attractant of the gypsy moth, *Porthetria dispar*, was 10-acetoxy-*cis*-7-hexadecen-1-ol (2).

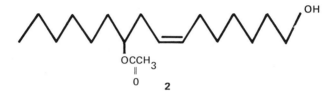

2

Most of the work through the decade of the 1960s was concerned with the sex attractants of Lepidoptera, and single compounds were reported to be *the* attractant for a number of species, mainly on the basis of a laboratory bioassay that had been used appropriately by Butenandt et al. (1) to measure short-range sexual excitation in a unique "domesticated" insect that simply did not exist in the field. Field trials based on these single compounds were almost invariably disappointing, and the purer the compound, the more disappointing were the results.

In 1966, Silverstein et al. (5) reported that the pheromone produced by the male bark beetle, *Ips paraconfusus* Lanier [formerly *Ips confusus* (LeConte)] (Scolytidae), which attracts both males and females, was a mixture of three compounds, none of which was appreciably active in the field by itself. The compounds were terpene alcohols: (−)-2-methyl-6-methylene-7-octen-4-ol (ipsenol, 3), (+)-*cis*-verbenol (4), and (+)-2-methyl-6-methylene-2,7-octadien-4-ol (ipsdienol, 5).

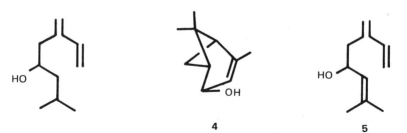

4                                                5

This report did not go completely unchallenged, and a brisk polemical exchange ensued (6) (7). However, one or more of these terpene alcohols have since been found to be present in a number of other *Ips* species, although in most cases the components of the actual chemical message have not been delineated. Cross-attraction in the genus *Ips* has been extensively studied, and probably results largely from these compounds (8)-(14).

This was the first demonstration of a multicomponent attractant pheromone and of the phenomenon of synergistic effects in pheromone components (15). The specificity of synergistic effects was demonstrated by the fact that *Ips latidens* responded in the field to a mixture of compounds **3** and **4**, but not to a mixture of **3**, **4**, and **5**; this "blocking" effect was confirmed in the laboratory (15). Furthermore, these pheromone components also attracted the predator, the black-bellied clerid, *Enocleris lecontei* (Cleridae) (16).

Here we have a nice example of the "parsimony" of nature: the same compounds that regulate *intra*specific aggregation also enable the predators to find their prey. Such interspecific compounds have been termed kairomones (17), and are described by Blum (18) as "pheromones and allomones which have evolutionarily backfired."

Another variation on the theme of multicomponent pheromones is the aggregation pheromone of the western pine beetle, *Dendroctonus brevicomis* (Scolytidae), which consists of three components, one each contributed by the female, the male, and the host tree: *exo*-7-ethyl-5-methyl-6,8-dioxabicyclo[3.2.1]octane (*exo*-brevicomin, **6**), 1,5-dimethyl-6,8-dioxabicyclo [3.2.1]octane (frontalin, **7**), and myrcene (**8**) (19)-(24).

The component contributed by the female also serves as a kairomone for the predator, *Temnochila chlorodia* (Trogositidae). The isomeric *endo*-brevicomin (**9**) is also contributed by the female, but it seems less important than the *exo* compound. Large scale field tests have been carried out and are described by Wood in Chapter 22.

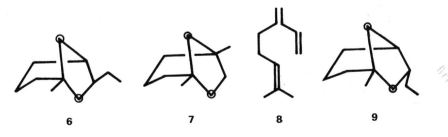

|   |   |   |   |
|---|---|---|---|
| **6** | **7** | **8** | **9** |

Frontalin, *exo*-brevicomin, and *endo*-brevicomin have been found in several other *Dendroctonus* species, and 3-methyl-2-cyclohexen-1-ol (seudenol, **10**) has been identified as an attractant component from *D. pseudo-*

*tsugae* (13) (14) (25). Verbenone (**11**) and *trans*-verbenol (**12**) occur in a number of species, but their functions are not clear. Host-produced terpenes also contribute to the effectiveness of the many mixtures tested.

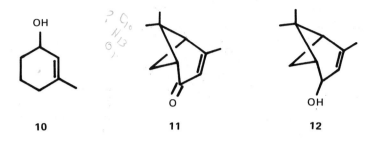

10                          11                          12

One of the spectacular examples of a complex pheromone is afforded by the four-component mixture that was identified from the fecal pellets of the male boll weevil, *Anthonomus grandis* (Curculionidae) (26). All of the compounds are required for effective attraction in the field: (+) *cis*-2-isopropenyl-1-methylcyclobutaneethanol (**13**), *cis*-3,3-dimethyl-$\Delta^{1,\beta}$-cyclohexaneethanol (**14**), *cis*-3,3-dimethyl-$\Delta^{1,\alpha}$-cyclohexaneacetaldehyde (**15**), and *trans*-3, 3-dimethyl-$\Delta^{1,\alpha}$-cyclohexaneacetaldehyde (**16**).

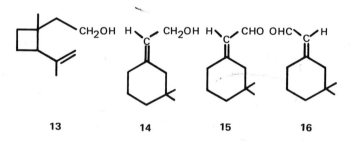

13                  14                  15                  16

A semantic problem may be raised here. Should the term *pheromone* be applied to the total active mixture (and the individual compounds designated pheromone components), or should each compound, albeit individually inactive, be termed a *pheromone*? My own preference is for the former usage because it seems to honor the meaning of the phrase "cause a specific reaction" in the original definition of *pheromone* (27).

Furthermore, it avoids the problem of what to call an inactive compound such as myrcene (**8**) which is contributed by the host rather than the insect; admittedly this usage does some violence to the phrase "secreted by the animal" in the original definition. [See also the discussion by Birch (28)].

It is interesting to note that as early as 1964, Wright (29) suggested that multicomponent pheromones would convey more information than single-component systems. Despite this suggestion, and despite the apparent validity of the multicomponent pheromone in several coleopteran species, the "magic bullet" concept held sway among the investigators of moth pheromones. It is impossible to avoid speculating on causes.

As in many rapidly developing fields, pheromone studies have been plagued with partial, erroneous, and conflicting results and conclusions. Two areas of weakness are apparent: there is frequently a gross lack of knowledge of the behavior of the organism under study; and the biologic and chemical methodology selected are sometimes inappropriate. Wood et al. (7) affirmed that a laboratory bioassay must reflect as nearly as possible the behavior of the natural population. Silverstein et al. (30) (31) formulated a sequence of six steps considered essential to a rigorous study of insect pheromones; a key step was the isolation of each compound in a state of high purity, each stage of the isolation being monitored by an *appropriate* bioassay. Furthermore, no fraction was to be discarded "until it had been tested in combination with other fractions"; this was to detect synergistic effects. (For a review of biologic and chemical methodology, see Young and Silverstein (32).) There are two difficulties here. First, development of a bioassay that is a realistic simulation of a field response, as Wood et al. (7) prescribed, requires considerable ingenuity and experimentation, and some risk. Second, Silverstein's precepts (30) (31) demand a large commitment of biologic and chemical talents, and time to decipher in detail the often complex chemical message. Bioassays of numerous combinations of fractions on a quantitative basis are especially tedious. Some of the difficulties and pitfalls in the biological and chemical methodology have been pointed out (32).

On the other side, as recently as 1970, H.H. Shorey, one of the more persuasive writers in the field, firmly stated his conviction that "although a female moth may release a medley of compounds, only one of them has been selected behaviorally as the sex pheromone" (33). And in the same chapter, Shorey also postulated: "It is not necessary to observe the whole sequence of behavioral steps in a pheromone bioassay. In fact, it is advisable not to base the bioassay on the sequence, but instead to select only one step, such as the forward movement of the antennae, wing extension or vibration, upwind flight, or copulatory attempts." These concepts had indeed held sway generally throughout the 1960s, and the message was welcomed in the real world of strictures on time and funds. However, if one reads the entire chapter, one is struck by the almost evangelical plea for more intensive and incisive basic studies of insect behavior and by the strong case made for the need for *quantitative* bioassays; the author's own research is a testament.

It is interesting retrospectively to discern the development of a pronounced dichotomy in approaches to pheromone studies. This was remarked in 1972 by Wood:

"A dichotomy in research goals appears to be developing in the field of insect pheromones in which one goal is to isolate and/or synthesize compounds based on known molecular configurations that have been shown to be active as attractants, synergists or masking agents [(34)-(36)], and the second is to explain the precise molecular bases of certain olfactory behavior patterns. Such dichotomy is not new to science. . . ."

After it became apparent that the compounds eliciting the "excitation" response from male moths were nearly all straight-chain alcohols and acetates with one or two double bonds, mixtures of these compounds and convenient homologues and analogues of them were exposed in field trials, and very interesting and useful results were obtained. In some cases, greatly improved catches resulted from empirically chosen additives. But for many species for which attractant pheromones have been designated, it remains unclear that the designated compound or compounds have actually been isolated from that species and actually *comprise the true chemical message* for that species. An advocate of this approach may defend it on the grounds that: "The synthetic mixtures found by field tests to be the best attractant for some species may not always be the natural pheromone mixture, but they must generally bear close resemblance to it" (37). "Probably true in many cases," the purist would yield, "but what of elegance in scientific inquiry?" The riposte from the real world of limited funding based on "productivity" could be devastating. But at the same time, proponents of the "run it up the flagpole and see who salutes" methodology have not been too successful in rationalizing biologic activity from chemical structure of homologues and analogues. Shortcuts and interpretations based on them have resulted in a literature that taxes the analytical abilities of anyone who attempts a critical review. Despite the inability to correlate activity and structure, however, the empirical "screening" approach has developed several very useful insect attractants and repellants (38). Compounds that do not occur in the natural system but elicit a response similar to that produced by the natural components have been termed *parapheromones* by Payne et al. (39). Hexalure (40), for example, is a parapheromone for the pink bollworm and has been a useful tool in field studies; however the attractant produced by the female is from 100 to 1000 times more effective (41) (42).

In 1970, Jacobson et al. (43) identified two compounds, *cis*-9-tetradecen-1-ol acetate and *cis*-9-*trans*-12-tetradecadien-1-ol acetate from the southern army worm, *Spodoptera eridania* (Noctuidae). Each compound excited the males in a laboratory bioassay, but both compounds were needed for attraction in the field. Brady et al. (44) reported that extracts of abdominal tips of *Cadra cautella* (Pyralididae) females contained an unidentified compound that was inactive by itself but caused an increase in the excitatory and attractant response elicited by the active identified compound. These may be the first reports of synergism by individually inactive compounds in moths.

In 1971, Silverstein (45) stated that, on the whole, single-component pheromone systems are probably much less common than multicomponent systems, even among moths. The first few years of the decade of the 1970s have seen a number of reports of multicomponent pheromones in moths. Just a few examples will suffice.

Ritter (46) and Tamaki et al. (47) isolated and identified a mixture of *cis*-9- and *cis*-11-tetradecen-1-ol acetate as the attractant pheromone of *Adoxophyes orana* (Tortricidae); Minks et al. (48) showed that two tortricid moths, *Adoxophyes orana* and *Clepsis spectrana*, are reproductively isolated by different blends of these two compounds. In 1973, Klum et al. (49) demonstrated that the reported field responses of the red-banded leaf roller, *Argyrotaenia velutinana* (Tortricidae) (50), and of the European corn borer, *Ostrinia nubilalis* (Tortricidae) (51) to *cis*-11-tetradecen-1-ol acetate were in fact due to mixtures containing small amounts of the *trans* isomer; each isomer tested by itself showed little, if any, activity. Roelofs et al. (52) reinvestigated the red-banded leaf roller and showed that, indeed, a small amount of the *trans* isomer was present; furthermore, a compound, dodecyl acetate, that had been found empirically to enhance field catches was also found to be present.

I have gone to some length to gut the "magic bullet" concept of pheromones. But nothing I have said should be taken to mean that there are no single-component pheromones whatever. In fact, the second report of a coleopteran pheromone (53) described a single component secreted by the female black carpet beetle, *Attagenus megatoma* (Dermestidae) that excited and attracted the male: *trans*-3, *cis*-5-tetradecadienoic acid (**17**). In view of their experience with the bark beetle, *Ips paraconfusus*, the investigators made a thorough but unsuccessful search for other compounds that might be involved in the response; even so, they hedged a bit in their report. However, several dermestid beetles in the genus *Trogoderma* secrete multicomponent pheromones, each component showing its own kind and level of activity. Five of the six active components of the excitatory and attractant pheromone secreted by the female *Trogoderma glabrum* have been identified (54): methyl *trans*-14-methyl-8-hexadecenoate (**18**), *trans*-14-methyl-8-hexadecen-1-ol (**19**), methyl *cis*-7-hexadecenoate (**20**), hexanoic acid (**21**), and γ-caprolactone (**22**).

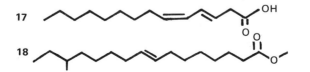

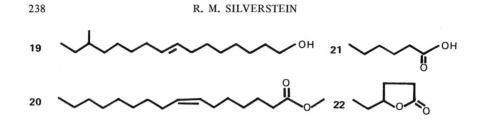

A high degree of cross-attraction occurs among a number of *Trogoderma* species (55). Presumably these compounds will be found widely distributed through the genus in various combinations and ratios. Chapter 20 by Burkholder contains further discussions of pheromones of dermestid beetles, and describes the field tests. See note p. 251.

In contrast with the bark beetles described above, which construct galleries in the phloem tissue, the ambrosia beetles bore into the xylem. One of the serious pests of timber in the Northwest is the ambrosia beetle, *Gnathotrichus sulcatus* (Scolytidae). The isolation and identification of the aggregation pheromone, produced by the male, followed the protocol developed in the bark beetle investigations: boring dust (from the male) was extracted and fractionated, and each fraction was monitored alone and in combination with other fractions by laboratory bioassay with beetles of both sexes. No synergism was apparent; at each stage in the isolation process, the single attractive fraction was competitive with all fractions combined. The attractive compound was identified as 6-methyl-5-hepten-2-ol (sulcatol **23**) and was found to be very effective in field tests (56) (57). Once again in the Coleoptera, we are confronted with what appears to be a single-component pheromone (see below for further explanation).

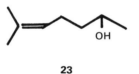

**23**

For a final example of an aggregating pheromone from the Coleoptera, I shall briefly summarize the recent investigation of the smaller European elm bark beetle, *Scolytus multistriatus* (Scolytidae), the principal vector for the Dutch elm disease pathogen, *Ceratocystus ulmi*. Uninfested elm bolts are weakly attractive to the beetle, but when females bore into the phloem-cambium region of elm trees, they produce a potent aggregation pheromone that attracts both males and females (58). The aggregation pheromone consists of two compounds produced by the female (**24** and **25**) and one compound produced by the tree (26): 4-methyl-3-heptanol (24), 2,4-dimethyl-5-ethyl-6,8-dioxabicyclo (3.2.1) octane (α-multistriatin, 25), and α-cubebene (26) (59).

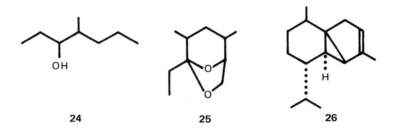

24            25            26

## II. EXAMPLES TAKEN FROM HYMENOPTERA

It is difficult to top Blum et al.'s reference (60) to Wray (61) who, in 1670, reported "Some uncommon observations and experiments made with an acid juyce to be found in ants." Wray described experiments carried out by Mr. Samuel Fisher of Sheffield who had steam-distilled a volatile acid from ant bodies and was obviously dealing with formic acid (which he described as "like Spirit of Vinegar"), an alarm and defensive secretion of some formicine ants. Although Fisher may have been the first recorded experimentalist in the field of chemicals affecting the behavior of Hymenoptera, his report was antedated by the observation of Charles Butler, "who in 1609 described how honeybees are attracted and provoked to mass attack by a substance released as a result of a single initial sting" [quotation taken from MacConnell and Silverstein (62)].

Much of the early work in Hymenoptera was done on defensive and alarm secretions of ants. It was obvious from the beginning that these glandular secretions were complex mixtures, and so the "magic bullet" concept never developed into a motif in hymenopteran studies; the ability of single compounds to serve several functions was generally recognized (18) (60) (63)-(66). Callow et al. (67), for example, showed that at least 32 compounds were present in the mandibular gland of the honeybee queen, *Apis mellifera* (Apidae). Bergström's group in Sweden has identified most of the components of highly complex mixtures by gas chromatography—mass-spectrometry. The biologic meanings of these message components—of a variety of chemical classes—and their combinations have not been completely deciphered, but comparative studies on related species indicate that distinctive messages play a part in species recognition (68).

In contrast with the attractive and excitatory compounds with which the studies of moths, beetles, and weevils were concerned, the alarm and defense compounds of ants and bees often are produced in large amounts, and most of the early work was directed toward chemical analysis of these secretions and interpretation of their striking effects. As recently as 1972, Blum and Brand (60) commented that "our total knowledge of sex attractants in female social insects is predicated on the identification of the sex pheromone produced by the queen honeybee, *Apis mellifera*"; Gary (69) had demonstrated that the queen substance *trans*-9-oxo-2-decenoic acid (27) was an attractant for airborne drones.

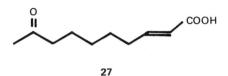

**27**

It is the same compound that had been designated "queen substance" (70) and identified (71) (72) as a major component in maintaining hive "order," including inhibition of queen cell construction, which occurs only on removal of the queen. Thus the queen substance functions as both a primer and a releaser pheromone.

Blum and Brand (60) attempted to demonstrate a chemosystematic distribution of alarm pheromones in three subfamilies of ants, but there are too many inconsistencies to warrant any statement stronger than "the distribution appears to be chemosystematically significant." The general occurrence of 3-alkanones in the subfamily Myrmicinae is striking. The compound 4-methyl-3-heptanone (28) has been identified as an alarm pheromone in species of four genera in the subfamily Myrmicinae, and careful quantitative studies were carried out on the response by workers of *Atta texana* and *Atta cephalotes* (73)-(76). At a concentration of about 33 million molecules per cubic centimeter; the compound elicited an aggregation response from workers in a laboratory bioassay; at a tenfold higher concentration, the typical alarm response was obtained. Both species responded to the other compounds identified in the mandibular gland only at much higher concentrations. Thus, in these cases a single compound from the mixture appears to function as the alarm pheromone; the presence of the other compounds may serve to afford species recognition.

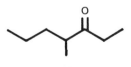

**28**

One of the most fascinating aspects of behavior in some ants, wasps, and bees is trail-following. Moser and Blum (77) described the trail-following behavior of the myrmicine ant, *Atta texana*, in response to a pheromone secreted in the poison gland, stored in the poison sac, and deposited through the sting apparatus, which has been converted to serve a social function. The trail made by streaking an extract of poison sacs on cardboard could be detected by workers for a brief period at a distance of about 2 cm above the trail; workers placed on the trail could still follow it on the sixth day after its preparation; apparently at least two compounds of very different orders of volatility (78) were present. The major volatile compound was identified as methyl-4-methylpyrrole-2-carboxylate (29) (79) (80). Several other active fractions were noted, but the small amounts of the compounds precluded identification. In the laboratory bioassay, the synthesized compound (29) was detectable by minor workers at 0.08 pg/cm ($3.48 \times 10^8$ molecules/cm); at this level, only 0.33 mg would be required to draw a detectable trail around the world. In the field, medium and large workers could be induced to leave a natural trail and follow another trail made by dribbling a solution of 4.0 pg/$\mu$l of the synthesized compound on the sand leading off at a 45-degree angle from the natural trail. They readily followed a 2.7 pg/cm trail on a 15-cm strip of cardboard placed across an "erased" portion of the natural trail.

Methyl 4-methylpyrrole-2-carboxylate (29) appears to be a common component in a number of attine species in the genera *Cyphomyrmex,* *Apterostigma, Trachymyrmex, Acromyrmex,* and *Atta.* A number of species in each genus respond to it (81); it has been identified in *Atta cephalotes* (75), which follows a trail of the synthesized compound, and in *Atta sexdens rubropilosa* (82), which does not. It is interesting, however, that the latter ant follows laboratory trails prepared from the poison sac of *Atta texana* (81), as well as trails made of a mixture of 29 and methyl phenylacetate, which is also present in *Atta sexdens rubropilosa* (82). Robinson and Cherrett (83), reporting preliminary studies,

**29**

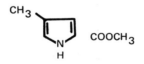

suggested that the pyrrole compound (29) may increase the effectiveness of bait for several myrmicine species.

Recently, Ritter et al. (84) identified 5-methyl-3-butyloctahydroindolizine (30) as one of the attractant components of Pharaoh's ant, *Monomorium pharaonis* (Formicidae), which is also a trail-following ant. Ritter et al. noted that other biologically active components were present. Finally, Huwyler et al.

(85) identified six components of the trail pheromone of the ant, *Lasius fuliginosus* (Formicidae); these are the $C_6$, $C_7$, $C_8$, $C_9$, $C_{10}$, and $C_{12}$ straight-chain carboxylic acids in the rectal fluid. Each compound is individually active; the authors note that other unidentified components may also be present.

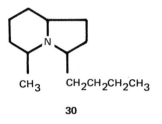

CH$_3$     CH$_2$CH$_2$CH$_2$CH$_3$

**30**

Finally, Huwyler et al. (85) identified six components of the trail pheromone of the ant, *Lasius fuliginosus* (Formicidae); these are the $C_6$, $C_7$, $C_8$, $C_9$, $C_{10}$, and $C_{12}$ straight-chain carboxylic acids in the rectal fluid. Each compound is individually active; the authors note that other unidentified components may also be present.

The very limited amount of work done on identification of trail pheromones of ants precludes fruitful speculation, but the evidence to date points to general use of multicomponent pheromones. Although ants of various species generally do follow each other's trail in the laboratory, they readily distinguish and only follow their own trails in the field. It even seems likely that various components perform distinct functions. Hölldobler (86) and Hölldobler and Wilson (87) showed that the myrmicine harvester ant (*Pogonomyrmex badius*) utilizes primarily material from the Dufour's gland for nest-finding (orientation trails) and material from the poison gland to lay recruitment trails to food sources.

The ability of several unrelated animals to follow ant trails has been reported by several investigators. Most of these reports deal with army ant trails, which are followed by several coleopteran species (88), by mites (87), by millipedes (89), and by blind snakes (90) (91). Several cockroaches follow trails of *Atta texana* (92) or of *Acromyrmex coronatus* (93). The compounds responsible for these interspecific responses have not been identified.

Stingless bees of several species lay trails by depositing droplets from mandibular glands at intervals as they return to the nest from a food source (94). Bees orient aerially to the vapor trail from these deposits. Blum et al. (95) identified neral and geranial as the dominant components in the mandibular-gland secretion of the stingless bee, *Trigona subterranea* (Apidae), and benzaldehyde and a number of straight-chain alcohols, ketones, and acetates in the mandibular secretions of three other *Trigona* species.

## III. THESIS

The extraordinary diversity, complexity, and specificity of behavior-modifying chemicals are evident from the examples selected. We have noted examples in which a pheromone consisted of a single active compound; two or more active compounds whose combined activity is the sum of the parts; and two or more compounds, some active and some inactive, whose combined activity is greater than the sum of the parts (synergism).

There are truly pheromone languages that range from a grunt (single component) to complex statements with a vocabulary and syntax. The meaning of an individual word (chemical component) depends on context, and even a grunt requires interpretation. It seems safe to say that multicomponent pheromones predominate. It is also safe to say that, in most cases reported, the chemical message has been only partially deciphered.

Although chemosystematics can hardly claim to be a developed branch of science, some very nice studies on cross-attraction in related species have been reported [e.g., (8)-(10) (12) (55)]. Many of the chemical components responsible for the cross-attraction in these studies of dermestid beetles and bark beetles have been identified. Several examples of reproductive isolation of sympatric species by mixtures of different identified components or different proportions of the same components have been cited above. A very nice example of mutual inhibition by two sympatric species of bark beetles has been reported recently (96); again, the two compounds responsible have been identified. By and large, closely related species share the same or closely related compounds. But one cannot argue a converse effect from this. The occurrence of the same compound in different orders of animals means nothing except that evolution in this particular aspect took a similar course and produced similar enzymes.

We have placed behavioral responses on a molecular basis—that is, we have identified chemical compounds that evoke behavioral responses. Now let us examine the levels of specificity of the response relevant to subtleties of chemical structure. When we do so, we observe—as we learn to expect—the widest diversity.

At the grossest level we find that receptor systems generally distinguish one chemical compound from another—say, ethyl alcohol from LSD. But even at this level, an alarm response can be elicited from an ant by two different classes of compounds with different molecular formulas, provided they are grossly similar in size and shape (97). Some insects, such as termites, respond to large numbers of unrelated chemical stimuli (98).

At the next level is the distinction between *isomers*—compounds that have the same molecular formula. These are of two kinds: *positional isomers* (*cis*-9-tetradecen-1-ol acetate and *cis*-11-tetradecen-1-ol acetate; 2-heptanone and 3-heptanone) and *functional group isomers* (ethyl ether and *n*-butyl alcohol). Again the distinction is easy for most organisms (73) (97).

The difference between *stereoisomers* is somewhat more subtle. These are isomers that differ from one another only in the spatial orientation of the atoms; there is no difference in the order in which the atoms are joined. Stereoisomers are either related to one another as nonsuperimposable mirror images (*enantiomers*) or they are not (*diastereomers*). In the latter category are *geometric* stereoisomers, which owe their existence to the presence of a rigid structure, which can be a double bond or a ring system. *cis*-11-Tetradecen-1-ol acetate and *trans*-11-tetradecen-1-ol acetate are a pair of geometric stereoisomers, and the red-banded leaf roller can distinguish one from the other; in fact, it is highly sensitive to the precise ratio (49). The bark beetle, *Dendroctonus brevicomis*, can distinguish between *exo*-brevicomin (**6**) and *endo*-brevicomin (**9**) (98).

An object and its nonsuperimposable mirror image constitute a pair of enantiomers (also called *optical isomers*). In terms of symmetry elements, an object has a nonsuperimposable mirror image only if it is a chiral object—that is, it lacks an alternating axis of symmetry; it may or may not have a simple axis of symmetry. The ability to distinguish between a pair of enantiomers requires a chiral agent; a glove is a chiral agent, which can distinguish between the right hand and the left hand, which are enantiomers. Enzymes have chiral centers which should be able to function as chiral agents; and, in fact, enzymes generally synthesize or degrade only one of a pair of enantiomers. Nonetheless, the question of whether the human nose can distinguish between a pair of enantiomers was not satisfactorily resolved until Friedman and Miller (100) and Russell and Hills (101) carried out their rigorous experiments. Kafka et al. (102) succeeded in conditioning honeybees to discriminate between the optical isomers of 4-methylhexanoic acid (**31a** and **31b**).

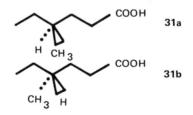

However, these optical isomers must be considered to be artifacts since they are not normally utilized by the insects tested. Similarly, Lensky and Blum (103) conditioned honeybees to discriminate between the enantiomers of carvone (**32a** and **32b**) and of 2-octanol (**33a** and **33b**). Neither compound is utilized by the bee under natural conditions.

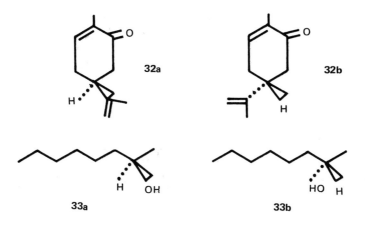

Riley and colleagues (74) tested *Atta texana* and *Atta cephalotes* against the naturally occurring (in both species) alarm pheromone, (S)-(+)-4-methyl-3-hep-tanone (28a), and its enantiomer (28b), which must be considered to be an artifact. Both insects, without conditioning, were more responsive to the naturally occurring enantiomer. Thus, chiral sites are implicated at the receptor, and in some insects at least, these sites can make the distinction between enantiomers, which can be observed at the behavioral level.

Usually one expects an enzyme to produce only one enantiomer—in our bodies, for example, we have only the L series of amino acids and the D series of sugars. Only a few pheromones have been examined with regard to chirality; and, predictably, the results have been diverse. Again, we were conditioned by our experience of having detected only a single enantiomer of the alarm component (28a), and we were quite unprepared for the discovery that sulcatol (23) occurs as a 65/35 mixture of the (S)-(+) enantiomer (23a) and the (R)-(−) enantiomer (23b), respectively (56). Thus the description earlier in this chapter of sulcatol as a single-compound pheromone is strictly true only in a world devoid of chirality. A complete identification of a chiral pheromone should therefore include a statement of enantiometric purity and a description of the absolute configuration(s) of the chiral center(s) (R and S designations).

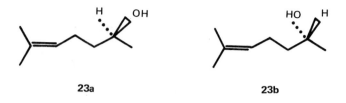

23a                                                    23b

Chemical interaction among organisms—in broad terms, chemical ecology—is now recognized as a vast, diverse, and fruitful area for collaborative investigations. With the increased sophistication being brought to bear in this area, many more surprises will be uncovered. Nature indeed "speaks a various language."

## REFERENCES

(1)   A. Butenandt, R. Beckman, D. Stamm and E. Hecker. 1959. Über den Sexuallockstoff des Seidenspinners *Bombyx mori*, Reindarstellung und Konstitution. *Z. Naturforsch.* **14b**:283-284.

(2)   P. DeKruif. 1926. *Microbe Hunters*. Harcourt Brace, New York, p. 334.

(3)   M. Jacobson, M. Beroza and W.A. Jones. 1960. Isolation, identification, and synthesis of the sex attractant of the gypsy moth. *Science* **132**:1011.

(4)   M. Jacobson, M. Schwarz and R.M. Waters. 1970. Gypsy moth sex attractants: A reinvestigation. *J. Econ. Entomol.* **63**:943.

(5)   R.M. Silverstein, J.O. Rodin and D.L. Wood. 1966. Sex attractants in frass produced by male *Ips confusus* in ponderosa pine. *Science* **154**:509-510.

(6)   J.P. Vité. 1967. Sex attractants in frass from bark beetles. *Science* **156**:105.

(7)   D.L. Wood, R.M. Silverstein and J.O. Rodin. 1967. Sex attractants in frass from bark beetles. *Science* **156**:105.

(8)   J.P. Vité and R.I. Gara. 1962. Volatile attractants from ponderosa pine attacked by bark beetles (Coleoptera: Scolytidae). *Contrib. Boyce Thompson Inst.* **21**:251-273.

(9)   J.P. Vité, R.I. Gara and H.D. van Scheller. 1964. Field observations on the response to attractants of bark beetles infesting southern pines. *Contrib. Boyce Thompson Inst.* **22**:461-470.

(10)  D.L. Wood. 1970. Pheromones of bark beetles. In *Control of Insect Behavior by Natural Products*. D.L. Wood, R.M. Silverstein and M. Nakajima, Eds. Academic, New York, pp. 301-316.

(11)  J.P. Vité and J.A.A. Renwick. 1971. Population aggregating pheromones in the bark beetle, *Ips grandicollis. J. Insect Physiol.* **17**:1699-1704.

(12)  G.N. Lanier and D.L. Wood. 1975. Specificity of population aggregating pheromones in the genus *Ips* (Coleoptera: Scolytidae). *J. Chem. Ecol.* **1**:9-23.

(13)  G.N. Lanier and W.E. Burkholder. 1974. Pheromones in speciation of Coleaptera. In *Pheromones*. M.C. Birch, Ed. North-Holland, Amsterdam, pp. 161-189.

(14)  J.H. Borden. 1974. Aggregation pheromones in the Scolytidae. In *Pheromones*. M.C. Birch, Ed. North-Holland, Amsterdam, pp. 135-160.

(15)  D.L. Wood, R.W. Stark, R.M. Silverstein, and J.O. Rodin. 1967. Unique synergistic effects produced by the principal sex attractant compounds of *Ips confusus* (Le Conte) (Coleaptera: Scolytidae). *Nature* **215**:206.

(16)  D.L. Wood, L.E. Browne, W.D. Bedard, P.E. Tilden, R.M. Silverstein, and J.O. Rodin. 1968. Response of *Ips confusus* to synthetic sex pheromones in nature. *Science* **159**:1373-1374.

(17)  W.L. Brown, Jr., T. Eisner, and R.H. Whittaker. 1970. Allomones and kairomones, transspecific chemical messengers. *Bioscience* **20**:21-22.

(18)  M.S. Blum. 1974. Deciphering the communicative Rosetta Stone. *Bull. Entomol. Soc. Am.* **20**:30-35.

(19)  R.M. Silverstein, R.G. Brownlee, T.E. Bellas, D.L. Wood, and L.E. Browne. 1968. Brevicomin: Principal sex attractant in the frass of the female western pine beetle. *Science* **159**:889-891.

(20)  W.D. Bedard, P.E. Tilden, D.L. Wood, R.M. Silverstein, R.G. Brownlee, and J.O. Rodin. 1969. Western pine beetle: Field response to its sex pheromone and a synergistic host terpene, myrcene. *Science* **164**:1284-1285.

(21)  G.W. Kinzer, A.F. Fentiman, T.F. Page, R.L. Foltz, J.P. Vité, and G.B. Pitman. 1969. Bark beetle attractants: Identification, synthesis, and field bioassay of a new compound isolated from *Dendroctonus*. *Nature* **221**:477-478.

(22)  J.P. Vité and G.B. Pitman. 1969. Aggregation behavior of *Dendroctonus brevicomis* in response to synthetic pheromones. *J. Insect Physiol.* **15**:1617-1622.

(23)  W.D. Bedard, R.M. Silverstein, and D.L. Wood. 1970. Bark Beetle pheromones. *Science* **167**:1638-1639.

(24)  D.L. Wood. 1972. Selection and colonization of ponderosa pine by bark beetles. In: Insect/Plant Relationship Symp. Roy. Entomol. Soc. Lond. No. 6, pp. 101-117.

(25)  J.P. Vité, G.B. Pitman, A.F. Fentiman, and G.W. Kinzer. 1972. 3-Methyl-2-cyclo-hexen-1-ol isolated from *Dendroctonus*. *Naturwissenschaften* **59**:469.

(26)  J.H. Tumlinson, D.D. Hardee, R.C. Gueldner, A.C. Thompson, P.A. Hedin, and J.P. Minyard. 1969. Sex pheromone produced by male boll weevil: Isolation, identification, and synthesis. *Science* **166**:1010-1012.

(27)  P. Karlson and A. Butenandt. 1959. Pheromones (ectohormones) in insects. *Ann. Rev. Entomol.* **4**:39-58.

(28)  M.E. Birch. 1974. Introduction. In *Pheromones.* M.C. Birch, Ed. North-Holland, Amsterdam, pp. 1-7.

(29)  R.H. Wright. 1964. After pesticides—what?*Nature* **204**:121-125.

(30)  R.M. Silverstein, J.O. Rodin and D.L. Wood. 1967. Methodology for isolation and identification of insect pheromones with reference to studies on the California five-spined *Ips. J. Econ. Entomol.* **60**:944-949.

(31)  R.M. Silverstein. 1970. Attractant pheromones of Coleoptera. In *Chemicals Controlling Insect Behavior.* M. Beroza, Ed. Academic, New York, pp. 21-40.

(32)  J.C. Young and R.M. Silverstein. 1976. Biological and chemical methodology in the study of animal communication systems. In *Methods in Olfactory Research.* D.G. Moulton, J.M. Johnston, Jr. and A. Turk, Eds. Academic, London.

(33)  H.H. Shorey. 1970. Sex pheromones of Lepidoptera. In *Control of Insect Behavior by Natural Products.* D.L. Wood, R.M. Silverstein, and M. Nakajima, Eds. Academic, New York, pp. 249-284.

(34)  N. Green, M. Jacobson and J.C. Keller. 1969. Hexalure, an insect attractant discovered by empirical screening. *Experientia* **25**:682-683.

(35)  W.L. Roelofs and A. Comeau. 1970. Lepidopteran sex attractants discovered by field screening tests. *J. Econ. Entomol.* **63**:969-974.

(36)  W.L. Roelofs and A. Comeau. 1971. Sex pheromone perception: Synergists and inhibitors for the red-banded leafroller attractant. *J. Insect Physiol.* **17**:435-438.

(37) W.E. Roelofs and R.T. Cardé. 1974. Sex pheromones in reproductive isolation of lepidopteran species. In *Pheromones*. M. Birch, Ed. North-Holland, Amsterdam, pp. 96-114.

(38) M. Beroza. 1970. Current usage and some recent developments with insect attractants and repellants in the USA. In *Chemicals Controlling Insect Behavior*. M. Beroza, Ed. Academic, New York, pp. 145-163.

(39) T.L. Payne, H.H. Shorey and L.K. Gaston. 1973. Sex pheromones of Lepidoptera. XXXVIII. Electroantennogram responses in *Autographa californica* to *cis*-7-dodecenyl acetate and related compounds. *Ann. Entomol. Soc. Am.* **66**:703-704.

(40) J.C. Keller, L.W. Sheets, N. Green and M. Jacobson. 1969. *cis*-7-Hexadecen-1-ol acetate (hexalure), a synthetic sex attractant for pink bollworm males. *J. Econ. Entomol.* **62**:1520-1521.

(41) H.H. Shorey. Unpublished.

(42) H.E. Hummel, L.K. Gaston, H.H. Shorey, R.S. Kaae, K.J. Byrne and R.M. Silverstein. 1973. Clarification of the chemical status of the pink bollworm sex pheromone. *Science* **181**:873-875.

(43) M. Jacobson, R.E. Redfern, W.A. Jones and M.H. Aldridge. 1970. Sex pheromones of the southern armyworm moth: Isolation, identification, and synthesis. *Science* **170**:542-543.

(44) U.E. Brady, J.H. Tumlinson, R.G. Brownlee and R.M. Silverstein. 1971. Sex stimulant and attractant in the Indian meal moth and in the almond moth. *Science* **171**:802-804.

(45) R.M. Silverstein. 1971. Recent and current collaborative studies of insect pheromones. In *Chemical Releasers in Insects*. A.S. Tahori, Ed. Gordon and Breach, New York, pp. 69-89.

(46)* F.J. Ritter. 1971. Some recent developments in the field of insect pheromones. *Meded. Ryksfak Landbouwwetensch.* (Gent) **3b**:874-882.

(47) Y. Tamaki, H. Noguchi, T. Yushima and C. Hirano. 1971. Two sex pheromones of the smaller tea tortrix: Isolation, identification and synthesis. *Appl. Entomol. Zool.* **6**:139-141.

(48) A.K. Minks, W.L. Roelofs, F.J. Ritter and C.J. Persoons. 1973. Reproductive isolation of two tortricid moth species by different ratios of a two-component sex attractant. *Science* **180**:1073-1074.

(49) J.A. Klun, O.L. Chapman, K.C. Mattes, P.W. Wojtkowski, M. Beroza and P.E. Sonnet. 1973. Insect sex attractants: Minor amounts of opposite geometric isomer critical to attraction. *Science.* **181**:661.

(50) W.L. Roelofs and H. Arn. 1968. Sex attractant of the red-banded leaf roller moth. *Nature* **219**:513.

(51) J.A. Klun and T.A. Brindley. 1970. *cis*-11-Tetradecenyl acetate, a sex stimulant of the European corn borer. *J. Econ. Entomol.* **63**:779-780.

(52) W.L. Roelofs, A. Hill and R. Cardé. 1975. Sex pheromone components of the redbanded leafroller, *Argyrotaenia velutinana* (Lepidoptera: Tortricidae) *J. Chem. Ecol.* **1**:83-89.

(53) R.M. Silverstein, J.O. Rodin, W.E. Burkholder and J.E. Gorman. 1967. Sex attractant of the black carpet beetle. *Science* **157**:85-87.

(54) R.G. Yarger, R.M. Silverstein and W.E. Burkholder. 1975. Sex pheromone of the female dermestid beetle, *Trogoderma glabrum* (Herbst). *J. Chem. Ecol.* **1**:323–334.

(55) K.W. Vick, W.E. Burkholder and J.E. Gorman. 1970. Interspecific response to sex pheromones of *Trogoderma* species (Coleoptera: Dermestidae). *Ann. Entomol. Soc. Am.* **63**:379-381.

(56)  K.J. Byrne, A.A. Swigar, R.M. Silverstein, J.H. Borden and E. Stokkink. 1974. Sulcatol: Population aggregating pheromone in the scolytid beetle, *Gnathotrichus sulcatus*. *J. Insect Physiol.* **20**:1895-1900.

(57)  J.A. McLean and J.H. Borden. 1975. Survey of a sawmill population of *Gnathotrichus sulcatus* (Coleoptera: Scolytidae) using the pheromone, sulcatol. *Can. J. Forest Res.* **5**:586-591.

(58)  J.W. Peacock, A.C. Lincoln, J.B. Simeone and R.M. Silverstein. 1971. Attraction of *Scolytus multistriatus* (Coleoptera: Scolytidae) to a virgin female-produced pheromone in the field. *Ann. Entomol. Soc. Am.* **64**:1143-1149.

(59)  G.T. Pearce, W.E. Gore, R.M. Silverstein, J.W. Peacock, R. Cuthbert, G.N. Lanier and J.B. Simeone. 1975. Chemical attractants for the smaller European elm bark beetle, *Scolytus multistriatus* (Coleoptera: Scolytidae). *J. Chem. Ecol.* **1**:115-124.

(60)  M.S. Blum and J.M. Brand. 1972. Social insect pheromones: Their chemistry and function. *Am. Zool.* **12**:553-576.

(61)  J. Wray. 1670. Some uncommon observations and experiments made with acid juyce to be found in ants. *Phil. Trans. Roy. Soc.* (Lond.) 2063-2069.

(62)  J.G. MacConnell and R.M. Silverstein. 1973. Recent results in insect pheromone chemistry. *Angew. Chem.* (Int. ed.) **12**:644-654.

(63)  M.S. Blum. 1974. Pheromonal sociality in Hymenoptera. In *Pheromones*. M.C. Birch, Ed. North-Holland, Amsterdam, pp. 222-249.

(64)  E.O. Wilson. 1970. Chemical communication within animal species. In *Chemical Ecology*. E. Sondheimer and J.B. Simeone, Eds. Academic, New York, pp. 133-155.

(65)  E.O. Wilson. 1971. *The Insect Societies*. Harvard University Press, Cambridge.

(66)  N.E. Gary. 1974. Pheromones that affect the behavior and physiology of honeybees. In *Pheromones*. M.C. Birch, Ed. North-Holland, Amsterdam, pp. 200-221.

(67)  R.K. Callow, J.R. Chapman and P.N. Patton. 1964. Pheromones of the honeybee: Chemical studies of the mandibular secretion of the queen. *J. Apicult. Res.* **3**:77-89.

(68)  G. Bergström and J. Löfquist. 1971. *Camponotus ligniperda* Latr.–A model for the composite volatile secretions of Dufour's gland in formicine ants. In *Chemical Releasers in Insects*. A.S. Tahori, Ed. Proc. 2nd Int. IUPAC Congr., Tel-Aviv, Vol. 3, pp. 195-223.

(69)  N.E. Gary. 1962. Chemical mating attractants in the queen honeybee. *Science* **136**:773-774.

(70)  C.G. Butler. 1954. *The World of the Honeybee*. Collins, London, pp. 97-110.

(71)  R.K. Callow and N.G. Johnson. 1960. The chemical constitution and synthesis of queen substance of honeybees (*Apis mellifera*). *Bee World* **41**:152-153.

(72)  J. Barbier and E. Lederer. 1960. Structure chimique de la substance royale de la rein d'abeille (*Apis mellifica* L.) *C. R. Acad. Sci.* (Paris) **250**:4467-4469.

(73)  J.C. Moser, R.G. Brownlee, and R.M. Silverstein. 1968. Alarm pheromones of the ant, *Atta texana*. *J. Insect Physiol.* **14**:529-535.

(74)  R.G. Riley, R.M. Silverstein, and J.C. Moser. 1974. Biological responses of *Atta texana* to its alarm pheromone and the enantiomer of the pheromone. *Science* **183**:760-762.

(75)  R.G. Riley, R.M. Silverstein and J.C. Moser. 1974. Isolation, identification, synthesis, and biological activity of volatile compounds from the heads of *Atta* ants. *J. Insect Physiol.* **20**:1629-1637.

(76)  J.C. Moser. 1970. Pheromones of social insects. In *Control of Insect Behavior by Natural Products*. D.L. Wood, R.M. Silverstein and M. Nakajima, Eds. Academic,

New York, pp. 161-178.

(77)   J.C. Moser and M.S. Blum. 1963. Trail-marking substance of the Texas leaf-cutting ant: Source and potency. *Science* **140**:1228.

(78)   J.C. Moser and R.M. Silverstein. 1967. Volatility of trail marking substance of the town ant. *Nature* **215**:206-207.

(79)   J.H. Tumlinson, R.M. Silverstein, J.C. Moser, R.G. Brownlee and J.M. Ruth. 1971. Identification of the trail pheromone of a leaf-cutting ant, *Atta texana*. *Nature* **234**:348-349.

(80)   J.H. Tumlinson, J.C. Moser, R.M. Silverstein, R.G. Brownlee and J.M. Ruth. 1972. A volatile trail pheromone of the leaf-cutting ant, *Atta texana*. *J. Insect Physiol.* **18**:809-814.

(81)   S.W. Robinson, J.C. Moser, M.S. Blum and E. Amante. 1974. Laboratory investigation of the trail-following response of leaf-cutting ants with notes on the specificity of a trail pheromone of *Atta texana* (Buckley). *Insect Sociaux* **21**:87-94.

(82)   J.C.Moser and R.M. Silverstein, Unpublished.

(83)   S.W. Robinson and J.M. Cherrett. 1973. Studies on the use of leaf-cutting ant scent trail pheromones as attractants in baits. Proc. VII Congr. IUSSI, London, pp. 332-338.

(84)   F.J. Ritter, I.E.M. Rotgans, E. Talman, P.E.J. Verwiel and F. Stein. 1973. 5-Methyl-3-butyl-octahydroindolizine, a novel type of pheromone attractive to Pharaoh's ants (*Monomorium pharaonis* (L.)). *Experientia* **29**:530.

(85)   S. Huwyler, K. Grob and M. Viscontini. 1975. The trail pheromone of the ant, *Lasius fuliginosus:* Identification of six components. *J. Insect Physiol.* **21**:299-304.

(86)   B. Hölldobler. 1971. Homing in the harvester ant *Pogonomyrmex badius*. *Science* **171**:1149-1151.

(87)   B. Hölldobler and E.O. Wilson. 1970. Recruitment trails in the harvester ant *Pogonomyrmex badius*. *Psyche* **77**:385-399.

(88)   R.D. Akre and C.W. Rettenmyer. 1968. Trail-following by guests of army ants (Hymenoptera: Formicidae: Ecitonini). *J. Kansas Entomol. Soc.* **41**:165-174.

(89)   C.W. Rettenmeyer. 1962. The behavior of millipedes found with Neotropical army ants. *J. Kansas Entomol. Soc.* **35**:377-384.

(90)   J.F. Watkins, T.W. Cole and R.S. Baldridge. 1967. Laboratory studies on interspecies trail following and trail preference of army ants (Dorylinae). *J. Kansas Entomol. Soc.* **40**:146-151.

(91)   F.R. Gehlbach, J.F. Watkins and J.C. Kroll. 1971. Pheromone trail-following studies of typhlopid, leptotyphlopid, and colubrid snakes. *Behavior* **40**:282-294.

(92)   J.C. Moser. 1964. Inquiline roach responds to trail-marking substance of leaf-cutting ants. *Science* **143**:1048-1049.

(93)   M.S. Blum and C.A. Portocarrero. 1966. Chemical releasers of social behavior. X. An attine trail substance in the venom of a non-trail laying myrmicine, *Daceton armigerum* (Lattreille). *Psyche* **73**:150-155.

(94)   M. Lindauer and W.E. Kerr. 1958. Die gegenseitige Verständigung bei den stachellosen Bienen. *Z. vgl. Physiol.* **41**:405-434.

(95)   M.S. Blum, R.M. Crewe, W.E. Kerr, L.H. Keith, A.W. Garrison and M.M. Walker. 1970. Citral in stingless bees: Isolation and functions in trail laying and robbing. *J. Insect Physiol.* **16**:1637-1648.

(96)   M.C. Birch and D.L. Wood. 1975. Mutual inhibition of the attractant pheromone response by two species of *Ips* (Coleoptera: Scolytidae). *J. Chem. Ecol.* **1**:101-113.

(97)    J.E. Amoore, G. Palmieri and M.S. Blum. 1969. Ant alarm pheromone activity: Correlation with molecular shape by scanning computer. *Science* 165:1266-1269.
(98)    B.P. Moore. 1974. Pheromones in the termite societies. In *Pheromones*. M.C. Birch, Ed. North-Holland, Amsterdam, pp. 250-266.
(99)    D.L. Wood. Unpublished.
(100)   L. Friedman and J.G. Miller. 1971. Odor incongruity and chirality. *Science* 172:1044-1046.
(101)   G.F. Russell and J.I. Hills. 1971. Odor differences between enantiomeric isomers. *Science* 172:1043-1044.
(102)   W.A. Kafka, G. Ohloff, D. Schneider and E. Vareschi. 1973. Olfactory discrimination of two enantiomers of 4-methylhexanoic acid by the migratory locust and the honeybee. *J. Comp. Physiol.* 87:277-284.
(103)   Y. Lensky and M.S. Blum. 1974. Chirality in insect chemoreceptors. *Life Sci.* 14:2045-2049.

Note added in proof:

The most important pheromone component in these species, 14-methyl-8-hexadecenal, was not present in the beetle extracts but was found by aeration of the beetles and collection on Porapak. (J.H. Cross, R.C. Byler, R.F. Cassidy, Jr., R.M. Silverstein, R.E. Greenblatt, W.E. Burkholder, A.R. Levinson, and H.Z. Levinson. 1976. Porapak-Q collection of pheromone components and isolation of (Z) and (E) -14-methyl-8-hexadecenal, potent sex attracting components, from the females of four species of *Trogoderma* (Coleoptera:Dermestidae). *J. Chem. Ecol.* 2:457-468.

## CHAPTER 15

# Complexity, Diversity, and Specificity of Behavior-Modifying Chemicals in Lepidoptera and Diptera

Yoshio Tamaki

*Division of Entomology,*
*National Institute of Agricultural Sciences*
*Nishigahara, Tokyo, Japan*

Insect behavior is under the control of various chemical messengers, most of which fall into one of the following three groups: chemical messengers within an insect species (pheromones); chemical messengers between insect species (allomones and kairomones); and chemical messengers from hosts (e.g., attractants and stimulants for oviposition or feeding). Of these, pheromones, with their high biologic activity, have attracted the attention of many researchers, such as chemists, biologists, and applied entomologists. During the present decade many investigations have been conducted on insect sex pheromones, mainly because of their potential usefulness as tools in pest management. But to employ them effectively, we must have detailed knowledge of their behavior-modifying properties.

## I. STRUCTURAL DIVERSITY

### A. Sex Pheromones of Female Lepidoptera

Sex pheromones secreted by female moths have been characterized for 50 species, involving 30 different types of identified compounds (Table 1). Table

TABLE I
Female Sex-Attractant Pheromones of Lepidopera

| Compound[a] | Species | Family | Reference |
|---|---|---|---|
| Hydrocarbon | | | |
| 2-Methyl-heptadecane | *Homomelia nigricans* | Arctiidae | (1) |
| | *H. ferruginos* | Arctiidae | (1) |
| | *H. fragilis* | Arctiidae | (1) |
| | *H. imaculata* | Arctiidae | (1) |
| | *H. lamae* | Arctiidae | (1) |
| | *H. aurantica* | Arctiidae | (1) |
| | *Pyrrharctia isabella* | Arctiidae | (1) |
| Epoxy-hydrocarbon | | | |
| $c$-7, 8-Epoxy-2-methyl-octadecane | *Porthetria dispar* | Lymantriidae | (2) |
| Alcohol | | | |
| $t$8,$t$10-Dodecadienol | *Laspeyresia pomonella* | Tortricidae | (3) (4) |
| $c$11-Tetradecenol (+ $t$11-14:OH, $c$11-14:Ac, $t$11-14:Ac) | *Platynota stultana* | Tortricidae | (5) |
| $t$11-Tetradecenol (+ $c$11-14:OH, $c$11-14:Ac,$t$11-14:Ac) | *P. stultana* | Tortricidae | (5) |
| (+ $t$11-14:Ac) | *P. idaeusalis* | Tortricidae | (6) |
| $t$10,$c$12-Hexa-decadienol | *Bombyx mori* | Bombycidae | (7) |
| Aldehyde | | | |
| $c$9-Tetradecenal (+ $c$11-16:Ald) | *Heliothis virescens* | Noctuidae | (8) |
| $c$11-Tetradecenal (+ $c$11-14:Ac) | *Argyrotaenia citrana* | Tortricidae | (9) |
| $t$11-Tetradecenal | *Choristoneura fumiferana* | Tortricidae | (10) |
| $c$11-Hexadecenal | *Heliothis zea* | Noctuidae | (8) |
| (+ $c$9-14:Ald) | *H. virescens* | Noctuidae | (8) |
| Ketone | | | |
| $c$6-Heneicosen-11-one | *Orgyia pseudotsugata* | Lymantriidae | (11) |
| Isovalerate | | | |
| $c$5-Decenyl isovalerate | *Nudauletia cytherea cytherea* | Saturniidae | (12) |
| Acetate | | | |
| Dodecyl acetate (+ $c$11-14:Ac, $t$11-14:Ac) | *Archips argyrospilus* | Tortricidae | (13) |
| (+ $c$11-14:Ac, $t$11-14:Ac) | *Argyrotaenia velutinana* | Tortricidae | (14) |
| $c$7-Dodecenyl acetate | *Autographa bioloba* | Noctuidae | (15) |
| | *A. californica* | Noctuidae | (16) |
| | *Pseudoplusia includens* | Noctuidae | (17) |
| | *Rachiplusia ou* | Noctuidae | (15) |

TABLE I. *(continued)*

| Compound | Species | Family | Reference |
|---|---|---|---|
| | *Trichoplusia ni* | Noctuidae | (15) (131) |
| t7-Dodecenyl acetate | *Argyroploce leucotreta* | Tortricidae | (18) (19) |
| c8-Dodecenyl acetate | *Grapholitha molesta* | Tortricidae | (20) |
| c9-Dodecenyl acetate | *Paralobesia biteana* | Tortricidae | (21) |
| t9-Dodecenyl acetate | *Rhyacionia buoliana* | Tortricidae | (22) |
| (+ t9,11-12:Ac, | | | |
| 11-12:Ac) | *Diparopsis castanea* | Noctuidae | (23) |
| 11-Dodecenyl acetate | | | |
| (+ t9-12;Ac, | | | |
| t9,11-12:Ac) | *D. castanea* | Noctuidae | (23) |
| t9, 11-Dodecadienyl acetate | | | |
| (+ t9-12:Ac, 11-12:Ac) | *D. castanea* | Noctuidae | (23) |
| c9-Tetradecenyl acetate | | | |
| (+ c11-14:Ac) | *Adoxophyes fasciata* | Tortricidae | (24) |
| (+ c11-14:Ac) | *A. orana* | Tortricidae | (25) (117) |
| (+ c9,t12-14:Ac) | *Cadra cautella* | Pyralidae | (26) |
| (+ c9,t12-14:Ac) | *Spodoptera eridania* | Noctuidae | (27) |
| | *S. frugiperda* | Noctuidae | (28) |
| c10-Tetradecenyl acetate | *Archips semiferanus* | Tortricidae | (29) |
| c11-Tetradecenyl acetate | | | |
| (+ c9-14:Ac) | *Adoxophyes fasciata* | Tortricidae | (24) |
| (+ c9-14:Ac) | *A. orana* | Tortricidae | (25) (117) |
| (+ t11-14:Ac, 12:Ac) | *Archips argyrospilus* | Tortricidae | (13) |
| (+ t11-14:Ac) | *Archips podana* | Tortricidae | (30) |
| (+ c11-14:Ald) | *Argyrotaenia citrana* | Tortricidae | (9) |
| (+ t11-14:Ac, 12:Ac) | *A. velutinana* | Tortricidae | (14) (31) |
| | *Choristoneura rosaceana* | Tortricidae | (32) |
| (+ t11-14:Ac) | *Ostrinia nubilalis* | Pyralidae | (33) (34) |
| (+ t11-14:Ac, c11-14; OH, t11-14:OH) | *Platynota stultana* | Tortricidae | (5) |
| t11-Tetradecenyl-acetate | | | |
| (+ c11-14:Ac, 12:Ac) | *Archips argyrospilus* | Tortricidae | (13) |
| (+ c11-14:Ac) | *Archips podana* | Tortricidae | (30) |
| (+ c11-14:Ac) | *Argyrotaenia velutinana* | Tortricidae | (14) |
| (+ c11-14:Ac) | *Ostrinia nubilalis* | Pyralidae | (34) |
| (+ c11-14:Ac, t11-14: OH, c11-14:OH) | *Platynota stultana* | Tortricidae | (5) |
| c9,t11-Tetradecadienyl acetate | | | |
| (+ c9,t12-14:Ac) | *Spodoptera littoralis* | Noctuidae | (23) (35) |
| (+ c9,t12-14:Ac) | *S. litura* | Noctuidae | (36) |
| c9,t12-Tetradecadienyl acetate | *Anagasta kuehniella* | Pyralidae | (37) |
| (+c9-14:Ac) | *Cadra cautella* | Pyralidae | (26) (38) (39) |
| | *C. figulilella* | Pyralidae | (40) |
| | *Ephestia elutella* | Pyralidae | (41) |

TABLE I. *(continued)*

| Compound | Species | Family | Reference |
|---|---|---|---|
| | *Plodia interpunctella* | Pyralidae | (38) (39) |
| (+ $c$9-14:Ac) | *Spodoptera eridania* | Noctuidae | (27) |
| | *S. exigua* | Noctuidae | (42) |
| (+ $c$9,$t$11-14:Ac) | *S. littoralis* | Noctuidae | (23) (35) (36) |
| (+ $c$9,$t$11-14:Ac) | *S. litura* | Noctuidae | (36) |
| $c$7,$c$11-Hexadecadienyl acetate (+ $c$7,$t$11-16:Ac) | *Pectinophora gossypiella* | Gelechiidae | (43) (44) |
| $c$7,$t$11-Hexadecadienyl acetate (+ $c$7,$c$11-16:Ac) | *P. gossypiella* | Gelechiidae | (43) (44) |
| | *Sitotroga cerealella* | Gelechiidae | (45) |
| $c$3,$c$13-Octadecadienyl acetate | *Sanninoidea exitiosa* | Sesiidae | (46) |
| $t$3,$c$13-Octadecadienyl acetate | *Synanthedon pictipes* | Sesiidae | (46) |

$^a$Key to abbreviated notation of chemical formulae: $c$ or $t$ = *cis* or *trans*; the number following $c$ or $t$ indicates the position of unsaturation; the number following the hyphen indicates the length of the carbon chain; Ac = acetate; Ald = aldehyde.

1 includes only those insect species for which sex pheromones were isolated and identified. All the compounds are aliphatic, and their structural diversity is indicated by differences in number of carbon atoms, in the number and position of unsaturation, and in their functionality. Of the 30 compounds, 18 (60%) are acetates having one or two double bonds, although other types of compounds—such as hydrocarbons, an epoxy-hydrocarbon, alcohols, aldehydes, a ketone, and an isovalerate—have been characterized as natural sex pheromones.

The carbon numbers of lepidopterous sex pheromones range from 12 to 21. The smallest compound is dodecadienol of the codling moth (3) (4), and the largest is heneicosenone of the Douglas-fir tussock moth, *Orgyia pseudotsugata* (11). Those compounds having carbon numbers 14, 16, and 18 comprise about 93% of the total compounds so far identified. These three are acetates with carbon chain lengths of 12, 14, and 16 carbons in their alcohol moiety. The molecular weights of the 30 female attractant pheromones range from 182 to 308, and 78.8% of these compounds have molecular weights between 226 and 254. The mean molecular weight of the 30 pheromones is 246.3. The carbon-carbon double bond in the pheromones is distributed from the third to the thirteenth position. Positions 7, 9, and 11 are the most frequent, and these three comprise 73.9% of the total. Double bonds in positions 3, 12, and 13 have been found only in doubly unsaturated compounds. Geometry of the double bond is also an important character of pheromonal compounds, and sometimes only the *cis*- or *trans*-isomer shows biologic activity as the sex pheromone for a particular species. Sixty percent of the double bonds found in the natural-attractant pheromones are of the *cis*-configuration.

Relationships between insect taxonomy and the structure of pheromonal compounds are not yet clear because of the small number of species so far investigated. The 50 species for which female sex attractant pheromones have been identified belong to only 9 families of Lepidoptera, compared with the total number of more than 100 families in this order. In both the Tortricidae and Noctuidae, compounds of different chemical classes are utilized by different species within a family. Acetates are the major lipid class for sex pheromones of 5 different families. An interesting relationship is found in the Tortricidae. The subfamily Tortricinae utilizes compounds having 14 carbon atoms, and the subfamily Olethreutinae utilizes compounds having 12 carbon atoms (47) (48).

The difference between 30 (the number of compounds) and 50 (the number of insect species) indicates that many compounds are utilized as common sex pheromones by different insect species.

The biologic activity of synthetic sex pheromones, especially their competitiveness with live virgin females in the field, is of great interest from the viewpoint of their field application. Typical data comparing synthetic pheromones with virgin females are summarized in Table 2. Most of the synthetic sex pheromones are very competitive with virgin females in the field. However, there is surprisingly little information on the comparison between the release rates of synthetic compounds from evaporation substrates and natural pheromones from virgin females.

## B. Synthetic Sex Attractants for Male Lepidoptera

During field tests of various compounds proposed to be sex pheromones or the empirical screening of pheromone analogues, male moths of a variety of insect species were attracted to a number of the synthetic compounds. At the present date, 35 compounds function as attractants or attractant components for 107 insect species of 14 families (Table 3). Twenty-four acetates, individually or in combination with other classes of compounds, attracted 92 species. It is not yet certain whether these synthetic compounds are the sex pheromones of each attracted species, but they act as sex attractants for males of these species. All the attracted individuals were males, and each species was attracted by only one compound or by a specific mixture of two compounds.

## TABLE II

### Comparison of Biologic Activity between Synthetic Pheromonal Compounds and Virgin Females in the Field

| Species | Compound | Number of Male Moths Captured by | | Reference |
|---|---|---|---|---|
| | | Synthetic Compound (mg/evaporation substrate) | Virgin Female (No. females used) | |
| *Adoxophyes orana* | $c9$-14:Ac + $c11$-14:Ac = 9:1 | 40.9 (0.1/plastic) | 14.9 (1) | (49) |
| *Archips podana* | $c11$-14:Ac + $t11$-14:Ac = 1:1 | 131 (0.8/plastic) | 77 | (30) |
| *Argyrotaenia velutinana* | $c11$-14:Ac + 12:Ac = 4:6 | 825 (10 µl/plastic) | 1805 (5) | (50) |
| *Choristoneura fumiferana* | $t11$-14:Ald | 53.7 (0.1/plastic) | 45.8 | (10) |
| *Ostrinia nubilalis* | $c11$-14:Ac | 168 (0.014[a]/filter paper) | 32 | (51) |
| *Pectinophora gossypiella* | $c7,c11$-16:Ac + $c7,t11$-16:Ac = 1:1 | 22.2 (25 mm planchet) | 6.4 (10) | (44) |
| *Pseudoplusia includens* | $c7$-12:Ac | 424 (0.05 ml/plastic) | 530 (100) | (17) |
| *Rhyacionia buoliana* | $t9$-12:Ac | 27.8 (1.4/plastic) | 2.5 (2) | (22) |
| *Spodoptera litura* | $c9,t11$-14:Ac + $c9,t12$-14:Ac = 10:1 | 4842 (1/rubber) | 4860 (10) | (52) |
| *Trichoplusia ni* | $c7$-12:Ac | 359 (0.05 ml/plastic) | 1112 (100) | (17) |

[a] 0.35 ml of olive oil was added.

TABLE III
Synthetic Sex Attractants Found by Field Tests

| Compound | Species | Family | Reference |
|---|---|---|---|
| Hydrocarbon | | | |
| 2-Methyl-heptadecane | *Homomelina laeta* | Arctiidae | (1) |
| | *H. rubicundaria* | Archiidae | (1) |
| Epoxy-hydrocarbon | | | |
| $c$7,8-Epoxy-2-methyl-octadecane | *Lymantria obfuscata* | Lymantriidae | (53) |
| | *L. fumida* | Lymantriidae | (54) |
| Alcohol | | | |
| $c$7-Dodecenol (+$c$7-12:Ac) | *Harpipteryx "xylostella" auct.* | Ypomoneutidae | (48) |
| $c$7-Dodecenol | *Shizura semirufescens* | Notodontidae | (48) |
| | *Raphia frater* | Noctuidae | (55) |
| | *Exartema* sp. | Tortricidae | (55) |
| $c$8-Dodecenol | *Epiblema scudderiana* | Tortricidae | (48) |
| $t$8-Dodecenol (+$t$8-12:Ac) | *Hedia chinosema* | Tortricidae | (48) |
| $t$9-Dodecenol (+$t$9-12:Ac) | *Dichrorampha* sp. | Tortricidae | (48) |
| $t$8,$t$10-Dodecadienol | *Thiodia* sp. | Tortricidae | (56) |
| $c$11-Tetradecenol (+$c$11-14:Ac) | *Archips rosannus* | Tortricidae | (56) |
| | *Choristoneura fractivittana* | Tortricidae | (47) |
| (+ $c$11-14:Ac) | *Clepsis melaleucana* | Tortricidae | (48) |
| | *Nedra ramosula* | Noctuidae | (47) |
| | *Sparganothis niveana groteana* | Tortricidae | (48) |
| Aldehyde | | | |
| $t$11-Tetradecenal | *Acleris emargana* | Tortricidae | (55) |
| | *Choristoneura biennis* | Tortricidae | (57) |
| | *C. occidentalis* | Tortricidae | (10) |
| Acetate | | | |
| $c$7-Decenyl acetate | *Battaristis* sp. | Gelechiidae | (47) |
| | *Lobesia aelopa* | Tortricidae | (59) |
| $c$5-Dodecenyl acetate | *Autoplusia egena* | Noctuidae | (58) |
| | *Chionodes fuscomaculella* | Gelechiidae | (47) |
| | *Scrobipalpa atriplicella* | Gelechiidae | (47) |
| $c$7-Dodecenyl acetate | *Autographa ampla* | Noctuidae | (47) |
| | *A. falcifera* | Noctuidae | (47) |
| | *A. precationis* | Noctuidae | (48) |
| | *Chrysaspidia contexta* | Noctuidae | (47) |
| | *Epinotia atistriga* | Tortricidae | (48) |
| | *E. zandana* | Tortricidae | (48) |
| | *Filatima* sp. | Tortricidae | (47) |
| (+ $c$7-12:OH) | *Harpipteryx "xylostella" auct.* | Ypomoneutidae | (48) |

TABLE III. *(continued)*

| Compound | Species | Family | Reference |
|---|---|---|---|
| | *Nippoptila issikiii* | Pterophoridae | (59) |
| | *Orthogonia sera* | Noctuidae | (59) |
| | *Phlyctaenia terrealis* | Pyralidae | (47) |
| | *Plusia aereoides* | Noctuidae | (47) |
| | *Pterophorus tenuidactylus* | Pterophoridae | (48) |
| *c*8-Dodecenyl acetate | *Aphania infida* | Tortricidae | (48) |
| | *Cryptophlebia ombrodelta* | Tortricidae | (59) |
| | *Epiblema desertana* | Tortricidae | (48) |
| | *Grapholitha funebrana* | Tortricidae | (56) |
| | *G. prunivora* | Tortricidae | (20)(47) |
| | *Laspeyresia* sp. | Tortricidae | (48) |
| | *Pammene nemorosa* | Tortricidae | (59) |
| | *Pseudexentera maracana* | Tortricidae | (48) |
| *t*8-Dodecenyl acetate | *Ecdytolopha insiticiana* | Tortricidae | (48) |
| | *Grapholitha packardii* | Tortricidae | (20)(47) |
| (+ *t*8-12:OH) | *Hedia chinosema* | Tortricidae | (48) |
| *c*9-Dodecenyl acetate | *Episimus argutanus* | Tortricidae | (48) |
| | *Cosmopterix* sp. | Cosmopterigidae | (59) |
| *t*9-Dodecenyl acetate | *Loxostege chortalis* | Pyralidae | (48) |
| (+ *t*9-12:OH) | *Dichrorampha* sp. | Tortricidae | (48) |
| *c*10-Dodecenyl acetate | *Argyrotaenia quadrifasciana* | Tortricidae | (56) |
| *t*10-Dodecenyl acetate | *Thiodia alterana* | Tortricidae | (56) |
| *t*7,*c*9-Dodecadienyl acetate | *Lobesia botrana* | Tortricidae | (60) |
| *c*7-Tetradecenyl acetate | *Amathes c-nigrum* (large) | Noctuidae | (47) |
| | *Lacinipolia lorea* | Noctuidae | (48) |
| *t*7-Tetradecenyl acetate | *Amathes c-nigrum* (small) | Noctuidae | (47) |
| *c*9-Tetradecenyl acetate | *Apamea interoceanica* | Noctuidae | (47) |
| (+ *c*11-14:Ac) | *Aphania laeteifacies* | Tortricidae | (61) |
| | *A. auricrigtana* | Tortricidae | (59) |
| (+ *c*11-14:Ac) | *Archippus ingentanus* | Tortricidae | (61) |
| (+ *c*11-14:Ac) | *Clepsis spectrana* | Tortricidae | (62) |
| | *Bryotropha similis* | Gelechiidae | (47) |
| | *Cucullia intermedia* | Noctuidae | (48) |
| (+ *c*11-14:Ac) | *Dadica lineosa* | Noctuidae | (61) |
| | *Hermonassa cecilia* | Noctuidae | (59) |
| | *Leucania phragmitidicola* | Noctuidae | (48) |
| | *Nemapogon apicisignatellus* | Tineidae | (48) |
| | *Pyreferra citrombra* | Noctuidae | (48) |

TABLE III. *(continued)*

| Compound | Species | Family | Reference |
|---|---|---|---|
| *t*9-Tetradecenyl acetate | *Bryotropha* sp. | Gelechiidae | (47) |
| | *Loxostege neobliteralis* | Pyralidae | (47) |
| | *Polia grandis* | Noctuidae | (47) |
| *c*10-Tetradecenyl acetate | *Exartema* sp. | Tortricidae | (63) |
| *c*11-Tetradecenyl acetate | | | |
| (+ *c*9-14:Ac) | *Aphania leteifacies* | Tortricidae | (61) |
| (+ *c*9-14:Ac) | *Archippus ingentanus* | Tortricidae | (61) |
| (+ *c*11-14:OH) | *Archips rosanus* | Tortricidae | (56) |
| | *Ceramica picta* | Noctuidae | (48) |
| (+ *c*11-14:OH) | *Clepsis melaleucana* | Tortricidae | (48) |
| (+ *c*9-14:Ac) | *C. spectrana* | Tortricidae | (62) |
| (+ *c*9-14:Ac) | *Dadica lineosa* | Noctuidae | (61) |
| | *Hoshinoa longicellanus* | Tortricidae | (59) |
| (+ *t*11-14:Ac) | *Ostrinia obumbratalis* | Pyralidae | (64) |
| | *Thyris maculata* | Thyrididae | (48) |
| *t*11-Tetradecenyl acetate | *Choristoneura viridis* | Tortricidae | (57) |
| | *Cnephasia* sp. | Tortricidae | (64) |
| | *Dichomeris ligulella* | Gelechiidae | (48) |
| (+ *c*11-14:Ac) | *Ostrinia obumbratalis* | Pyralidae | (64) |
| | *Phalonia* sp. | Phaloniidae | (47) |
| | *Pyrausta ochosalis* | Pyralidae | (48) |
| | *Sparganothis albicaudana* | Tortricidae | (48) |
| | *S. sulfureana* | Tortricidae | (47) |
| | *Zeiraphera diniana* | Tortricidae | (65) |
| *c*9,*t*12-Tetradecadienyl acetate | *Rusidrina depravata* | Noctuidae | (59) |
| *c*7-Hexadecenyl acetate | *Pectinophora gossypiella* | Gelechiidae | (66) |
| | *Exoa fessellata* | Noctuidae | (47) |
| *t*9-Hexadecenyl acetate | *Brachmia macroscopa* | Gelechiidae | (67) |
| *c*11-Hexadecenyl acetate | *Apamea velata* | Noctuidae | (48) |
| | *Amphipyra monolitha* | Noctuidae | (59) |
| | *Epiplema moza* | Epiplemidae | (59) |
| | *Morrisonia confusa* | Noctuidae | (48) |
| | *Orthodes crenulata* | Noctuidae | (47) |
| | *O. vecors* | Noctuidae | (47) |
| | *Scotogramma trifolii* | Noctuidae | (47) |
| | *Telorta divergens* | Noctuidae | (59) |
| *t*11-Hexadecenyl acetate | *T. edentata* | Noctuidae | (59) |
| *c*3,*c*13-Octadecadienyl acetate | *Carmenta teta* | Sesiidae | (68) (69) |
| | *Paranthrene simulans* | Sesiidae | (68) (69) |
| | *Podosesia syringae* | Sesiidae | (68) (69) |
| | *Sylovora acerni* | Sesiidae | (68) (69) |
| | *Synanthedon geleformis* | Sesiidae | (68) (69) |
| | *S. fatifera* | Sesiidae | (68) (69) |
| | *S. proxima* | Sesiidae | (68) (69) |
| (+ *t*3, *c*13-18:Ac) | *S. rubrofascia* | Sesiidae | (68) (69) |

TABLE III. *(continued)*

| Compound | Species | Family | Reference |
|---|---|---|---|
| | *S. sapygaeformis* | Sesiidae | (68) (69) |
| | *S. scitula* | Sesiidae | (68) (69) |
| (+ *t*3, *c*13-18:Ac) | *Vespamime sequoiae* | Sesiidae | (68) (69) |
| *t*3,*c*13-Octadecadienyl acetate | | | |
| (+ *c*3,*c*13-18:Ac) | *Synanthedon rubrofascia* | Sesiidae | (68) (69) |
| (+ *c*3,*c*13-18:Ac) | *Vespamime sequoiae* | Sesiidae | (68) (69) |

Males of no species were attracted by two different individual compounds presented separately. These facts suggest that each synthetic compound is very close to, or more possibly a major component of, the sex pheromone of each attracted species.

The difference between 35, (the number of tested chemicals) and 107 (the number of attracted insect species) again indicates that one chemical attracts several insect species.

## C. Sex Pheromones of Male Lepidoptera

Male moths of at least two lepidopterous species secrete pheromones that excite female moths sexually. *n*-Nonanal (1) and *n*-undecanal (2) were identified as the male pheromone of the greater waxmoth, *Galleria mellonella* (70) (71). Both compounds individually elicit typical sexual behavior from females. *n*-Undecanal is also a male pheromonal component, with *cis*-11-octadecenal (3) in the lesser waxmoth, *Achroia grisella* (72). In the case of the lesser waxmoth, a mixture of the two components is essential for the sexual excitement of females; the individual components elicit no response from females. The combination of the multicomponent stimulant pheromone and an auditory signal from males mediates orientation behavior in the lesser waxmoth females (72) (Fig. 1).

Male moths of many noctuid species possess various types of brush organs that affect copulatory behavior. Arrestant odors are released by noctuid male moths of subfamilies Cucullinae (73), Hadeninae (74) (76), and Amphipyrinae (74). The compounds shown in Fig. 2 are generally present as mixtures in the brush organs. Benzaldehyde (8) acted as an arrestant for females during the courtship of *Pseudaletia separata* (75). However, experimental demonstration of the function of this compound is still inadequate. The biologic functions of the other components (4-7, 9-12) are also undetermined.

A heterocyclic ketone, 2,3-dihydro-7-methyl-1*H*-pyrrolizin-1-one (13) has been isolated and identified from the hair-pencil organ of queen butterfly

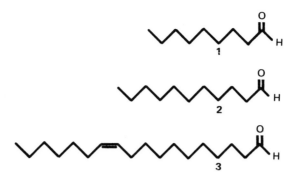

Fig. 1. Sex-Stimulant Pheromones of Male Lepidoptera (70) (72): *n*-Nonanal (**1**), *n*-undecanal (**2**), *cis*-11-octadecenal (**3**)

males, *Danaus gillippus berenice* (77) (Fig. 3). The same ketone was also found in other danaiid butterflies, such as *Lycorea ceres ceres* (79) (83), *Danaus chrisippus* (82), *D. limniace ptiverana* (84), *D. affinis affinis* (78), *D. hamatus hamatus*(78), *Amauris albimaculata* (84), *A. echeria* (84), *A. niavius* (84), and *A. ochlea* (84). The biologic function of this ketone has only been confirmed in the queen butterfly, in which it acts as an aphrodisiac (85) (86). If the aphrodisiacs possess species specificity in danaiid

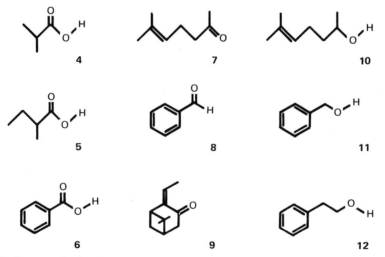

Fig. 2. Compounds Identified from the Brush Organs of Male Noctuid Moths of Subfamilies Cuculliinae, Hadeninae, and Amphipyrinae (73) (76): 2-methyl-propionic acid (**4**), 2-methyl butanoic acid (**5**), benzoic acid (**6**), 6-methyl-5-hepten-2-one (**7**), benzaldehyde (**8**), pinocarvone (**9**), 6-methyl-5-hepten-2-ol (**10**), benzyl alcohol (**11**), 2-phenyl-ethanol (**12**)

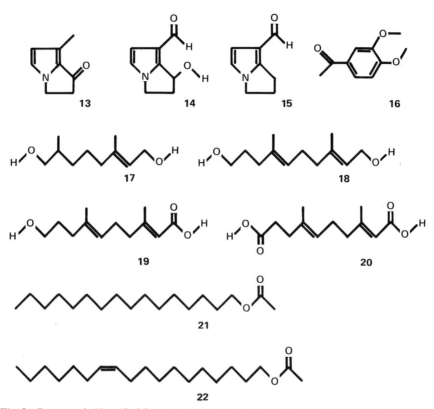

Fig. 3. Compounds Identified from the Hair-Pencils of Danaid Butterflies (77)-(84):
2,3-dihydro-7-methyl-1*H*-pyrrolizin-1-one (**13**) and its homologues (**14,15**), 3,4-dimethoxy-
acetophenone (**16**), 3,7-dimethyl-*trans*-2-octen-1,8-diol (**17**), 3,7-dimethyl-*trans,trans*-2,6-
decadien-1,10-diol (**18**), 3,7-dimethyl-10-hydroxy-*trans,trans* -2,6-decadienoic acid (**19**),
3,7-dimethyl-*trans,trans*-2,6-decadien-1,10-dioic acid (**20**), *n*-hexadecnyl acetate (**21**),
*cis*-11-octadecenyl acetate (**22**)

butterflies, it is reasonable to predict the presence of some additional active
components in the pheromone of each species, because the ketone is the
common compound in several danaiid species. The function of other
compounds, such as heterocyclic aldehydes (**14,15**), an acetophenone (**16**),
and aliphatic compounds (**17-22**), is not clear.

### D. Sex Pheromones of Diptera

Sex pheromones of Diptera have been identified for only two species.
These pheromones are aliphatic compounds (Fig. 4). *cis*-9-Tricosene (**23**), the
major component of the female sex pheromone of the housefly, *Musca*

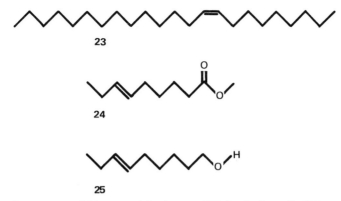

Fig. 4. Sex Pheromones of Diptera: *cis*-9-tricosene (23) for the housefly (87); methyl *trans*-6-nonenoate (24) and *trans*-6-nonenol (25) for the Mediterranean fruitfly (88)

*domestica* (87), has been isolated from the cuticular wax of mature females, and is a component of the cuticular wax of the cockroaches *Periplaneta australiasiae, P. brunnea,* and *P. fulginosa* (88). Recent studies on related unsaturated hydrocarbons revealed that *cis*-9-heneicosene has a potent synergistic effect on the activity of *cis*-9-tricosene (89). In the Mediterranean fruitfly, *Ceratitis capitata,* methyl *trans*-6-nonenoate (24) and *trans*-6-nonen-1-ol (25) are pheromonal components secreted by males (90). In addition to these two compounds, a third component, an unknown acidic compound, is a natural synergist (90). A sex pheromone blend of four components has recently been isolated from Caribbean fruitfly males. The active ingredients are mono- and diunsaturated 9-carbon-alcohols and two lactone esters having a molecular weight of 196. A combination of all four components is significantly more active than combinations of two or three components (91). The sex pheromone of the Queensland fruitfly, *Dacus tryoni,* is a mixture of at least six aliphatic amines (92). Thus we can recognize multicomponent sex pheromone systems in fruitflies.

Potent synthetic attractants—trimedlure (26), cure-lure (27), and methyl eugenol (28)—for the Mediterranean fruitfly, the melon fly (*Dacus cucurbitae*), and the oriental fruitfly (*Dacus dorsalis*) have been widely used for detection of infestations and direct control of these insects (93) (Fig. 5).

### E. Structure-Activity Relationships in Sex-Attractant Pheromones

Because of the relative simplicity of the chemical structure of female sex-attractant pheromones, a considerable number of pheromone analogues or congeners have been synthesized and their biologic activity evaluated with various insect species. Generally, any slight changes in the chemical structure

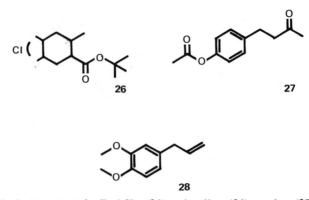

C, 10

H, 7

O, 2

Cl, 1

**26**                                          **27**

**28**

Fig. 5. Synthetic Attractants for Fruitflies (93): trimedlure (26), cue-lure (27), and methyl eugenol (28)

of pheromones result in a pronounced reduction in activity. The apparent biologic activity of pheromone analogues or congeners differs according to which assay method is used. The following methods, either singly or in combination, will determine biologic activity: field tests for male attraction; laboratory tests for sexual excitement or short-range orientation behavior; and electroantennogram (EAG) responses. Of these three methods, the highest specificity in structure-activity relationships is usually observed by field tests; and the lowest by EAG response as shown in the cabbage looper, *Trichoplusia ni* (94), and the gypsy moth, *Porthetria dispar* (95). The low specificity of the EAG method for determining structure-activity relationships is shown with doubly unsaturated sex pheromones and their congeners. For example, in the case of the sex pheromone of the almond moth (*Cadra cautella*), *cis*-9,*trans*-12-tetradecadienyl acetate, there were strong EAG responses with the mono-unsaturated tetradecenyl acetates having the *cis*-9 or the *trans*-12 double bond (48). The low specificity in the EAG method has been successfully utilized for the structural elucidation of the sex pheromone of the codling moth, *trans*-8,*trans*-10-dodecadien-1-ol (4).

Laboratory assays on a considerable number of pheromone analogues or congeners revealed that some of these synthetic compounds elicit sexual behavior from male moths such as the cabbage looper moth (94) (96) (97), the southern armyworm moth (*Spodoptera eridania*) (97), the codling moth (98) (99), the gypsy moth (100), and the almond moth (101). Most of the compounds found to be active in laboratory bioassays, however, were inactive in the field for attraction of males.

Field tests with various pheromone congeners show that a minor structural change of a pheromone results in a considerable decrease in attractive potency. From 80 to 90% decrease in the male catch was brought about by changing the carbon chain length or by moving the position of the double bond in the sex pheromone, cis-7-dodecenyl acetate, of the cabbage looper moth (94) (96). There was a 90% decrease in male catch when additional branches were added to the carbon skeleton of the sex pheromone, 2-methyl-heptadecane, of the tiger moth (1). Most of the 55 analogues of cis-7,8-epoxy-2-methyl-octadecane, the sex pheromone of the gypsy moth, were inactive in the field; of these, the several compounds most closely related to the sex pheromone in structure were 40- to 100-fold less active than the pheromone (100). Hexalure (cis-7-hexadecenyl acetate), a potent synthetic sex attractant of the pink bollworm moth, possesses only 18% activity of the natural sex pheromone (1:1 mixture of cis-7,cis-11- and cis-7,trans-11-hexadecadienyl acetates) when tested in the field (44).

These structure-activity relationships indicate that a slight change in the structure of a pheromonal compound results in a great decrease in biologic activity and that the most active analogues are those most closely related to the pheromonal compounds.


## II. SYNERGISTS AND INHIBITORS OF SEX-ATTRACTANT PHEROMONES

Mixing the pheromone with particular synthetic compounds often enhances or inhibits biologic activity of a lepidopterous sex-attractant pheromone. The activity-modifiers for sex-attractant pheromones, such as synergists or inhibitors, are generally very close in structure to the sex pheromones. This is reasonable when we consider the structural similarity among the lepidopterous sex pheromones so far identified and when we consider that some sex pheromones or pheromonal components function as allomones or kairomones. Synergists and inhibitors (Tables 4 and 5, respectively) of sex-attractant pheromones show either one or two of the following structural characteristics:

1. Geometric isomer of the sex pheromone
2. Positional isomer in unsaturation
3. Same carbon skeleton, having different function
4. Different carbon skeleton, having the same function

Typical examples of artificial synergists are the geometric isomers and the saturated homologues of the sex pheromones of several tortricid and noctuid moths. In geometric isomers the amount of synergist is very important for biologic activity. There is an optimum ratio in the mixture of a pheromone and its geometric isomer. Biologic activities of the sex pheromone of the oriental fruit moth, Grapholitha molesta, are increased 23-fold when 6 to 7%

## TABLE IV

### Synergists of the Sex-Attractant Pheromones of Several Lepidopterous Species

| Species and Attractants | Synergist | Amount[a] | Increase in Male Catch (fold) |
|---|---|---|---|
| *Argyrotaenia velutinana* (104) (105) | | | |
| $c11$-14:Ac + $t11$-14:Ac = 93:3 | 12:Ac[b] | — | 2.7 |
| $c11$-14:Ac + $t11$-14:Ac = 93:3 | $c7$-12:Ac | — | 8.5 |
| $c11$-14:Ac pure | $t11$-14:Ac[b] | 6-8% | 20-40 |
| *Grapholitha molesta* (102) (103) (106) (107) | | | |
| $c8$-12:Ac pure | $t8$-12:Ac | 6-7% | 23 |
| $c8$-12:Ac + $t8$-12:Ac = 93.3:6.7 | 12:OH | 3-fold | 2 |
| *Ostrinia nubilalis* (104) | | | |
| $c11$-14:Ac pure | $t11$-14:Ac | 4% | 4 |
| *Paralobesia viteana* (21) | | | |
| $c9$-12:Ac | 9:Ac | 10-fold | 20 |

[a]Relative to sex pheromone.
[b]Recently verified as natural pheromonal components.

of the geometric isomer is added (102) (103). Similarly, about four-fold increase in male catch of the European corn borer, *Ostrinia nubilalis* (in Iowa), occurred with 4% of the *trans*-isomer was added to its sex pheromone, *cis*-11-tetradecenyl acetate (104). There was a synergistic effect of saturated homologues in the grape berry moth, *Paralobesia viteana*. The activity of *cis*-9-dodecenyl acetate, the sex pheromone of this tortricid moth, increased about 20-fold when a 10-fold amount of lauryl acetate or nonyl acetate was added to the pheromone (21).

The geometric isomers of a sex pheromone often serve as potent inhibitors of the pheromone. An example is the lesser peach-tree borer moth, *Synanthedon pictipes* (46). Addition of 1% *cis,cis*-isomer to the sex pheromone *trans*-3,*cis*-13-octadecadienyl acetate caused a 95% decrease in the male catch. Inhibitory effects of geometrical isomers were also noticed with the tufted apple bud moth (*Platynota idaeusalis*) (6), the pink bollworm moth (*Pectinophora gossypiella*) (43), the European pine shoot moth (*Rhyacionia buoliana*) (22), and the summerfruit tortrix (*Adoxophyes orana*) (49) by concentrations less than 10% of the known pheromonal compounds. On the other hand, the addition of 10% of the geometric isomers did not affect male catches for the false codling moth (*Argyroploce leucotreta*) (19), the codling moth (4), and the tobacco budworm moth (*Heliothis virescens*) (8). Inhibitory effects of the saturated homologues of sex pheromones are found in the European pine shoot moth (22) and the oriental fruit moth (103). In addition to these compounds, various analogues of each sex pheromone showed potent inhibitory effects on male attraction in field tests. The analogues include the compounds that belong to either one of the four groups mentioned earlier. It is noteworthy that the discovery of a pheromone inhibitor for a particular insect species is easier than the discovery of a synergist. In other words, the olfactory communication system in insects is delicate, and the addition of various chemicals easily disturbs it.

## III. MULTICOMPONENT PHEROMONE SYSTEMS

Recent studies on insect sex pheromones, especially of Lepidoptera, indicate that a particular pheromonal component is often utilized by more than one species. cis-7-Dodecenyl acetate, for example, is the sex pheromone of five noctuid moths and attracts 11 other species in the field. cis-11-Tetradecenyl acetate is the sex pheromone or the pheromonal component of five species and also attracts males of six other species in the field. Similarly, cis-9,trans-12-tetradecadienyl acetate is the sex pheromone or the pheromonal component of nine different species. These facts indicate that specificity of lepidopterous sex pheromones is relatively low when a single compound is considered as the pheromone of each species.

Dethier (115) pointed out the importance of a mixture in odor quality: "... one of the most important principles relating to attractants is that mixtures are superior to any one of the component parts." Wright (116) also presumed that insect pheromones are composed of multiple components. Jacobson et al. (27) reported the first example of a multicomponent sex pheromone in Lepidoptera in the southern armyworm moth. In this case, however, a possible difference of the two compounds in biologic functions was indicated, and no data were presented on the possible combined effect and importance of the mixture of the two components, cis-9-tetradecenyl acetate and cis-9,trans-12-tetradecadienyl acetate. Strongly suggestive of the importance of a mixture are the data of Roelofs and Comeau (105), which indicate that the addition of a particular compound to a sex pheromone causes a drastic change in the biologic activity. cis-11-Tetradecenyl acetate is the common sex pheromone of both the red-banded leaf roller moth and the oblique-banded leaf roller moth (Choristoneura rosaceana). Lauryl acetate is an artificial synergist for the former species, but is an inhibitor for the latter species.

The importance of admixture and the ratio of two pheromonal compounds in lepidopterous sex pheromones was experimentally proved for the first time on the two sibling tortricid moths, the smaller tea tortix (Adoxophyes fasciata) and the summerfruit tortrix (A.orana) (24) (25) (117). Their sex

## TABLE V

### Inhibitors of the Sex Attractant Pheromones of Several Lepidopterous Species

| Species and Sex Pheromone | Inhibitor | Amount | Inhibition of Male Catch (%)[a] |
|---|---|---|---|
| *Adoxophyes orana* (49) (108) (109) | | | |
| $c9$-14:Ac + $c11$-14:Ac = 9:1 | $t9$-14:Ac + $t11$-14:Ac = 9:1 | 2-10% | 67-94 |
| $c9$-14:Ac + $c11$-14:Ac = 9:1 | 9-yne- or 11-yne-14:Ac | 10-50% | 59-98 |
| $c9$-14:Ac + $c11$-14:Ac = 9:1 | 11-methoxy- or 11-ethoxy-11:Ac | 40% | 92-96 |
| $c9$-14:Ac + $c11$-14:Ac = 9:1 | $c7$- or $c10$-14:Ac | 40% | 77-96 |
| *Argyrotaenia velutinana* (105) | | | |
| $c11$-14:Ac | $t11$-13:Ac, 11-yne-13:Ac, 11-yne-14:Ac, 11-yne-14:Pr | 50% | 80-100 |
| *Cadra cautella* (110) | | | |
| $c9,t12$-14:Ac | $c9,t12$-14:OH | 50% | 75[b] |
| *Grapholitha molesta* (103) | | | |
| $c8$-12:Ac | 12:Ac | 10-50 fold | 100 |
| *Laspeyresia pomonella* (111) (112) | | | |
| $t8,t10$-12:OH | $c8$-12:Ac, $c10$-12:Ac | 50% | 85 |
| $t8,t10$-12:OH | $t8,t10$-12:Ac | 25% | 95-100 |
| 10 virgin females | $t8,t10$-12:Ac | 4 mg | 95-100 |
| *Ostrinia nubilalis* (51) | | | |
| $c11$-14:Ac | $t11$-13:Ac, $c9$-14:Ac, $t11$-14:Ac, $c11$-15:Ac,$t11$-15:Ac | 50% | 97-100 |

270

| | | | |
|---|---|---|---|
| *Pectinophora gossypiella* (43) | | | |
| c7,c11-16:Ac + c7,t11-16:Ac = 1:1 | t7,c11- or t7,t11-16:Ac | 5-10% | 51-78 |
| *Platynota idaeusalis* (6) | | | |
| t11-14:OH + t11-14:Ac = 2:1 | c11-14:OH | 10-20% | 50 |
| *Porthetria dispar* (113) | | | |
| virgin female | 2-Me-c7-18 | 1 mg | 97 |
| *Rhyacionia buoliana* (22) | | | |
| c9-12:Ac | 12:Ac, t9-12:Ac | 2% | 96 |
| *Synanthedon pictipes* (46) | | | |
| t3,c13-18:Ac | c3,c13-18:Ac | 1% | 95 |
| *Trichoplusia ni* (114) | | | |
| c7-12:Ac | c7-12:OH | 0.01-0.1% | 72-98 |
| 50 virgin females | c7-12:OH | 5 mg | 96 |

[a]Relative to sex pheromone.
[b]Data from laboratory bioassay

pheromones are mixtures of *cis*-9- and *cis*-11-tetradecenyl acetates. Neither compound elicited behavioral activity when presented individually in a laboratory bioassay and in field tests to either species. Mixture of the two compounds is essential for the release of overt sexual activity of male moths of both species. The ratio of *cis*-9-tetradecenyl acetate to *cis*-11-tetradecenyl acetate in the pheromonal mixture found in and released by female moths was different between the two species. The optimum response of male moths in the laboratory bioassay and maximum catch of males in the field occurred when the two different specific ratios of the pheromonal components found in the females of the two species were used (Table 6).

Similarly the importance of a specific ratio of pheromonal components has been demonstrated in the armyworm moth, *Spodoperta litura* (52). In this case *cis*-9,*trans*-11-tetradecadienyl acetate alone showed weak activity as an attractant in the field; the activity was greatly enhanced by adding *cis*-9,*trans*-12-tetradecadienyl acetate (Table 7). The latter compound is a natural synergist in the multicomponent sex-pheromone system of the armyworm moth.

At present, multicomponent sex-attractant pheromones have been found in 15 species of Lepidoptera (Table 8). Examples of multicomponent systems apparently show that the ratio of components is critical for biological activity, except for the almond moth and the southern armyworm moth. It is notable that the ratio of the pheromonal components found in female moths and the ratio needed for maximal male activity are fairly similar. However, the ratio of the two components in the released sex pheromone has not been determined for these species, except for the smaller tea tortrix (67), the summerfruit tortrix (67), the red-banded leaf roller (14), and the organe tortrix moth (*Argyrotaenia citrana*) (9).

In addition to attraction to known natural pheromones, several species are attracted by specific mixtures of two different synthetic compounds in the

TABLE VI

Differences in Relative Quantities of Pheromonal
Components in Two *Adoxophyes* spp. (24) (62) (67) (117)

| | Ratio of Two Components $(c9\text{-}14\!:\!Ac/c11\text{-}14\!:\!Ac)$ | |
| --- | --- | --- |
| | *A. fasciata* | *A. orana* |
| Found in female body | 64/36 | 76/24 |
| Released by female | 63/37 | 82/18 |
| Optimum response of males in laboratory bioassay | 60-70/40-30 | 80-90/20-10 |
| Maximum catch of males in field | 60-70/40-30 | 90/10 |

TABLE VII

Attraction of Male *Spodoptera litura* to
Different Mixtures of Pheromonal
Components (52)

| Ratio of components[a] (*c*9,*t*11-14:Ac/*c*9,*t*12-14:Ac) | No. Males Caught by 3 Traps for 6 Nights |
|---|---|
| 0/10 | 0 |
| 5/5 | 90 |
| 8/2 | 253 |
| 9/1 | 238 |
| 19/1 | 305 |
| 39/1 | 318 |
| 10/0 | 20 |

[a]Amount of the mixture applied on a rubber septum was 1 mg.

field. Examples are found in *Aphania laeteifacies* (61), *Archippus ingentanus* (61), *Archips rosanus* (56), *Clepsis spectrana* (62), *Dadica lineosa* (61), *Dichrorumpha* sp. (48), *Harpipteryx "xylostella"* auct. (48), *Hedia chinosema* (48), *Synanthedon rubrofaocia* (68), and *Vespamime sequoiae* (68). The mixtures of compounds to which these moths are attracted have not been proved to be the sex pheromone of the particular species. However, these species probably utilize similar mixtures as their true sex pheromones. These data again indicate the importance of the multicomponent idea of sex-attractant pheromones in Lepidoptera.

Generally, the nature of the pheromonal signal of a particular insect species can be expressed by a sum of the pheromonal components, as follows:

$$X_1 + X_2 + \ldots + X_n = \sum_{i=1}^{n} X_i$$

where, $X_1$, $X_2$, and $X_n$ are the relative quantities of components $1, 2, \ldots n$ in the multicomponent pheromone. Thus a pheromonal signal is defined by both the type of components and the relative quantity of each component. Most of the multicomponent attractant pheromone systems so far identified are $n = 2$ systems. Lauryl acetate, *trans*-9-dodecenyl acetate, 11-dodecenyl acetate, and *trans*-9, 11-dodecadienyl acetate were reported to be pheromonal components of *Diparopsis castanea*; and myristyl acetate, *cis*-9-tetradecenyl acetate, *trans*-11-tetradecenyl acetate, and *cis*-9,*trans*-11-tetradecadienyl acetate were similarly proposed as pheromonal components of the cotton leaf worm (23). However, the biologic activity of these compounds, especially the effect of the combinations of the four compounds, has not been proved with these two species. Attractive potency of *cis*-9,*trans*-11-tetradecadienyl acetate for males

**YOSHIO TAMAKI**

TABLE VIII

Multi Component Sex-Attractant Pheromones of Lepidoptera

| Species and Compound[a] | Ratio Found in Female | Optimum Ratio for Male Catch | Reference |
|---|---|---|---|
| *Adoxophyes fasciata* | | | |
| $c9$-14:Ac/$c11$-14:Ac | 64/36 | 70/30 | (24) |
| *A. orana* | | | |
| $c9$-14:Ac/$c11$-14:Ac | 76/24 | 90/10 | (62) (117) |
| *Archips podana* | | | |
| $c11$-14:Ac/$t11$-14:Ac | 58/42 | 60/40 | (30) |
| *A. argyrospilus* | | | |
| $c11$-14:Ac/$t11$-14:Ac/12:Ac | 60/40/? | 12/8/80 | (13) |
| *Argyrotaenia citrana* | | | |
| $c11$-14:Ald/$c11$-14:Ac | 1/100 | 66-20/33-80 | (9) |
| *A. velutinana* | | | |
| $c11$-14:Ac/$t11$-14:Ac/12:Ac | 15/2/83 | 30-38/2-4/67-50 | (14) |
| *Heliothis virescens* | | | |
| $c11$-16:Ald/$c9$-14:Ald | 94-57/6-25 | 99.5-67/0.5-33 | (8) |
| *Ostrinia nubilalis* (N.Y.) | | | |
| $t11$-14:Ac/$c11$-14:Ac | 96/4 | 98/2 | (34) |
| *Pectinophora gossypiella* | | | |
| $c7,c11$-16:Ac/$c7,t11$-16:Ac | 50/50 | 50/50 | (43) (44) |
| *Platynota idaeusalis* | | | |
| $t11$-14:OH/$t11$-14:Ac | 67.5/32.5 | 50-33/50-67 | (6) |
| *Platynota stultana* | | | |
| $t11$-14:Ac/$c11$-14:Ac/$t11$-14:OH/ | | | |
| $c11$-14:OH | 82/11/6/1 | 92/6/2/1 | (5) |
| *Spodoptera eridania* | | | |
| $c9$-14:Ac/$c9,t12$-14:Ac | 83/17 | | (27) |
| *S. littoralis* | | | |
| $c9,t11$-14:Ac/$c9,t12$-14:Ac | 95/5 | 98/2[b] | (35) |
| *S. litura* | | | |
| $c9,t11$-14:Ac/$c9,t12$-14:Ac | 88/12 | 98/2 | (36) (52) |

[a]$c9$-14:Ac has been proposed as an additional component of the sex pheromone, $c9,t12$-14:Ac, of *Cadra cautella* (26).
[b]Laboratory assay only.

of the cotton leaf worm in the field did not improve by the addition of the other three compounds (118). The four-component pheromone systems proposed for *D. castanea* and for the cotton leaf worm are questionable. Hendry et al. (119) reported that the oak leaf roller moth (*Archips semiferanus*) possesses a sex-attractant pheromone system composed of more than 10 components (119), but further evaluation on the biologic activity of these possible components is needed. Recent studies, however, indicate the presence of $n = 3$ systems in the fruittree leaf roller moth (*Archips argyrospilus*) (13) and in the red-banded leaf roller moth (14), and of a $n = 4$

system in the omnivorous leaf roller moth (*Platynota stultana*) (5). All the insect species for which the sex pheromone was identified as a single component are now interesting targets for reinvestigation. When we suppose that insect sex-attractant pheromones are principally multicomponent systems, we can explain the gradual speciation of insects from the viewpoint of continuous divergence in pheromonal signals (120).

## IV. BEHAVIOR CONTROL BY PHEROMONAL COMPONENTS

The response patterns of male moths that are exposed to two-component sex-attractant pheromones can be classified in three groups. In the first, no overt behavior is elicited by single components, and admixture of the two is essential. Examples are found in the smaller tea tortrix (24), the summerfruit tortrix (25) (117), *Archips podana* (30), and the tobacco budworm moth (8). In the second, one component is slightly active and the other component is synergistic. Examples are found in *Spodoptera litura* (52) and the tufted apple bud moth (6). In the third, both components are individually active but the activity is enhanced by mixing the two. Examples are the southern armyworm (121) and the pink bollworm (43) (44).

A potentially fruitful approach for manipulating the chemical communication systems of insects may be based on an analysis of the detailed function of each component of multicomponent sex pheromones. Recent data on *S. litura* (122) suggest that the minor component, *cis*-9,*trans*-12-tetradecadienyl acetate, of its sex pheromone acts as an arrestant, which suppresses the flight behavior of male moths. The major component, *cis*-9,*trans*-11-tetradecadienyl acetate, on the other hand, acts as a long-distance attractant. Thus the presence of a small amount of the *cis*-9,*trans*-12-isomer with a large amount of the *cis*-9,*trans*-11-isomer causes male moths to find the exact location of females. Artificial distribution of an excess amount of the minor component, therefore, might disturb the orientation flight of male moths to natural females in the field.

A sex pheromone is, in the strict sense, an intraspecific chemical signal. However, several sex pheromones also function as interspecific chemical messengers such as allomones and kairomones. The gelechiid moths serve as an example (123). *cis*-9-Tetradecenyl acetate attracts male moths of *Bryotropha similis,* and the *trans*-isomer attracts another *Bryotropha* sp. The mixture of the two isomers attracts males of neither species. Thus an attractant of one species actually inhibits attraction of males of the other species. If each isomer is the sex pheromone of one of the species, the inhibitory function of the sex pheromone to the other species apparently effects complete sexual isolation and is beneficial, as a result, to both the releasing and receiving species. Thus these compounds show the properties of both allomones and kairomones. Recent studies on the sex pheromones of the lesser peach tree borer moth and the peach tree borer moth indicate a typical example of the allomone-kairomone properties of a true sex pheromone. Addition of a small amount of the peach tree borer sex pheromone, *cis*-3,*cis*-13-octadecadienyl acetate, to the lesser peach tree borer sex pheromone, brought about a great repression of the attractive potency of the latter pheromone (46). Also, permeation of the air with *cis*-7-dodecenyl acetate, the sex pheromone of the cabbage looper moth, *Autographa californica,* and *Pseudoplusia includens,* disrupts the communication between males and females of several insect species other than the above three noctuid moths (124).

Each component of a multicomponent sex-attractant pheromone may also have a function as an interspecific chemical signal. Sexual excitement and mating behavior of male moths of the smaller tea tortrix are greatly suppressed by exposure to either one of the two pheromonal components, *cis*-9- or *cis*-11-tetradecenyl acetate (125) (126). Therefore, if there is a sympatric species whose sex pheromone is one of these two compounds, this pheromone secondarily serves as an interspecific signal in addition to the primary pheromonal function. We can presume that sexual isolation through chemical communication systems between sympatric insect species is performed not only by differences in the multicomponent pheromone signals of the two species but also by the interspecific chemical signals provided by the pheromonal components. The term *pheromone* alone is inadequate to describe the total system of the complex chemical communication during the mating behavior of insects.

Several artificial inhibitors presented in Table 4 have been proposed to be potent communication-disruption agents. Recent results on some of these inhibitors, however, are disappointing. Attraction of the oriental fruit moth to traps baited with virgin females or the synthetic sex pheromone, *cis*-8-dodecenyl acetate, was greatly repressed when lauryl acetate was simultaneously evaporated from the traps (103). However, distribution of this inhibitor over a wide area in the field increased the number of male moths caught in pheromone traps, compared to control traps (127). Similarly, 2-methyl-*cis*-7-octadecene, a potent inhibitor of male attraction for the gypsy moth (113), was not effective for suppressing mating in the field (128). Furthermore, *cis*-7-dodecen-1-ol, a potent inhibitor of the cabbage looper sex pheromone, appeared to have little potential for behavioral control of this insect in the field (114) (129).

Next, we consider examples of the behavior-modifying effects of the individual components of multicomponent sex-pheromone systems. The first finding of the potential usefulness of single components of multicomponent sex pheromones as communication-disruption agents arose from observations of the behavior of the male of the smaller tea tortrix moths under the presence of or preexposure to the individual components (125). Preexposure of males to either of the two components remarkably suppressed their sexual excitment in subsequent exposures to the sex pheromone. There was remarkable suppression of mating also observed when the male moths were exposed to either of the two pheromonal components or their *trans*-isomers. However, the *trans*-isomers were far less effective as mating inhibitors (126). A similar effect of individual pheromonal components on sexual behavior has been demonstrated in the armyworm moth, *S. litura* (130). Both the major and the minor component of the sex pheromone of the armyworm moth effectively repressed the attractiveness of virgin female moths in the field when the component was individually attached to virgin-female-baited traps. Furthermore, both components remarkably suppressed the mating of virgin female moths in the field when the components were individually evaporated over the area in which the females were tethered. We should pay attention to the fact that the pheromonal components were much more effective than the total sex pheromone for suppression of mating, and the minor component was more effective than the major one.

Evaporation of one pheromonal component, especially the minor one, must bring about a great change in the pheromonal signal to the male antennae from that typical of the pheromone released by natural females. The greater disruptive effect of each pheromonal component compared to the pheromone indicates that the multicomponent system is relatively susceptible to our manipulation. Considerable future effort is necessary to explore these possible weak points in multicomponent pheromone systems in insect pests. Utilization of a pheromonal component instead of the sex pheromone itself has opened a new path in the practical application of insect sex pheromones in pest-management strategies.

## V.  CONCLUSIONS

We should recognize the importance of mixtures in the chemical messengers controlling insect behavior. Field screening of various synthetic compounds is an interesting approach for the discovery of substances, because all the sex-attractant pheromones of Lepidoptera and Diptera are of a simple structure, having aliphatic skeletons. Special attention, however, should be directed to various mixtures of the synthetic compounds. Multicomponent systems in sex pheromones are rather beneficial for us from the viewpoint of behavioral control of insect pests, because the pheromonal signals depend not only on the type of compounds but also on the relative quantity of each component, and can be modified relatively easily. Artificial modification of the chemical signal of a multicomponent pheromone system will become an important component of new tactics for pest management. Further effort should be concentrated to find all the components of multicomponent chemical messengers and to clarify the detailed functions of each component and the total mixture. From these efforts, we can find possible weak points in the chemical communication systems in insect behavior. It is very important to know the details of each chemical signal and to simulate it exactly, because an insect species can develop an ability to discriminate minor differences, if any, between an artificial signal and a natural pheromone. Continuous use of an incorrect synthetic signal might bring about a development of insect "resistance" to the synthetic signal.

## REFERENCES

(1)  W.L. Roelofs and R.T. Cardé, 1971. Hydrocarbon sex pheromone in tiger moths (Arctiidae). *Science* **171**:684-686.

(2)  B.A. Bierl, M. Beroza, and C.W. Collier. 1970. Potent sex attractant of the gypsy moth: Its isolation, identification, and synthesis. *Science* **170**:87-89.

(3)  M. Beroza, B.A. Bierl, and H.R. Moffit. 1974. Sex pheromones: (E,E)-8, 10-dodecadien-1-ol in the codling moth. *Science* **183**:89-90.

(4)  W. Roelofs, A. Comeau, A. Hill, and G. Milicevic. 1971. Sex attractant of the codling moth: Characterization with electroantennogram technique. *Science* **174**:297-299.

(5)  A.S. Hill and W.L. Roelofs. 1975. Sex pheromone components of the omnivorous leafroller moth, *Platynota stultana*. *J. Chem. Ecol.* **1**:91-99.

(6)  A. Hill, R.T. Cardé, A. Comeau, W. Bode, and W. Roelofs. 1974. Sex pheromones of the tufted apple bud moth (*Platynota idaeusalis*). *Environ. Entomol.* **3**:249-252.

(7)  A. Butenandt, R. Beckman, D. Stamm, and E. Hecker. 1959. Über den Sexual-lockstoff des Seidenspinners *Bombyx mori*. Reidarstellung und Konstitution. *Z. Naturforsch.* **B14**:283-284.

(8)  W.L. Roelofs, A.S. Hill, R.T. Cardé, and T.C. Baker. 1974. Two sex pheromone components of the tobacco budworm moth, *Heliothis virescens*. *Life Sci.* **14**:1555-1562.

(9)  A.S. Hill, R.T. Cardé, H. Kido, and W.L. Roelofs. 1975. Sex pheromone of the orange tortrix moth, *Argyrotaenia citrana*. *J. Chem. Ecol.* **1**:215-224.

(10)  J. Weatherston, W. Roelofs, A. Comeau, and C. Sanders. 1971. Studies of physiologically active arthropod secretions. X. Sex pheromone of the eastern spruce budworm, *Choristoneura fumiferana* (Lepidoptera: Tortricidae). *Can. Entomol.* **103**:1741-1747.

(11)  R.G. Smith, G.E. Daterman, and G.D. Daves, Jr. 1975. Douglas-fir tussock moth: Sex pheromone identification and synthesis. *Science* **188**:63-64.

(12)  H.E. Henderson and F.L. Warren. 1972. Sex pheromones. *cis*-Dec-5-en-1-yl 3-methyl butanoate as the pheromone from the pine emperor moth *(Nudaurelia cytherea cytherea* Fabr.). *Chem. Comm.* **18**:686-687.

(13)  W. Roelofs, A. Hill, R. Cardé, J. Tette, H. Madsen, and J. Vakenti. 1974. Sex pheromones of the fruitree leafroller moth, *Archips argyrospilus*. *Environ. Entomol.* **3**:747-751.

(14)  W. Roelofs, A. Hill, and R. Cardé. 1975. Sex pheromone components of the redbanded leafroller, *Argyrotaenia velutinana* (Lepidoptera: Tortricidae). *J. Chem. Ecol.* **1**:83-89.

(15) R.S. Berger and T.D. Canerday. 1968. Specificity of the cabbage looper sex attractant. *J. Econ. Entomol.* **61**:452-454.

(16) H.H. Shorey, L.K. Gaston, and J.S. Roberts. 1965. Sex pheromones of noctuid moths. VI. Absence of behavioral specificity for the female sex pheromones of *Trichoplusia ni* versus *Autographa californica*, and *Heliothis zea* versus *H. virescens* (Lepidoptera: Noctuidae). *Ann. Entomol. Soc. Am.* **58**:600-603.

(17) J.H. Tumlinson, E.R. Mitchell, S.M. Browner, and D.A. Lindquist. 1972. A sex pheromone for the soybean looper. *Environ. Entomol.* **1**:466-468.

(18) J.S. Read, F.L. Warren and P.H. Hewitt. 1968. Identification of the sex pheromone of the false codling moth *(Argyroploce leucotreta)*. *Chem. Comm.* **14**:792-793.

(19) J.S. Read, P.H. Hewitt, F.L. Warren and A.C. Myberg. 1974. Isolation of the sex pheromone of the moth *Argyroploce leucotreta*. *J. Insect Physiol.* **20**:441-450.

(20) W.L. Roelofs, A. Comeau and R. Selle. 1969. Sex pheromone of the oriental fruit moth. *Nature* **224**:723.

(21) W.L. Roelofs, J.P. Tette, E.F. Taschenberg and A. Comeau. 1971. Sex pheromone of the grape berry moth: Identification by classical and electroantennogram methods, and field tests. *J. Insect Physiol.* **17**:2235-2243.

(22) G.G. Smith, G.E. Daterman, G.D. Daves, K.D. McMurtrey and W.L. Roelofs. 1974. Sex pheromone of the European pine shoot moth: Chemical identification and field tests. *J. Insect Physiol.* **20**:661-668.

(23) B.F. Nesbitt, P.S. Beevor, R.A. Cole, R. Lester and R.G. Poppi. 1973. Sex pheromones of two noctuid moths. *Nature (New Biol.)* **244**:208-209.

(24) Y. Tamaki, H. Noguchi, T. Yushima and C. Hirano. 1971. Two sex pheromones of the smaller tea tortrix: Isolation, identification, and synthesis. *Appl. Entomol. Zool.* **6**:139-141.

(25) G.M. Meijer, F.J. Ritter, C.J. Persoons, A.K. Minks and S. Voerman. 1972. Sex pheromones of the summer fruit tortrix moth *Adoxophyes orana:* Two synergistic isomers. *Science* **175**:1469-1470.

(26) U.E. Brady. 1973. Isolation, identification and stimulatory activity of a second component of the sex pheromone system (complex) of the female almond moth, *Cadra cautella* (Walker). *Life Sci.* **13**:227-235.

(27) M. Jacobson, R.D. Redfern, W.A. Jones and M.H. Aldridge. 1970. Sex pheromones of the southern armyworm moth: Isolation identification, and synthesis. *Science* **170**:542-544.

(28) A.A. Sekul and A.N. Sparks. 1967. Sex pheromone of the fall armyworm moth: Isolation, identification and synthesis. *J. Econ. Entomol.* **60**:1270-1272.

(29) L.B. Hendry, J. Jugovich, L. Roman, M.E. Anderson and R.O. Mumma. 1974. *cis*-10-Tetradecenyl acetate, an attractant component in the sex pheromone of the oak leaf roller moth *(Archips semiferanus* Walker). *Experientia* **30**:886-888.

(30) C.J. Persoons, A.K. Minks, S. Voerman, W.L. Roelofs and F.J. Ritter. 1974. Sex pheromones of the moth, *Archips podana*: Isolation, identification and field evaluation of two synergistic geometrical isomers. *J. Insect Physiol.* **20**:1181-1188.

(31) W.L. Roelofs and H. Arn. 1968. Sex attractant of the red-banded leafroller moth. *Nature* **219**:513.

(32) W.L. Roelofs and J. Tette. 1970. Sex pheromone of the oblique-banded leaf roller. *Nature* **226**:1172.

(33) J.A. Klun and T.A. Brindley. 1970. *cis*-11-Tetradecenyl acetate, a sex stimulant of the European corn borer. *J. Econ. Entomol.* **63**:779-780.

(34) J. Kochansky, R.T. Cardé, J. Liebherr and W.L. Roelofs. 1975. Sex pheromones

of the European corn borer, *Ostrinia nubilalis,* in New York. *J. Chem. Ecol.* 1:225-231.

(35)  Y. Tamaki and T. Yushima. 1974. Sex pheromone of the cotton leafworm, *Spodoptera littoralis. J. Insect Physiol.* 20:1005-1014.

(36)  Y. Tamaki, H. Noguchi and T. Yushima. 1973. Sex pheromone of *Spodoptera litura* (F.) (Lepidoptera: Noctuidae): Isolation, identification, and synthesis. *Appl. Entomol. Zool.* 8:200-203.

(37)  Y. Kuwahara, H. Hara, S. Ishii and H. Fukami. 1971. The sex pheromone of the Mediterranean flour moth. *Agric. Biol. Chem.* 35:447-448.

(38)  U.E. Brady, J.H. Tumlinson, III, R.G. Brownlee and R.M. Silverstein. 1971. Sex stimulant and attractant in the Indian meal moth and in the almond moth. *Science* 171:802-804.

(39)  Y. Kuwahara, C. Kitamura, S. Takahashi, H. Hara, S. Ishii and H. Fukami. 1971. Sex pheromone of the almond moth and Indian meal moth: *cis*-9, *trans*-12-tetradecadienyl acetate. *Science* 171:801-802.

(40)  U.E. Brady and R.C. Caley. 1972. Identification of a sex pheromone from the female raisin moth, *Cadra figulilella. Ann. Entomol. Soc. Am.* 65:1356-1358.

(41)  U.E. Brady and D.A. Nordlund. 1971. *cis*-9, *trans*-12-Tetradecadien-1-y1 acetate in the female tobacco moth, *Ephestia elutella* (Hübner) and evidence for an additional component of the sex pheromone. *Life Sci.* 10:797-801.

(42)  U.E. Brady and M.C. Ganyard. Identification of a sex pheromone of the female beet armyworm, *Spodoptera exigua. Ann. Entomol. Soc. Am.* 65:898-899.

(43)  B.A. Bierl, M. Beroza, R.T. Staten, P.E. Sonnet and V.E. Adler. 1974. The pink bollworm sex attractant. *J. Econ. Entomol.* 67:211-216.

(44)  H.E. Hummel, L.K. Gaston, H.H. Shorey, R.S. Kaae, K.J. Byrne and R.M. Silverstein. 1973. Clarification of the chemical status of the pink bollworm sex pheromone. *Science* 181:873-875.

(45)  K.W. Vick, H.C.F. Su, L.L. Sower, P.G. Mahany and P.C. Drummond. 1974. (Z,E)-7,11-Hexadecadien-1-o1 acetate: The sex pheromone of the angoumois grain moth, *Sitotroga cerealella. Experientia* 30:17-18.

(46)  J.H. Tumlinson, C.G. Yonce, R.E. Doolittle and R.R. Heath. 1974. Sex pheromones and reproductive isolation of the lesser peachtree borer and the peachtree borer. *Science* 185:614-616.

(47)  W.L. Roelofs and A. Comeau. 1970. Lepidopterous sex attractants discovered by field screening tests. *J. Econ. Entomol.* 63:969-974.

(48)  W.L. Roelofs and A. Comeau. 1971. Sex attractants in Lepidoptera. *Proc. 2nd Int. Congr. Pesticide Chem., IUPAC,* Tel Aviv, pp. 91-114.

(49)  A.K. Minks and S. Voerman. 1973. Sex pheromones of the summerfruit tortrix moth, *Adoxophyes orana:* Trapping performance in the field, *Entomol. exp. appl.* 16:541-549.

(50)  E.O. Hamstead, R.E. Dolphin and W. Roelofs. 1972. Field test of virgin female and synthetic sex-lure traps for male red-banded leaf rollers. *Environ. Entomol.* 1:488-489.

(51)  J.A. Klun and J.F. Robinson. 1971. European corn borer moth: Sex attractant and sex attraction inhibitors. *Ann. Entomol. Soc. Am.* 64:1083-1086.

(52)  T. Yushima, Y. Tamaki, S. Kamano and M. Oyama. 1974. Field evaluation of a synthetic sex pheromone, "litlure", as an attractant for males of *Spodoptera litura*(F.) (Lepidoptera: Noctuidae). *Appl. Entomol. Zool.* 9:147-152.

(53)  M. Beroza, A.A. Punjapi and B.A. Bierl. 1973. Disparlure and analogues as attractants for *Lymantria obfuscata. J. Econ. Entomol.* 66:1215-1216.

(54)    M. Beroza, K. Katagiri, Z. Iwata, H. Ishizuka, S. Suzuki and B.A. Bierl. 1973.
        Disparlure and analogues as attractants for two Japanese Lymantriid moths.
        *Environ. Entomol.* 2:966.

(55)    J. Weatherston, L.M. Davidson and D. Simonini. 1974. Attractants for several
        male forest Lepidoptera. *Can. Entomol.* 107:781-782.

(56)    W.L. Roelofs and R.T. Cardé. 1974. Sex pheromones in the reproductive
        isolation of lepidopterous species. In *Pheromones*. M.C. Birch, Ed. North-
        Holland, Amsterdam, pp. 96-114.

(57)    C.J. Sanders, G.E. Daterman, R.F. Sheppard and H. Cerezke. 1973. Sex
        attractants for two species of western spruce budworm, *Choristoneura biennis*
        and *C. viridis* (Lepidoptera: Tortricidae). *Can. Entomol.* 106:157-159.

(58)    R.H. Kaae, H.H. Shorey, S.U. McFarland and L.K. Gaston. 1973. Sex
        pheromones of Lepidoptera. XXXVII. Role of sex pheromones and other factors
        in reproductive isolation among ten species of Noctuidae. *Ann. Entomol. Soc.
        Am.* 66:444-448.

(59)    T. Ando, S. Yoshida, N. Takahashi and S. Tatsuki. 1975. Studies on sex
        attractants of Lepidoptera. Part 1. *Proc. Ann. Meeting Jap. Soc. Appl. Entomol.
        Zool.*, Machida, Tokyo, p. 109.

(60)    W.L. Roelofs, J. Kochansky, R. Cardé, H. Arn and S. Ranscher. 1973. Sex
        attractant of the grape vine moth, *Lobesia botrana. Mitteil. Schweiz. Entomol.
        Ges.* 46:71-73.

(61)    K. Honma. 1973. Personal communication.

(62)    A.K. Minks, W.L. Roelofs, F.J. Ritter and C.J. Persoons. 1973. Reproductive
        isolation of two tortricid moth species by different ratios of a two-component
        sex attractant. *Science* 180:1073-1074.

(63)    A. Comeau and W.L. Roelofs. 1973. Sex attraction specificity in the Tortricidae.
        *Entomol. exp. appl.* 16:191-200.

(64)    J.A. Klun and J.F. Robinson. 1972. Olfactory discrimination in the European
        corn borer and several pheromonally analogous moths. *Ann. Entomol. Soc. Am.*
        65:1337-1340.

(65)    W.L. Roelofs, R. Cardé, G. Benz and G. von Salis. 1971. Sex attractant of the
        larch    bud    moth    found    by    electroantennogram    method.    *Experientia*
        27:1438-1439.

(66)    N. Green, M. Jacobson and J.C. Keller. 1969. Hexalure, an insect sex attractant
        discovered by empirical screening. *Experientia* 25:682-683.

(67)    Y. Tamaki. Unpublished results.

(68)    J.H. Tumlinson. 1974. Personal communication.

(69)    D.G. Nielsen, F.F. Purrington, J.H. Tumlinson, R.E. Doolittle and C.E. Yonce.
        1975. Response of male clearwing moths to caged virgin females, female extracts,
        and synthetic sex attractants. *Environ. Entomol.* 4:451-454.

(70)    R.L. Leyrer and R.E. Monroe. 1973. Isolation and identification of the scent of
        the moth, *Galleria mellonella,* and a revaluation of its sex pheromone. *J. Insect
        Physiol.* 19:2267-2271.

(71)    H. Röller, K. Biemann, J.S. Bierke, D.W. Norgard and W.H. McShan. 1968. Sex
        pheromones of pyralid moths. I. Isolation and identification of the sex-attractant
        of *Galleria mellonella* L. (greater wax moth). *Acta Entomol. Bohemoslov.*
        65:208-211.

(72)    K.H. Dahm, D. Meyer, W.E. Finn, V. Reinhold and H. Röller. 1971. The
        olfactory and auditory mediated sex attraction in *Achrois grissella* (Fabr.).
        *Naturwissenschaften* 58:265-266.

(73)   M.C. Birch. 1972. Male abdominal brush-organs in British noctuid moths and their value as a taxonomic character. Part I: *Entomologist* 105:185-205; Part II: *Entomologist* 105:233-244.

(74)   R.T. Aplin and M.C. Birch. 1972. Identification of odorous compounds from male Lepidoptera. *Experientia* 26:1193-1194.

(75)   J.K. Clearwater. 1972. Chemistry and function of a pheromone produced by the male of the southerm armyworm *Pseudaletia separata*. *J. Insect Physiol.* 18:781-789.

(76)   G.G. Grant, U.E. Brady and J.M. Brand. 1972. Male army worm scent brush secretion: Identification and electroantennogram study of major components. *Ann. Entomol. Soc. Am.* 65:1224-1227.

(77)   J. Meinwald, Y.C. Meinwald and P.H. Mazzocchi. 1969. Sex pheromone of the queen butterfly: chemistry. *Science* 164:1174-1175.

(78)   J.A. Edgar, C.C.J. Culvenor and L.W. Smith. 1971. Dihydropyrrolizine derivatives in the "hair-pencil" secretions of danaiid butterflies. *Experientia* 27:761-762.

(79)   J. Meinwald and Y.C. Meinwald. 1966. Structure and synthesis of the major components in the hairpencil secretion of a male butterfly, *Lycorea ceres ceres* (Cramer). *J. Am. Chem. Soc.* 88:1305-1310.

(80)   J. Meinwald, A.M. Chalmer, T.E. Pliske and T. Eisner. 1968. Pheromones. III. Identification of *trans,trans*-10-hydroxy-3,7-dimethyl- 2,6-decadienoic acid as a major component in "hairpencil" secretion of the male monarch butterfly. *Tetrahedron Lett.* 4893-4896.

(81)   J. Meinwald, A.M. Chalmers, T.E. Pliske and T. Eisner, 1969. Identification and synthesis of *trans,trans*-3,7-dimethyl-2,6-decadien-1,10-dioic acid, a component of the pheromonal secretion of the male monarch butterfly. *Chem. Comm.* 15:86-87.

(82)   J. Meinwald, W.R. Thompson, T. Eisner and D.F. Owen. 1971. Pheromones. VII. African monarch: Major components of the hairpencil secretion. *Tetrahedron Lett.* 38:3485-3488.

(83)   J. Meinwald, Y.C. Meinwald, J.W. Wheeler, T. Eisner and L.P. Brower. 1966. Major components in the exocrine secretion of a male butterfly *(Lycorea)*. *Science* 151:583-585.

(84)   J. Meinwald, C.J. Boriak, D. Schneider, M. Boppre, W.F. Wood and T. Eisner. 1974. Volatile ketones in the hairpencil secretion of danaiid butterflies *(Amauris* and *Danaus)*. *Experientia* 30:721-723.

(85)   T.E. Pliske and T. Eisner. 1969. Sex pheromone of the queen butterfly: biology. *Science* 164:1170-1172.

(86)   D. Schneider and U. Seibt. 1969. Sex pheromone of the queen butterfly: Electroantennogram responses. *Science* 164:1173-1174.

(87)   D.A. Carlson, M.S. Mayer, D.L. Silhacek, J.P. James, M. Beroza and B.A. Bierl. 1971. Sex attractant pheromone of the house fly: Isolation, identification and synthesis. *Science* 174:76-78.

(88)   L.L. Jackson. 1970. Cuticular lipids of insects: II. Hydrocarbons of the cockroaches *Periplaneta australiasiae, Periplaneta brunea* and *Periplaneta fuliginosa*. *Lipids* 5:38-41.

(89)   A. Mansingh, R.W. Steele, B.N. Smallman, O. Meresz and C. Mozogai. 1972. Pheromone effects of *cis*-9 long chain alkenes on the common housefly—An improved sex attractant combination. *Can. Entomol.* 104:1963-1965.

(90)   M. Jacobson, K. Ohinata, D.L. Chambers, W.A. Jones and M.S. Fujimoto. 1973. Insect sex attractants. 13. Isolation, identification, and synthesis of sex pheromones of the male Mediterranean fruit fly. *J. Med. Chem.* 16:248-251.

(91)   J.L. Nation. 1975. The sex pheromone blend of Caribbean fruit fly males: Isolation, biological activity, and partial chemical characterization. *Environ. Entomol.* 4:27-30.

(92)   B.S. Fletcher. 1974. Personal communication.

(93)   M. Beroza. 1970. Current usage and some recent developments with insect attractants and repellents in the USDA. In *Chemicals Controlling Insect Behavior.* M. Beroza, Ed. Academic, New York, pp. 145-163.

(94)   L.K. Gaston, T.L. Payne, S. Takahashi and H.H. Shorey. 1971. Correlation of chemical structure and sex pheromone activity in *Trichoplusia ni* (Noctuidae). *Olfaction and Taste* IV:167-173.

(95)   V.E. Adler, M. Beroza, B.A. Bierl and R. Sarmiento. 1972. Electroantennograms and field attraction of the gypsy moth sex attractant disparlure and related compounds. *J. Econ. Entomol.* 65:679-681.

(96)   M. Jacobson, H.H. Toba, J. DeBolt and A.N. Kishaba. 1968. Insect sex attractants. VIII. Structure activity relationships in sex attractant for male cabbage loopers. *J. Econ. Entomol.* 61:84-85.

(97)   H.H. Toba, A.N. Kishaba and W.W. Wolf. 1968. Bioassay of the synthetic female sex pheromone of the cabbage looper. *J. Econ. Entomol.* 61:812-816.

(98)   B.A. Butt, M. Beroza, T.P. McGovern and S.K. Freeman. 1968. Synthetic chemical sex stimulants for the codling moth. *J. Econ. Entomol.* 61:570-572.

(99)   L.M. McDonough, D.A. George and B.A. Butt. 1969. Artificial sex pheromones for the codling moth. *J. Econ. Entomol.* 62:243-244.

(100)  R. Sarmiento, M. Beroza, B.A. Bierl and J.G.R. Tardiff. 1972. Activity of compounds related to disparlure, the sex attractant of the gypsy moth. *J. Econ. Entomol.* 65:665-667.

(101)  S. Takahashi, C. Kitamura and Y. Kuwahara. 1971. The comparative pheromone activity of acetates of unsaturated alcohol to males of the almond moth. *Botyu-Kagaku* 36:24-26.

(102)  W.L. Roelofs and R.T. Cardé. 1974. Oriental fruit moth and lesser apple worm attractant mixtures refined. *Environ. Entomol.* 3:586-588.

(103)  W.L. Roelofs, R.T. Cardé and J.P. Tette. 1973. Oriental fruit moth attractant synergists. *Environ. Entomol.* 2:252-254.

(104)  J.A. Klun, O.L. Chapman, K.C. Mattes, P.W. Wojrkowski, M. Beroza and P.E. Sonnet. 1973. Insect sex pheromones: Minor amount of opposite geometrical isomer critical to attraction. *Science* 181:661-663.

(105)  W.L. Roelofs and A. Comeau. 1971. Sex pheromone perception: Synergists and inhibitors for the red-banded leaf roller attractant. *J. Insect Physiol.* 17:435-448.

(106)  M. Beroza, G.M. Muschik and C.R. Gentry. 1973. Small proportion of opposite geometrical isomer increases potency of synthetic pheromone of oriental fruit moth. *Nature (New Biol.)* 244:149-150.

(107)  M. Beroza, C.R. Gentry, J.L. Blythe and G.M. Muschik. 1973. Isomer content and other factors influencing captures of oriental fruit moth by synthetic pheromone traps. *J. Econ. Entomol.* 66:1307-1311.

(108)  S. Voerman and A.K. Minks. 1973. Sex pheromones of summerfruit tortrix moth, *Adoxophyes orana.* 2. Compounds influencing their attractant activity. *Environ. Entomol.* 2:751-756.

(109)  S. Voerman, A.K. Minks and N.W.H. Houx. 1974. Sex pheromones of summerfruit tortrix moth, *Adoxophyes orana:* Investigation on compounds modifying their attractancy. *Environ. Entomol.* 3:701-704.

(110)  L.L. Sower, K.W. Vick and J.H. Tumlinson. 1974. (Z,E)-9,12-Tetradecadien-1-ol: A chemical released by female *Plodia interpunctella* that inhibits the sex

pheromone response of male *Cadra cautella*. *Environ. Entomol.* 3:120-122.

(111)    H. Arn, C. Schwarz, H. Limacher and E. Mani. 1974. Sex attractant inhibitors of the codling moth, *Laspeyresia pomonella* L. *Experientia* 30:1142-1144.

(112)    D.O. Hathaway, T.P. McGovern, M. Beroza, H.R. Moffitt, L.M. McDonough and B.A. Butt. 1974. An inhibitor of sexual attraction of male codling moths to a synthetic sex pheromone and virgin females in traps. *Environ. Entomol.* 3:522-524.

(113)    R.T. Cardé, W.L. Roelofs and C.C. Doane. 1973. Natural inhibitor of the gypsy moth sex attractant. *Nature* 241:474-475.

(114)    J.H. Tumlinson *et al*. 1972. cis-7-Dodecen-1-ol, a potent inhibitor of the cabbage looper sex pheromone. *Environ. Entomol.* 3:354-358.

(115)    V.G. Dethier. 1947. *Chemical Insect Attractants and Repellents*. Blakiston, Philadelphia.

(116)    R.H. Wright. 1964. After pesticides – What? *Nature* 204:121-125.

(117)    Y. Tamaki, H. Noguchi, T. Yushima, C. Hirano, K. Honma and H. Sugawara. 1971. Sex pheromone of the summerfruit tortrix moth: Isolation and identification. *Kontyu* 39:338-340.

(118)    S. Neumark, M. Jacobson and I. Teich. 1970. Field evaluation of the four synthetic components of the sex pheromone of *Spodoptera littoralis* and their improvement with an antioxidant. *Environ. Lett.* 6:219-230.

(119)    L.B. Hendry, M.E. Anderson, J. Jugovich, R.O. Mumma, D. Robacker and Z. Kosarych. 1975. Sex pheromone of the oak leaf roller: A complex chemical messenger system identified by mass fragmentography. *Science* 187:355-357.

(120)    Y. Tamaki. 1972. Insect sex pheromones and species speciation. *Seibutsu-Kagaku* 24:119-129.

(121)    R.E. Redfern, E. Cantu, W.A. Jones and M. Jacobson. 1971. Response of the male southern armyworm in a field cage to prodenialure A and prodenialure B. *J. Econ. Entomol.* 64:1570-1571.

(122)    K. Nakamura. 1975. Functions of pheromonal components of *Spodoptera litura* in the orientation behavior of male moths. *Proc. Ann. Meeting Jap. Soc. Appl. Entomol. Zool.,* Machida, Tokyo, p. 114.

(123)    W.L. Roelofs and A. Comeau. 1969. Sex pheromone specificity: Taxonomic and evolutionary aspects in Lepidoptera. *Science* 165:398-400.

(124)    R.S. Kaae, J.R. McLaughlin, H.H. Shorey and L.K. Gaston. 1972. Sex pheromones of Lepidoptera. XXXII. Disruption of intraspecific pheromonal communication in various species of Lepidoptera by permeation of the air with looplure or hexalure. *Environ. Entomol.* 1:651-653.

(125)    Y. Hirai, Y. Tamaki and T. Yushima. 1974. Inhibitory effects of individual components of the sex pheromone of a tortricid moth on the sexual stimulation of males. *Nature* 247:231-232.

(126)    Y. Tamaki, T. Ishiwatari and M. Osakabe. 1975. Inhibition of sexual behavior of the smaller tea tortrix moth by the sex pheromone and its components. *Jap. J. Appl. Entomol. Zool.* 19:187-192.

(127)    G.H.L. Rothschild. 1974. Problems in defining synergists and inhibitors of the oriental fruit moth pheromone by field experimentation. *Entomol. exp. appl.* 17:294-302.

(128)    M. Beroza. 1973. Pheromones and related chemicals in some USDA insect control programs. *III Int. Symp. Chem. Toxicol. Aspects on Environ. Quality,* Tokyo.

(129)    R.S. Kaae, H.H. Shorey, L.K. Gaston and H.H. Hummel. 1974. Sex pheromones

of Lepidoptera: Disruption of pheromone communication in *Trichoplusia ni* and *Pectinophora gossypiella* by permeation of the air with nonpheromone chemicals. *Environ. Entomol.* 3:87-89.

(130)   T. Yushima, Y. Tamaki, S. Kamano and M. Oyama. 1975. Suppression of mating of the armyworm moth, *Spodoptera litura* (F.), by a component of its sex pheromone. *Appl. Entomol. Zool.* 10:237-239.

(131)   R.S. Berger. 1966. Isolation, identification, and synthesis of the sex attractant of the cabbage looper, *Trichoplusia ni. Ann. Entomol. Soc. Am.* 59:767-771.

## CHAPTER 16

# An Overview — The Evolving Philosophies and Methodologies of Pheromone Chemistry

Wendell L. Roelofs

*New York State Agricultural Experiment Station*
*Geneva, New York*

The general phenomenon of insects being attracted by the "smells" of their mates has been observed for more than a century. Until almost a decade ago studies on this behavior were limited because of a lack of information on the identity of the chemicals involved. The excellent discussions in Chapters 14 and 15 on the complexity and diversity of behavior-modifying chemicals reveal the tremendous amount of information that has accumulated on the chemistry of pheromones in the past decade. So many compounds have now been identified that the detailed information on insect behavior is found to be lagging. The ultimate goal, of course, is to describe the whole mating communication system qualitatively and quantitatively. Progress toward this goal has been interesting but slow because of the many variables and subtleties that exist in naturally occurring systems.

## I. AN HISTORICAL PERSPECTIVE

The differences in pheromone response patterns among insect species have allowed the development of many types of laboratory bioassays based on selected portions of the behavioral repertoire, such as stimulation, orientation, walking, flight, and antennal responses. Each method of assay has advantages that make it useful for certain insects or for assaying compounds effecting certain types of behavior, but each method is limited because all variables of nature and all subtleties of behavior are not perfectly duplicated in any single method.

What, then, constitutes a pheromone identification if one is never certain that the whole message is described and if the behavioral consequences are difficult to elucidate? A synthetic compound can be classified as a pheromone component if it is: isolated from the emitting organism or its effluvia; structurally defined; and shown to elicit behavioral responses in the receiving organism similar to those observed with naturally occurring pheromone components. As shown in Chapters 14 and 15, many compounds have properly been accepted as pheromone components, but many others are lacking in at least one of the criteria. The latter compounds are not labelled *pheromone components,* but in many instances they do provide key information about the chemical communication system. The importance of this information has been a matter of dispute throughout the past decade, and certain value judgements have been placed on some of the techniques used. Sometimes the sophistication of the method and the complexity of the chemical structures obscure the fact that for purposes of defining the communication system it is as important to know what behavioral role a very simple compound fulfills as it is to know the role of a complex chemical.

Knowledge of lepidopteran and coleopteran pheromones has apparently evolved to about the same levels, although some reports would lead one to believe that this knowledge accrued through different routes—the former through serendipity and arbitrary field screening, and the latter by ingenuity, experimentation, and elegance. It is my opinion that the goals were always the same and that the means to these goals were dealt with in ways that are philosophically similar. The next few sections will discuss my viewpoints on that conclusion.

## A. The One-Compound/One-Species Concept of the 1960s

Chapters 14 and 15 make it quite evident that pheromonal blends are a general phenomenon in the insect orders investigated. Many multicomponent pheromone systems have been described recently in the Lepidoptera and insect behavioral studies have progressed to the point where one can describe the behavior elicited by the individual components (1). However, it has been implied that during the 1960s the "moth" people were operating under the assumption that there was only one pheromone compound for each species. The literature might give the impression that this concept was generally accepted because the first two moth pheromone identifications, one from the silkworm, *Bombyx mori,* in 1961 (2) and the other from the cabbage looper, *Trichoplusia ni,* in 1966 (3), were found to be monocomponent pheromonal systems. Also, the pheromone of two major pest species—the gypsy moth, *Porthetria dispar,* in 1970 (4), and the codling moth, *Laspeyresia pomonella,* in 1971 (5)—subsequently were found to be single compounds. This does not necessarily mean that researchers subsequently wore blinders that influenced them to search for one compound only. Even today, when one expects to find blends, these four pheromones are still considered to be single-component systems.

The same implication could have been leveled at the "beetle" people if the single-component pheromones of the black carpet beetle, *Attagenus megatoma*, (6) and the ambrosia beetle, *Gnathotrichus sulcatus,* (7) had been the first of the pheromones identified for the Coleoptera.

To ascertain whether or not blinders were used, one must consider the intervening lepidopteran pheromone identifications that were reported as single compounds and later found to be mixtures. A critical reading of the literature shows that other compounds were implied in every case. The identification of a compound for the red-banded leaf roller moth, *Argyrotaenia velutinana*, (8) in 1968 was immediately followed by a report in 1968 (9) that revealed the "synergistic" effect of a second compound, dodecyl acetate. Although the second compound was not identified from the females at that time (10), the suggestion was made that other compounds could well be involved in the pheromone system to increase attractiveness of the single component. Anyone attending pheromone sessions of the national meetings of the Entomological Society of America in 1968 and thereafter should remember the vigorous support some investigators gave for multicomponent pheromonal systems in Lepidoptera. Further support for the use of other components was soon obtained when the oblique-banded leaf roller moth, *Choristoneura rosaceana,* pheromone was found to be the same as that of the red-banded leaf roller (11). Since dodecyl acetate in this case was inhibitory and the females of the two species were not cross-attractive, it was stated that "the emission of different secondary compounds by females of each species along with the sex attractant could effect reproductive isolation in the field for these two species". A similar conclusion was reached in 1969 (12) from the report of the Oriental fruit moth, *Grapholitha molesta,* pheromone. Although only one compound was found at that time, observations on this compound and on live females of the *G. molesta* and a sibling species "suggested the possible role of secondary chemicals." At the same time, research with a number of Arctiidae species revealed that some species were attracted to a single compound, but others were not (12). Again the suggestion was made that "specificity could be effected by the use of secondary chemicals or by differences in the release of the primary attractant, or both." The feeling for the use of pheromonal mixtures was strong enough that field-screening programs in 1969 and 1970 included many binary sets of chemicals (14). Strong support for the use of mixtures in Lepidoptera was also presented in 1969 by investigators working on the red bollworm (15).

## B. The Evolution of Methodology

A rigorous study of the pheromones with all combinations of fractions tested with an appropriate bioassay appears to be the best and perhaps the only way of elucidating the pheromone system of the *Ips* bark beetles (16).

Preset rigorous approaches applied to variable natural systems, however, are sometimes limiting. This was discovered when the same approach applied to a pheromone study of *Dendroctonus* bark beetles did not reveal all the necessary pheromone components. The mixture of components finally used in field studies utilized: only one of the two originally identified compounds (17), *exo*-brevicomin, from females; a compound, myrcene, incorporated in the beetles from the host tree (18); and a compound from the males, frontalin (19). Hindsight shows that perhaps the best approach originally to define this mixture would have been to take the total effluvia from a tree under attack and assay it in the natural environment.

The same assessment can be made on determining the pheromone blends of some lepidopteran species. With the red-banded leaf roller moth the first component, *cis*-11-tetradecenyl acetate, could be isolated and identified with the aid of any number of assay methods, such as electroantennograms or laboratory behavioral tests. The dodecyl acetate component, however, is interesting because it is extracted in extremely low quantities from the gland (10), and we have not been able to devise an appropriate laboratory behavioral assay to reveal activity for this compound—even though field observations have indicated the natural role of this compound in the communication system. Although the compound was first found by field testing various mixtures, hindsight now shows that the approach for defining this component in the female would be to trap female effluvia and assay by the electroantennogram technique, or possibly to test recombinations in the field. The electroantennogram can be extremely useful in assaying for some lepidopteran pheromone components without having to recombine fractions and with low numbers of males required, whereas it may not be useful at all when dealing with species in other insect orders in which the adults respond to a multitude of chemical stimuli for various behavioral responses.

The discovery (20) that chemicals previously found to be inhibitory at ratios of 1:1 to the attractant can be obligatory pheromone components at other ratios—such as 1:20—demonstrated the need to analyze pheromone components for small percentages of the geometric isomer. The knowledge that compounds can effect totally different behaviors at different ratios, coupled with the development of better gas chromatographic columns, has made this type of analysis a routine part of pheromone identifications.

As the methodology continues to improve, we find that our knowledge of communication systems continues to increase and new aspects of the system are revealed that require new methodologies for chemical definition.

## C. Field Tests

Simple techniques can sometimes be judged to be inferior to sophisticated ones, yet some of our great naturalists would have neither appreciated nor deserved such judgment on their simple techniques, which led to profound observations. In pheromone research such value judgements have sometimes depreciated the "field-screening" technique, which has been used for aims that some critics may not fully understand. One can find a sex attractant (Chapter 15) quite easily for some species if he uses the field-screening technique; thus, this method represents a comparatively simple route to obtaining active compounds. These compounds cannot be classified as pheromones because they have not been identified from the insects, but data obtained from good field studies can add valuable information about the chemical-communication system. It can be assumed that significant field responses similar to those produced by the natural mating signals are produced by chemicals affecting the same communication system. Although the method of obtaining the information may not be elegant, the importance of the information to the chemical-communication system may be much greater than information from compounds identified from the insects by sophisticated methods but without any knowledge of their effect on behavior. In the latter category would be the series of compounds identified from male moth hairpencils (Chapter 15) and the host of glandular mixtures identified in many hymenopteran species (Chapter 14). In some respects, much of the mammalian pheromone research falls into this category since typically the approach has been to identify the major compounds from a gland and then to test them to see if they exert any activity on the behavior of the subject. The final step is a form of "field-screening," although not recognized as such in the literature.

The *field-screening* term usually has been identified with lepidopteran programs, but an analysis of other pheromone-research programs reveals that this has been a common technique with the other programs as well. In comparison, varying mixtures of monounsaturated 14-carbon acetates, which were identified as pheromones in a lepidopteran species, and finding field attractancy with one of the mixtures for another lepidopteran species, is not much different from varying a mixture of *Ips* bark beetle pheromones and finding attractancy for another *Ips* species (21). Typically the results of the field-screening studies of coleopteran researchers have been described as interesting cases of species specificity or cross-attractancy. In other modifications of the field-screening technique, but not recognized as such, are the field assessments "of all compounds implicated as pheromones" (22).

Another aspect of "field-screening" that has incorrectly been considered to be similar to that just described is the "screening" of various isomers and analogues in conjunction with the attractant chemicals in order to obtain information on the receptor site structural requirements and how the input can be modulated. In these cases, the attractant system has been elucidated already, and the field studies are concentrated on the one species. Our efforts were initiated in 1968 and presented as studies on the "sequence of receptor site affinity-intrinsic activity relationships" of closely-related structures (23).

Information generated by these studies accumulated to give some behavioral significance to various molecular modifications. This information, then, became very useful in single-cell studies (23) designed to determine how many receptor cells are used for pheromone blends and the relative sensitivities of these cells to the various compounds that affect the communication system in the field. Additionally, a practical goal in this field testing has been to obtain compounds that can effect a decrease in attractiveness and thus be used in insect-control programs.

A general hypothesis on predicting structural modifications that would affect the sensory input seems to be valid with the moth species investigated to date. Briefly, the hypothesis (25) was that compounds possessing the potential to interact with at least two of the acceptor active sites (particularly for the functional group end, double-bond position, or the hydrocarbon end of the compound) could affect the sensory input in some way. With red-banded leaf roller moths, variations of the *cis*-11-tetradecenyl acetate attractant showed that trap-catch increases were effected by compounds that were acetates with a nucleophilic atom in the 11- or 12-position and possessed the same chain-length as the attractant, such as 10-propoxydecanyl acetate; and acetates that terminate their chain length in the 11- or 12-position, such as undecyl acetate or 11-bromoundecyl acetate. Trap catch decreases were effected by converting the double bond to an epoxide; changing the functional group from an acetate to an alcohol or formate; shifting the double bond to neighboring positions; or using a high ratio of the *trans*-isomer.

Extensive studies with the *Adoxophyes orana* moth (26) also showed that compounds with the greatest effect in decreasing trap catch with the attractants, *cis*-9- and *cis*-11-tetradecenyl acetates, were those that substantiate the general hypothesis. Thus, in a series of 70 compounds in which the chain length was varied and an oxygen atom was placed in the chain from positions 0 to 12, the only compounds with much inhibitory effect were those with the oxygen in the 9- or the 12-position with an overall chain length of 13 or 14 carbons, such as 11-ethoxyundecyl acetate.

Studies on the codling moth gave similar results (27). In a series of 34 monounsaturated 12- to 14-carbon acetates and alcohols, the compounds giving the most dramatic decreases in trap catch to the attractant, *trans*-8, *trans*-10-dodecadien-1-ol, were those with a double bond in common with the attractant and a chain of 12 or 13 carbons, such as *cis*-8 and *cis*-10-dodecenyl acetates. The *cis* isomers were more effective than the *trans,* and acetates more effective than the corresponding alcohols. In another study with codling moth (28), the investigators employed various functional group derivatives of the attractant. Ethers from methyl to *t*-butyl were prepared, as well as acylates from formate to butyrate. Trap-catch decreases were quite specifically effected by the ethyl ether and the acetate, again giving information about the receptor-site requirements for this sensory input.

A number of compounds fulfilling requirements of the hypothesis were tested (29) with the Oriental fruit moth attractant, *cis*-8-dodecenyl acetate, and trap-catch decreases were effected with dodecyl acetate, whereas trap-catch increases were found with dodecyl alcohol. Subsequent research with these two compounds has provided some interesting observations about the communication system. Communication-disruption experiments (30) with the inhibitor, dodecyl acetate, have revealed that its role is complex and that permeation of the atmosphere with this compound actually causes an increase in trap catch. The role of dodecyl alcohol was determined by field observations (1) and found to operate on the close-range behavior of the males, such as eliciting landing, fanning, and hairpenciling.

## II. PHEROMONE BLENDS AND INSECT CONTROL

The accumulated knowledge of the pheromone chemistry and behavioral roles of each pheromone component has provided a basis for mating-disruption experiments. The discussion in Chapter 15 on testing individual components of multicomponent pheromone systems in disruption studies underscores the need to know the structure of all pheromone components. The availability of synthetic components makes possible further important experiments on behavior, on single-cell receptors, and on the effect of a component or a particular mixture of components in adapting, habituating, or confusing the target males (or females).

## III. PHEROMONE BLENDS AND REPRODUCTIVE ISOLATION

The existence of multicomponent pheromone systems has been firmly established as a concept. Specificity with these systems can be obtained with the use of secondary pheromone components, and they provide a good mechanism for pheromonal changes when selective pressure demands better reproductive isolation. In this regard, it may be possible to find species utilizing only one pheromone compound if there has been no such pressure to produce a blend.

Any discussion of the possibility that variations in host-plant chemicals can cause pheromonal changes, and thus speciation (31), will immediately reveal the paucity of information available on pheromone biosynthesis. Data must still be obtained to show whether or not pheromonelike compounds in host plants are sequestered directly by the larvae and finally emitted as pheromone by the adult. Sequestering of pheromonelike compounds would differ from the ingestion and utilization of plant alkaloids of the pyrrolizidine group by danaid butterflies (32) (33). In the latter case the adult male butterflies are attracted to plants containing pyrrolizidine alkaloids; they modify these substances into the dihydropyrrolizines (34) found in their hairpencil secretions. Thus, the ingestion is not done by the larvae, which feed on a different foodplant lacking the pyrrolizidine alkaloids (33).

The conversion of related host-plant compounds to pheromone in the emitting insect, however, could be the route used by a number of other species. There is evidence that α-pinene, a major component of the monoterpenes in the phloem of ponderosa pine, can be transformed to *cis*- and *trans*-verbenol by microorganisms found in the gut of adult male and female *Ips paraconfusus* (35). Also, it has been postulated (36) that locusts can degrade lignin ingested in grass or shrubs to guaiacol in their crops, and then convert the guaiacol to 2-methoxy-5-ethylphenol, a gregarization pheromone excreted in the feces of the locust.

Obviously a plant-feeding insect must ingest all of the chemicals from plants that it needs for its life processes. The degree to which primary or secondary plant substances are modified is a moot question. For instance, the sequestering of trace compounds by the larval stage, and emission of them completely unaltered, later, by adults is hard to determine. Many factors can influence pheromone production that may not be related directly to pheromone precursor intake. Inbreeding in a laboratory culture can lower pheromone production (37). According to one unpublished study, bacteria in the intestines, usually controlled by antibiotics in cotton plants, can greatly reduce pheromone production in laboratory cultures of the cotton boll weevil, *Anthonomus grandis.* Experiments with stored-grain insect species have also shown that it is difficult to analyze accurately for pheromone content in the various life stages because of surface contamination of the eggs, larvae, and pupae with pheromone.

Experiments with radio-labeled compounds will need to be carried out to answer questions on the diversity of the biosynthetic routes insects use for pheromone production. Some insect species at least can carry out *de novo* biosynthesis at some stages in the production of pheromone. *Bombyx mori* female pupae evidently (38) are capable of dehydrogenating cetyl phosphate or sodium palmitate in a stereospecific way to the pheromone *trans*-10, *cis*-12-hexadecadien-1-ol, and female gypsy moth pupae can carry out the epoxidation of the precursor olefin to the pheromone, *cis*-7,8-epoxy-2-methyl-octadecane (39). Studies (40) have also revealed the *de novo* biosynthesis of the boll weevil pheromone from injected radio-labeled acetate, mevalonic acid, or glucose. The danaid butterflies, on the other hand, depend upon the exogenous alkaloid precursor and do not incorporate radioactivity from injected radio-labelled potential precursors (32).

Research with radio-labeled compounds in an effort to elucidate pheromone biosyntheses is still in its infancy, but the accumulated chemical information now available provides a good basis for these studies, just as it provides the basis for pertinent studies in a variety of other important areas, such as insect behavior, neurophysiology, reproductive isolation mechanism, and insect-control strategies.

## REFERENCES

(1)  R.R. Cardé, T.C. Baker and W.L. Roelofs. 1975. Behavioural role of individual components of a multichemical attractant system in the Oriental fruit moth. *Nature* 253:348-349.

(2)  A. Butenandt, R. Beckmann, D. Stamm and E. Hecker. 1959. Über den Sexuallockstoff des Seidenspinners *Bombyx mori*. Reindarstellung und Konstitution. *Z. Naturforsch.* B 14:283-284.

(3)  R.S. Berger. 1966. Isolation, identification, and synthesis of the sex attractant of the cabbage looper, *Trichoplusia ni. Ann. Entomol. Soc. Am.* 59:767-771.

(4)  B.A. Bierl, M. Beroza and C.W. Collier. 1970. Potent sex attractant of the gypsy moth: Its isolation, identification, and synthesis. *Science* 170:87-89.

(5)  W.L. Roelofs, A. Comeau, A. Hill and G. Milicevic. 1971. Sex attractant of the codling moth: characterization with electroantennogram technique. *Science* 174:297-299.

(6)  R.M. Silverstein, J.O. Rodin, W.E. Burkholder and J.E. Gorman. 1967. Sex attractant of the black carpet beetle. *Science* 157:85-87.

(7)  K.J. Byrne, A.A. Swigar, R.M. Silverstein, J.H. Borden and E. Stokkink. 1974. Sulcatol: Population aggregating pheromone in the scolytid beetle, *Gnathotrichus sulcatus. J. Insect Physiol.* 20:1895-1900.

(8)  W.L. Roelofs and H. Arn. 1968. Sex attractant of the redbanded leaf-roller moth. *Nature* 219:513.

(9)  W.L. Roelofs and A. Comeau. 1968. Sex pheromone perception. *Nature* 220:600-601.

(10)  W.L. Roelofs, A.S. Hill and R.T. Cardé. 1975. Sex pheromone components of the redbanded leafroller moth, *Argyrotaenia velutinana. J. Chem. Ecol.* 1:83-89.

(11)  W.L. Roelofs and J. Tette. 1970. Sex pheromone of the oblique-banded leafroller. *Nature* 226:1172.

(12)  W.L. Roelofs, A. Comeau and R. Selle. 1969. Sex pheromone of the Oriental fruit moth. *Nature* 224:723.

(13)  W.L. Roelofs and R.T. Cardé. 1971. Hydrocarbon sex pheromone in tiger moths (Arctiidae). *Science* 171:684-686.

(14)  W. Roelofs and A. Comeau. 1971. Sex attractants in Lepidoptera, In *Proc. 2nd Int. IUPAC Congr. Pesticide Chem.,* Tel Aviv. *Chemical Releasers in Insects.* A.S. Tahori, Ed. Gordon and Breach, New York, Vol. 3, pp. 91-112.

(15)  J.E. Moorhouse, R. Yeadon, P.S. Beevor and B.F. Nesbitt. 1969. Method for use in studies of insect chemical communication. *Nature* 223:1174-1175.

(16)  R.M. Silverstein, J.O. Rodin and D.L. Wood. 1966. Sex attractants in frass produced by male *Ips confusus* in ponderosa pine. *Science* 154:509-510.

(17)  R.M. Silverstein, R.G. Brownlee, T.E. Bellas, D.L. Wood and L.E. Browne. 1968. Brevicomin: Principal sex attractant in the frass of the female western pine beetle. *Science* 159:889-891.

(18)  W.D. Bedard, P.E. Tilden, D.L. Wood, R.M. Silverstein, R.G. Brownlee and J.O. Rodin. 1969. Western pine beetle: Field response to its sex pheromone and a synergistic host terpene, myrcene. *Science* 164:1284-1285.

(19)  G.W. Kinzer, A.F. Fentiman, T.F. Page, R.L. Foltz, J.P. Vité and G.B. Pitman.

1969. Bark beetle attractants: Identification, synthesis, and field bioassay of a new compound isolated from *Dendroctonus*. *Nature* 221:477-478.

(20) J.A. Klun, O.L. Chapman, K.C. Mattes, P.W. Wojkowski, M. Beroza and P.E. Sonnet. 1973. Insect sex pheromones: Minor amount of opposite geometrical isomer critical to attraction. *Science* 181:661-663.

(21) D.L. Wood, R.W. Stark, R.M. Silverstein and J.O. Rodin. 1967. Unique synergistic effects produced by the principal sex attractant compounds of *Ips confusus* (LeConte) (Coleoptera: Scolytidae). *Nature* 215:206.

(22) D.L. Wood. 1972. Selection and colonization of ponderosa pine by bark beetles. In *Insect/Plant Relationships* Sym. Roy. Entomol. Soc. Lond. No. 6, p. 106.

(23) W.L. Roelofs, H. Arn and A. Comeau. 1968. Sex pheromone activity: Redbanded leafroller. *Abstr. 156th ACS Nat 1. Meeting, Med. Chem.*, No. 46.

(24) R.J. O'Connell. 1975. Olfactory receptor responses to sex pheromone components in the redbanded leafroller moth. *J. Gen. Physiol.* 65:179-205.

(25) W.L. Roelofs and A. Comeau. 1971. Sex pheromone perception: Synergists and inhibitors for the redbanded leafroller attractant. *J. Insect Physiol.* 17:435-448.

(26) S. Voerman and A.K. Minks. 1973. Sex pheromones of summerfruit tortrix moth, *Adoxophyes orana*. 2. Compounds influencing their attractant activity. *Environ. Entomol.* 2:751-756.

(27) H. Arn, C. Schwarz, H. Limacher and E. Mani. 1974. Sex attractant inhibitors of the codling moth *Laspeyresia pomonella* L., *Experientia* 30:1142-1144.

(28) D.A. George, L.M. McDonough, D.O. Hathaway and H.R. Moffitt. 1975. Inhibitors of sexual attraction of male codling moths. *Environ. Entomol.* 4:606-608.

(29) W.L. Roelofs, R.T. Cardé and J. Tette. 1973. Oriental fruit moth attractant synergists. *Environ. Entomol.* 2:252-254.

(30) G.H.L. Rothschild. 1974. Problems in defining synergists and inhibitors of the Oriental fruit moth pheromone by field experimentation. *Entomol. exp. appl.* 17:294-302.

(31) L.B. Hendry, J.K. Wichmann, D.M. Hindenlang, R.O. Mumma and M.E. Anderson. 1975. Evidence for origin of insect sex pheromones: Presence in food plants. *Science* 188:59-64.

(32) J.A. Edgar and C.C.J. Culvenor. 1974. Pyrrolizidine ester alkaloid in danaid butterflies. *Nature* 248:614-615.

(33) D. Schneider, M. Boppré, H. Schneider, W.R. Thompson, C.J. Boriack, R.L. Petty and J. Meinwald. 1975. A pheromone precursor and its uptake in male *Danaus* butterflies. *J. Comp. Physiol.* 97:245-256.

(34) J. Meinwald, Y.C. Meinwald, J.W. Wheeler, T. Eisner and L.P. Brower. 1966. Major components in the exocrine secretion of a male butterfly (Lycorea). *Science* 151:583-585.

(35) J.M. Brand, J.W. Bracke, A.J. Markovetz, D.L. Wood and L.E. Browne. 1975. Production of verbenol pheromone by a bacterium isolated from bark beetles. *Nature* 254:136-137.

(36) D.J. Nolte, J.H. Eggers and I.R. May. 1973. A locust pheromone: Locustol. *J. Insect Physiol.* 19:1547-1554.

(37) A.K. Minks. 1971. Decreased sex pheromone production in an in-bred stock of the summerfruit tortrix moth, *Adoxophyes orana. Entomol. exp. appl.* 14:361-364.

(38) S. Inoue and Y. Hamamura. 1972. The biosynthesis of "Bombykol" sex pheromone of *Bombyx mori. Proc. Japan Acad.* 48:323-326.

(39)    G. Kasang and D. Schneider. 1974. Biosynthesis of the sex pheromone disparlure by olefin-epoxide conversion. *Naturwissenschaften* **61**:130-131.

(40)    N. Mitlin and P.A. Hedin. 1974. Biosynthesis of grandlure, the pheromone of the boll weevil, *Anthonomus grandis* from acetate, mevalonate, and glucose. *J. Insect Physiol.* **20**:1825-1831.

## CHAPTER 17

# Host-Plant Resistance to Insects—Chemical Relationships

Fowden G. Maxwell*

*Department of Entomalogy and Numatology*
*University of Florida*
*Gainesville, Florida*

## I. RESEARCH FACTORS

Plants that are inherently less severely damaged or less infested by a phytophagous pest than other plants under comparable environments in the field are termed *resistant* by Painter (1) (2). Considerable progress has been achieved, especially in recent years, in developing a great many resistant varieties. These achievements have been adequately reviewed by Painter (2), Beck (3), Maxwell et al. (4), and Maxwell (5). While considerable success has been achieved in breeding for resistance to certain key insect pests, basically little is known about the chemistry of resistance where these successes have been registered. In other words, we have been able to breed successfully for resistant mechanisms without knowing the chemical cause of resistance. Several reasons account for this. First, support, interest, and techniques have been lacking for exploring the field. Recently more support and basic interest have been expressed, and in several instances chemists and physiologists have become important members, of several host-plant resistance teams. Techniques for adequately detecting, analyzing, and identifying chemical compounds from plants, often in minute quantities, have also been fairly recent developments.

Second, resistance as expressed in the field is most often a very complex mechanism involving not only physical and biochemic factors, but frequently complex interrelationships among the insect, the plant, and the environment. Because of these complex relationships, it is often difficult if not impossible to determine the major chemical causes. In studies of the basic chemical nature of resistance, elaborate and often very expensive tests must be devised

---

Formerly coordinator, Environmental Activities, USDA, Washington, D.C.

to separate the major sources of variants from those that are contributing a significant effect to the total expression of resistance.

Third, we have not had the necessary basic behavioral and physiologic information on the insect pest. This is especially true in regard to chemoreception and other physiologic systems, which must be understood in greater detail so that research workers can devise and design adequate chemical and biologic assay techniques to assess properly the response of insects to various chemical fractions and compounds derived from plants. This has been a very important limiting factor, and currently constitutes one of the more serious problems facing most teams interested in determining causes of resistance. Recently, we have been making progress in this area primarily by bringing behavioral knowledge as well as basic electrophysiologic and electron microscopy techniques to bear on the problem.

Fourth and finally, insufficient pressure has existed in the past to investigate the chemical causes of resistance. The lack of pressure and the usual complex problems involved in investigating the area have caused entomologists and plant breeders intentionally to ignore the basic aspects and push ahead with "success-assured" applied aspects.

Recently, the interest in elucidating some of the chemical aspects of resistance has greatly increased. This has been brought about by several factors:

1. A greater interest in host-plant resistance as an alternative method to our current, almost unilateral, dependence upon pesticides; and the knowledge that resistance can and will play an important role in the development of effective integrated pest-management programs
2. The degree of success registered through past programs or host-plant resistance
3. Information derived from basic insect-plant interactions from host-plant resistance programs, which contribute materially to the field of insect behavior and control methodology
4. Greater interest shown by private and public funding agencies in supplying the necessary support to conduct more basic chemical studies
5. A greater awareness and interest on the part of natural-product chemists of the opportunities for research contribution in the field
6. The advent of better methodology and technology in microchemical techniques, especially gas chromatography and mass spectrometry
7. New Food and Drug Administration regulations, which will necessitate chemical research to document that changes (toxins, nutritional, etc.) made in the plant through the development of resistance to a pest will not constitute a hazard to the health or nutrition of humans and other animals that might consume the resistant crop.

This last factor is probably the most important one from the standpoint of the future. These regulations will have a tremendous impact, making necessary what we have largely ignored in the past—a thorough study and documentation of the chemical or physical causes of resistance wherever possible. This will greatly increase the cost of doing host-plant resistance work and may markedly delay the progress in releasing new resistant varieties. Although a serious disadvantage in this respect, many advantages will accrue from necessary chemical studies, which will enrich the general field of insect-plant interactions, and more specifically those of insect behavior, chemical ecology, and applied insect control.

## II. CHEMICAL MODIFIERS

To assess where we stand now in our knowledge of the chemical nature of resistance and the practical utilization of such knowledge to breeding programs is most difficult. Most of the chemical information accrued over the years is in the basic area of preference or nonpreference, dealing primarily with various chemicals in the host that affect behavior of the insect, specifically attractancy, repellency, feeding and oviposition stimulation, and deterrency. Hedin et al. (6) conducted an extensive survey of the existing literature and attempted to categorize the types of response that the insect and the host display to identify the compounds or classes of compounds that cause the responses if known, and to list the investigators who conducted the research. The survey showed that although quite a number of chemicals have been identified, only a few actually have been implicated as the causative agent per se in resistance of a host to a specific insect pest. Also, to my knowledge no practical chemical assay methods have been developed that are being used as a primary screening device for detecting resistant plants. This is not surprising; it reflects both the complexity of most cases of resistance and the general lack of suitable, economical chemical methods. As we learn more about the different components of behavior and the chemical and physical factors that control those segments of behavior, then and probably only then will we be able to develop successful, reliable, economical techniques for detecting the specific chemicals in host plants that initiate or control vital segments of behavior, such as attractancy, feeding, oviposition, and repellency, which so intimately relate to resistance of the plant.

There are numerous examples in the literature of known chemicals that directly affect the antibiosis component of resistance. These include an array of alkaloids, phenols, flavonoids, and more recently identified analogues of insect hormones. Well-known works include the identification of the benzoxazolinone compounds with European corn borer, *Ostrinia nubilalis*

(Hubner) in maize, *Zea mays* (3) (7)-(11); gossypol with *Heliothis zea* (Boddie) and *Heliothis virescens* (Fab.) in cotton *(Gossypium hirsutum* L.) (12)-(16); and saponins and protease inhibitors affecting several insects in legumes and stored legume seeds (17)-(23).

Suitable chemical methods for assay have been developed for 2,4-dihydroxy-7-methoxy-1,4-benzoxazin-3-one (DIMBOA) in corn and gossypol in cotton. The primary disadvantage at present for successful utilization of these methods in breeding programs is the expense of individual assays. A simple and inexpensive technique is badly needed for both of these compounds. Saponin chemistry is complicated, and the entire picture with many of the pests affected is still not clear; thus, application will be slow until more is known about the roles of various saponins.

The role of nutritional factors in resistance is also little understood. Although resistance mechanisms involving aspects of host nutrition may actually be prevalent, they are most difficult to prove because the net effect is likely to be more quantitative or subtle than are other resistance mechanisms. Also, genetic manipulation of the nutritional quality of the insect's host may be less of a feasible approach for the future because of the new FDA regulations restricting major changes in nutritional factors. A plant developed so that it would be nutritionally inadequate for the insect might also be equally inadequate for man or the animals that would utilize the crop. Maxwell (5) has reviewed the relationship of nutritional factors to resistance, particularly as it relates to antibiosis, preference, and overall pest management.

## III. SUMMARY AND CONCLUSIONS

In summary it is evident that currently little is known about the chemical nature of resistance in plants to insects. The current development of better chemical techniques is making it possible not only to explore the bases of plant resistance, but also the factors that make a plant attractive to an insect. The ability to breed successfully for lower concentrations of biologically active chemical compounds in plants depends largely upon recognizing the independent and collective effects of specific chemicals on the insect and the number of genes controlling the chemicals. Most of the problems involved to date with the preference mechanism of resistance indicate that a blend of chemicals is usually involved, rather than a single specific compound. If such is the case, breeding becomes very difficult because usually each chemical is under different genetic control.

In the immediate future it is unlikely that much change will take place in the way most host-plant resistance programs are conducted. Until more insect-behavioral and host-chemical studies can be conducted to develop the precision needed in a primary chemical approach, researchers will continue to use the classical applied screening techniques, utilizing chemical approaches where possible to help determine basic causes of resistance. Research must provide the necessary techniques in the future for a more positive chemical approach to the field. Increased support, evident now and likely to increase, will help to meet the challenges that are before the field and to meet the safety demands of concerned agencies such as FDA.

## REFERENCES

(1)  R.H. Painter. 1951. *Insect Resistance in Crop Plants.* Macmillan, New York.

(2)  R.H. Painter. 1958. Resistance of plants to insects. *Ann. Rev. Entomol.* 3:267-290.

(3)  S.D. Beck. 1965. Resistance of plants to insects. *Ann. Rev. Entomol.* 10:207-232.

(4)  F.G. Maxwell, J.N. Jenkins, and W.L. Parrott. 1972. Resistance of plants to insects. In *Advances in Agronomy* 24:187-265. Academic, New York.

(5)  F.G. Maxwell. 1972. Host plant resistance to insects—nutritional and pest management relationships. In *Insect and Mite Nutrition.* J.G. Rodriguez, Ed. North-Holland, Amsterdam.

(6)  P.A. Hedin, F.G. Maxwell, and J.N. Jenkins. 1974. Insect plant attractants, feeding stimulants, repellents, deterrents, and other related factors affecting insect behavior. In *Proceedings of the Summer Institute on Biological Control of Plant Insects and Diseases.* F.A. Harris, Eds. University Press of Mississippi, Jackson.

(7)  S.D. Beck and W. Hanec. 1958. Effect of amino acids on feeding behavior of the European corn borer, *Pyrausta nubilalis* (Hbn.). *J. Insect Physiol.* 2:85-96.

(8)  E.E. Smissman, J.B. LaPidus, and S.D. Beck. 1957. Isolation and synthesis of an insect resistance factor from corn plants. *J. Am. Chem. Soc.* 79:4697-4698.

(9)  J.A. Klun. 1965. Role of 6-methoxybenzoxazolinone (6MBOA) in maize inbred resistance to first brood European corn borer, *Ostrinia nubilalis. Diss. Abstr.* 29:(6):3544.

(10)  J. Klun, C. Tipton, and T. Brindley. 1967. 2,4-Dihydroxy-7-methoxy-1,4-benzo-xazin-3-one (DIMBOA), an active agent in the resistance of maize to the European corn borer. *J. Econ. Entomol.* 60:1529-1533.

(11)  J.A. Klun, W.D. Guthrie, A.R. Hallauer, and W. Russell. 1970. Genetic nature of concentration of 2,4-dihydroxy-7-methoxy-2H-1,4-benzoxazin-3(4H)-1 and resistance to European corn borer in a diallele set of 11 maize inbreds. *Crop Sci.* 10:87.

(12)  M.J. Lukefahr and D.F. Martin. 1966. Cotton plant pigments as a source of resistance to the bollworm and tobacco budworm. *J. Econ. Entomol.* 59:176-179.

(13)  M.J. Lukefahr and J.E. Houghtaling. 1969. Resistance of cotton strains with high gossypol content to *Heliothis* spp. *J. Econ. Entomol.* 62:588-591.

(14)  T.N. Shaver and M.J. Lukefahr. 1969. Effect of flavonoid pigments and gossypol on growth and development of the boll-worm, tobacco budworm and pink bollworm. *J. Econ. Entomol.* 62:643-646.

(15)  T.N. Shaver and W.L. Parrott. 1970. Relationship of larval age to toxicity of gossypol to bollworms, tobacco budworms and pink bollworms. *J. Econ. Entomol.* 63:1802-1804.

(16) B.F. Oliver, F.G. Maxwell and J.N. Jenkins. 1967. Measuring aspects of antibiosis in cotton lines to bollworm. *J. Econ. Entomol.* **60**:1459.

(17) S.W. Applebaum, B. Gestetner, and Y. Birk. 1965. Physiological aspects of host specificity in the Bruchidae. IV. Developmental incompatability of soybeans for *Callosobruchus. J. Insect Physiol.* **11**:611-616.

(18) S.W. Applebaum, S. Marco, and Y. Birk. 1969. Saponins as possible factors of resistance of legume seeds to the attack of insects. *J. Agric. Food Chem.* **17**:618-622.

(19) S.W. Applebaum and M. Guez. 1972. Comparative resistance of *Phaseolus vulgaris* beans to *Callosobruchus chinensis* and *Acanthoscelides obtectus* (Coleoptera: Bruchidae): The differential digestion of soluble heteropoly-saccharide. *Entomol. expt. appl.* 15:203-207.

(20) E. Horber. 1972. Alfalfa saponins significant in resistance to insects. In *Insect and Mite Nutrition.* J.G. Rodriguez, Ed. North-Holland, Amsterdam.

(21) M. Roof, E. Horber, and E.L. Sorensen. 1972. Bioassay techniques for the potato leafhopper, *Empoasca fabae* (Harris). *Proc. North Central Branch Entomol. Soc. Am.* 27:140-143.

(22) C.H. Hansen, M.W. Pedersen, B. Berang, M.E. Wall, and K.H. Davis. 1973. The saponins in alfalfa cultivars. In *Antiquality Components of Forages.* A.G. Matches, Ed. Special Publ. No. 4, Am. Soc. Agron., Madison, Wisconsin.

(23) M.W. Pedersen, E.L. Sorensen, and M.J. Anderson. 1976. A comparison of pea aphid-resistant and susceptible alfalfas for field performance, saponin concentration, digestibility, and insect resistance. *Crop Sci.* (in press).

## CHAPTER 18

# Mosquito Attractants and Repellents

A. A. Khan

*Department of Dermatology*
*University of California, School of Medicine*
*San Francisco, California*

## I. ATTRACTANTS

Chemicals play an important role in influencing insect vital activities such as finding food, the opposite sex, or an oviposition medium. Pheromones are chemicals secreted to influence communication between two or more members of the same species (1) (2). Their presence and importance are amply demonstrated in several insect orders, and the subject was reviewed by Jacobson (3) and recently by Shorey (4). Among mosquitoes, there are few reports documenting the importance of pheromones. They influence the mating behavior of *Culiseta inornata* (Williston) (5) and aggregation of females at a suitable oviposition site in some *Culex* and *Aedes* species (6)-(11).

Attraction and oviposition stimulation of gravid female mosquitoes by bacteria isolated from hay infusion has been reported also (12). The demonstration of oviposition attractants for some mosquitoes suggests that practical application of such substances for control may be possible, although chemical attractants for *Aedes aegypti* (L.) in "ovitraps" have not proved advantageous (13). Both chemical and physical stimuli play an important role in the blood-feeding drive of mosquitoes.

## A. Host-Seeking Behavior of Mosquitoes

The adult female mosquito seldom seeks a host soon after emergence. More often she mates and feeds on sugary rich material before she is in a proper physiologic state to feed on blood (14)-(16). Many species rest in situations of low light intensity and high humidity until ready to seek the host (17). After this initial rest period, the mosquitoes go through phases of random flight, exploratory flight, oriented flight toward the host, settling on the host, searching by walking, probing, and biting, and then take off. These states are documented in detail (17)-(19).

## B. Factors in Human Skin Attractive to Mosquitoes

Each stage in the hierarchy of response has a corresponding stimulus. The approach of a mosquito to its host in the field is generally upwind (19). This may be termed as a positive anemotaxis triggered by odor (17),(20). *Aedes sollicitans* (Walker) and *Aedes cantator* (Coq.) alight on the leeward side of a host and fly upwind toward the host even against a strong wind (20)-(22). Anopheline mosquitoes behave similarly (23). We studied the effect of several stimuli on the approach of *A. aegypti;* odor of the host attracted mosquitoes from the greatest distance (24). Gillies found odor effective at greater distance than $CO_2$ (25-27). Odor will be discussed in greater detail later.

### 1. Carbon Dioxide

The role of carbon dioxide in mosquito orientation has been interpreted differently by different workers. It has been used in mosquito traps for population sampling (28)-(32). Reeves (30) felt that $CO_2$ chemotropism of mosquitoes may be the principal factor in host selection and that different mosquito species react differently to varying $CO_2$ concentrations in the environment. Anderson and Olkowski (33) consider that $CO_2$ is an attractant for most groups of biting flies. On the other hand, Gillies (34) reports that a directional response is lacking on stimulation with a source of $CO_2$ and the behavior of an insect in its vicinity is often erratic and undirected. Most other workers consider $CO_2$ an activator (21) (24)-(27) (35)-(40). We also found $CO_2$ to be an activator (41)-(43). Smith (44) considers that some of the inconsistencies in the results of investigators regarding the action of $CO_2$ were caused by test techniques; also, some workers may have misinterpreted the behavioral responses. A true attractant induces an oriented approach to the source (45). The consensus regarding the effect of $CO_2$ is that its effect is as a locomotor stimulant (44) (46) (47).

## 2. Heat

The stimuli that elicit landing and probing by mosquitoes are heat and moisture. The effect of heat on mosquito attraction was studied by many workers. Convective rather than radiant heat is effective (48) (49). We found a heat source (34°C) presented at the bottom of a vertical tower effective in attracting mosquitoes from a height of 32 to 44 in (24). Its effect on inducing landing by mosquitoes was minimal (41) (50). Temperature may play a part in host selection. *A. aegypti* preferred a warmer over a cooler hand when both were offered simultaneously (51).

## 3. Moisture

The effect of moisture on mosquito attraction depends on the ambient conditions of temperature and humidity and on the hydration state of the mosquito. In a hot, dry environment moist air is attractive (47). In laboratory studies with *A. aegypti*, we found moisture minimally attractive, inducing some flying activity for distances up to 24 in. (24) (41). Few studies report on the effect of moisture alone. Most studies consider moisture in combination with heat or host odor (36)-(38) (51) (52).

## 4. Warm-Moist Convection Currents

The combination of heat and moisture has been reported as an effective short-range attractant (36) (37) (39) (53). To the contrary, Burgess (54) showed that *A. aegypti* did not respond to warm-moist convection currents in the absence of $CO_2$ or the host. We observed that a warm-moist surface was attractive to *A. aegypti*, and moisture in the presence of heat stimulated landing and probing. Introduction of $CO_2$ in the presence of heat and moisture attracted more mosquitoes (24) (41). The response also depends on the mosquito water balance. Dehydrated mosquitoes probed on a warm-moist surface more avidly than hydrated mosquitoes (55) (56). Diffusion of the convection currents and diffusing the odor gradient reduced the approach of mosquitoes to human skin by half; neutralizing the convection currents by raising the ambient temperature equal to that of the skin (34°C) eliminated the approach of mosquitoes to the human palm held below the bottom of the cage. When the hand was introduced inside the cage, half of the mosquitoes landed and bit (57). These studies emphasize the importance of warm-moist convection currents as short-range stimuli inducing landing and probing, and more importantly as vehicles of odor transport.

## 5. Odor

Host odor is a long-range attractant triggering positive anemotaxis in mosquitoes; its role in mosquito attraction has been demonstrated with different mosquito species. Using a dual-port olfactometer, odor was found attractive to *Anopheles quadrimaculatus* (Say) by Willis (35); *Anopheles labranchiae atroparvus* (van Thiel) by Laarman (38) (58); *Anopheles stephensi* (Liston) by Brouwer (59); and *A. aegypti* by Brown et al. (37), Thompson and Brown (60), and Rahm (61) (62). In experiments in a vertical tower, odor from the human palm was more attractive to *A. aegypti* than heat, moisture, and $CO_2$ individually or combined (24) (41). In tests of attractant sources placed side by side, most mosquitoes preferred the palm and few went to the warm-moist target with $CO_2$ (42). The difference between response to the skin and an artificial target depended on the physiologic and nutritional state of mosquitoes. The difference between response to the skin and the warm-moist surface in the presence of $CO_2$ observed with sugar-water-fed mosquitoes disappeared when mosquitoes were water-starved (55). Mosquitoes first fed on sugar water to repletion, and then on blood, and with continuous access to water thereafter did not respond to the skin until they oviposited. Those that had water first, then blood, but no water thereafter started responding to the human palm after 40 hr, while those that were fed on blood only and had no water, either before or after the blood meal, started probing on the palm after 24 hr (56). We observed the probing response to return in full 6 hr after the start of oviposition (63). We noted that 34°C, which is the temperature of human skin, was the optimum for probing by water-satiated mosquitoes; the probing decreased as the temperature of the warm-moist target increased, until at 40 to 42°C the mosquitoes ceased probing. With dehydrated mosquitoes the probing response was different. They probed avidly even when the temperature of the target surface reached 42°C (52).

## C. Varying Attractiveness of Hosts to Mosquitoes

Human beings differ sometimes markedly from one another in their attractiveness to mosquitoes (46) (64) (65). The difference in attractiveness may be due to differences in skin temperature (51) or skin color; darker skin being more attractive than lighter (51). Men attract more mosquitoes than do women (64). Newborn babies until 1 yr old were found to be less attractive to mosquitoes than children or adults (66) (67). Mosquitoes were attracted less to subjects in their fifties than to subjects younger or older up to 90 years of age (68). The main reason for differences in attractiveness is odor (65). We screened a large human population to establish a profile of its attractiveness to mosquitoes and to find a truly unattractive subject if one existed. The method used was that of exposing subjects to mosquitoes in a small cage and observing the number of mosquitoes biting in 3 min. Of the 838 persons tested, 17 were not bitten the first time. On retesting, all except one were bitten. Out of another 11 tests, this exceptional individual was bitten only three times. We substituted the assay response of mosquito biting with probing and developed techniques sophisticated enough to pick up small differences in attractiveness between any two individuals (69)-(71). The probing time 50 (PT 50) technique (69) measured time in seconds when three of six mosquitoes in a small cage placed 1 cm above the forearm probed simultaneously on the bottom toward the skin. Using this technique, we noted in a population of 100 subjects several that were more attractive and one or two relatively less attractive.

Differences among vertebrate hosts in attractiveness to mosquitoes may also depend on the quality of odor. Laarman (38) (58) observed A. *labranchiae atroparvus* more attracted to rabbit blood and less to bovine blood, and the laboratory mosquitoes reared on rabbit blood showed more attraction to rabbit blood than the wild mosquitoes of the same species. Similarly A. *stephensi* raised on human blood showed greater response to human odor than guinea pig odor (65). Based on similar observations of host specificity on *Anopheles maculipennis* (Meigen), the terms *zoophilous* and *anthropophilous* were coined in 1920 (72). These terms received credence by the results of precipitin tests done on captured wild caught mosquitoes (73) (74). Our A. *aegypti* strain raised for countless generations and years of feeding on guinea pig blood still shows preference for man (75). Bates and Hackett (76) criticized the terms *anthropophilous* and *zoophilous,* characterizing them as unfortunate. Hocking (20) remarked that techniques such as the precipitin test give an illusion of specificity. In our experiments with different vertebrate hosts (75), men were three times as attractive as guinea pigs and significantly more attractive than chickens or pigeons, although the skin temperature of the latter was 3 to 4°C higher. Guinea pigs were twice as attractive as birds (75). In another report (77) A. *aegypti* and *Culex tarsalis* (Coq.) were found to prefer mice over chicks. These preferences must rest on the quality of host odor perceived by mosquitoes when more than one host are available. In the absence of an alternate host they feed on the host present.

## D. Sources of Attractive Stimuli

### 1. Sweat

Sweat has been investigated as a likely source of odor. Some investigators found it unattractive to mosquitoes (21) (48) (78), but other obtained attractive responses. Parker (53) and Brown et al. (37) found armpit sweat attractive to *A. aegypti.* Rahm found odors from sweat obtained from the human hand similarly attractive (61) (62). Dummies soaked with sweat from the human armpits were more attractive to *Aedes* in the field than controls with equally moist clothing (36) (79). We examined two subjects with generalized eccrine anhidrosis due to an ectodermal defect; they were unattractive compared to normal subjects (80). The palm and forehead where sweating occurred in one subject were normally attractive. We measured the attractiveness of normal subjects by the PT 50 method. We induced eccrine sweating locally by injecting methacholine intradermally and by iontophoresis, and remeasured attractiveness. The area where sweating occurred became significantly more attractive (81). We collected sweat from the whole body and demonstrated its attractiveness to *A. aegypti* in a dual-port olfactometer (82). After lyophilization, the sweat remained attractive; the water removed by lyophilization was unattractive. On dialysis the attractive components were separated into the dialysable fraction. The lyophilized human sweat was extracted with different solvents (hexane, diethyl ether, acetone, isopropanol, and ethanol). The attractive component(s) were extracted only by diethyl ether and ethanol. This suggests that amino acids are not the major attractants in sweat. Liquid-liquid extraction with diethyl ether of whole sweat adjusted to pH 1 yielded a fraction repellent to mosquitoes (83).

### 2. Blood

Whole blood is attractive to mosquitoes (38) (58) (84) (85). Whole blood is more attractive than a suspension of corpuscles, and plasma is four to five times more attractive than water (85). Hemoglobin was noted to be attractive by Rudolfs (21). An unspecified factor from blood attractive to mosquitoes was reported by Schoerfenberg and Kupka (86).

### 3. Chemicals Attractive to Mosquitoes:

Several compounds are reported as mosquito attractants: lysine and alanine, arginine (87), methionine (88) (89), phenylalanine (21), tyrosine, benzoic acid, hydroquinone, oestradiol (90), L-lactic acid (91), and esters of aliphatic acids (92). None is as attractive as a natural host, and their value in the field for population studies or chemical control is doubtful.

## II. REPELLENTS

The use of repellents for personal protection against biting flies is not new. Smoke was used as an area or space repellent since antiquity. Until the 1940s materials such as pyrethrum, citronella, and other essential oils were the principal mosquito repellents. During and after World War II many effective repellents were developed. These included diethyl toluamide (deet), ethyl hexanediol, dimethyl phthalate (DMP), demethyl carbamate, and Indalone[R]. Test procedures for repellent evaluation were developed by Granett (93) (94). Deet was the most effective repellent against most mosquito species. Butyl ethyl propanediol, undecylenic acid, and benzyl benzoate were developed for use on clothing.

Despite all this progress, much room for improvement remains. Repellents are lost in a few hours after application. They also feel oily and are irritant to the mucous membrane. Since they are generally used on the skin, cosmetic acceptability and nontoxicity are important. They should not stain clothes or dissolve them on contact, and should be inexpensive. These requirements greatly limit the choice of compounds.

## A. Factors Affecting Repellent Loss

Insect repellents when applied on the skin are lost in time by evaporation, abrasion, and percutaneous absorption (95). The repellent will continue to protect if its amount on the skin does not fall below a certain level. This level below which mosquitoes would bite is termed the *minimum effective dose* (MED). To prevent repellents falling below the MED too soon they must be applied on the skin in relatively large amounts. The MED varies with repellents, requiring some to be applied more often than others. Smith et al. (95) reported the MED of ethyl hexanediol as two to five times higher than deet, and that of dimethyl phthalate 20 times higher than deet. We found the MED of dimethyl phthalate four times that of deet and equal to those of ethyl hexanediol and Indalone[R]. We observed that the MED is dependent on the size of a mosquito population. For some repellents, the MED was higher when determined in a high-density cage and lower with a low-density cage.

DMP and ethyl hexanediol evaporate about twice as fast as deet but are absorbed more slowly by the skin (95). Abrasion is an important source of repellent loss. Other than these factors, ambient conditions of wind velocity and temperature also determine the duration of protection by a repellent. An airflow of 192.3 m/min reduced the protection time of deet to one-third of the control (still air) and with every 10°C rise in temperature, the protection time was reduced by one-half. The protection times of deet, hexamethyleneimine butanesulfonamide, ethyl hexanediol, DMP, and Indalone[R] at room temperature (26°C) and minimal air movement were respectively eight, five, eleven, nine, and nine times higher than their protection times at 40°C and an airflow of 192.3 m/min (96).

## B. Factors Affecting Protection Time of Repellents

There are several other factors affecting protection time of mosquito repellents besides the depletion of the repellent on the skin. The relative effectiveness of repellents on human skin varies with the species of mosquitoes involved. In laboratory studies, Travis et al. (97) observed that DMP was more effective than ethyl hexanediol and *Indalone*[R] against *A. quadrimaculatus*. Against *Anopheles punctipennis* (Say) and *Anopheles freeborni* (Aitkin), Travis (98) found DMP and ethyl hexanediol equally effective. Against *Anopheles albimanus* (Wiedman) ethyl hexanediol was more effective than DMP (99). In field tests, Altman found deet more effective than DMP and other repellents tested against *A. albimanus* (100). Gilbert et al. (101) evaluated nine repellents against anopheline mosquitoes in Thailand and found both deet and DMP equally effective; ethyl hexanediol gave poor results. Against *A. aegypti*, deet is most effective, but against *Culex pipiens quinquefasciatus* (Say) the same dose lasts only half as long (95) (101). These intra- and intergeneric differences in response to the same repellent may perhaps be due to different threshold levels of perception.

The protection time of insect repellents also varies with the mosquito density it challenges. In a laboratory study we found protection with repellents significantly dependent on mosquito density (102). Deet applied at different dosages protected almost twice as long when tested against 100 *A. aegypti* per cage compared to 1000 to 2000 mosquitoes per cage. When three repellents—triethylene glycol monohexyl ether synthesized by Stanford Research Institute (SRI); 1,1'-carboxyl-bis-hexamethyleneimine (of Russian origin); and deet—were compared, the first two were superior to deet when tested in low- and medium-density cages (100 and 500 mosquitoes per cage), but deet became superior to the Russian repellent in a high-density cage (2000 per cage) (102).

Differences can be observed with the same dose of repellent applied on different subjects. This may be due to different rates of loss through absorption. The relative attractiveness of the host did not correlate with protection time (95). Mosquitoes recently gorged on sugar water took significantly longer to bite through repellent-treated skin than starved ones (102). Factors that do not seem to affect protection time of repellents are sex of the subject, amount of hair on the skin (95), and size and vigor of mosquitoes (102).

## C. Search for Better Topical Repellents

Most insect repellents were discovered through large-scale screening of organic compounds. About 11,000 compounds were screened in the USDA laboratories at Orlando, Florida (103). The efforts are now being continued at the USDA, Gainesville laboratories (104)-(106).

We have tested several hundred components synthesized in the laboratories at SRI. In these studies, deet was used as the standard. Compounds were selected on the basis of their performance relative to deet (equal or better). They were further tested in the laboratory and in the field by scientists at the Department of Dermatology, The Letterman Army Institute of Research, Presidio, San Francisco (directed by Col. W. Akers). The synthesis of these compounds was designed to enable us to understand the physiochemic factors necessary in a compound for repellent action. Only after such an understanding, could a rational approach to the development of better repellents be adapted. A series of ring-substituted diethylbenzamide analogues of deet was prepared for this purpose. The bioassay of the compounds for repellency indicated that repellency was proportional to volatility within the series (107). No direct relationship was evident with regard to partition coefficient among compounds of similar volatility. Five of the fluorine analogues of deet exhibited repellent activity comparable to deet. Little was gained by the addition of ring substituents in these series.

A series of ring-substituted aminoalkylbenzamide compounds was prepared in order to broaden the basis of structural comparison. The direct relationship of repellency to volatility seen in the case of diethylbenzamides was less obvious in this series, but the overall importance of volatility was still apparent (108). Pervomaisky et al. (109) reported certain sulfonamides derived from butanesulfonic acid as effective repellents to female *A. aegypti*. Exploration of the structure-activity relations revealed that the boiling point is again the most important factor for topical repellents. Quinoline-4-carboxylic acid derivatives were explored for repellency; amides were less active than esters perhaps because of their low volatility (110). A series of benzylic ethers with hydroxyl, epoxy, and unsaturated moieties was synthesized and evaluated for repellent action. None was as effective as deet (111). Normal and branched-chain aliphatic monoethers of triethylene glycol were synthesized and found effective topical mosquito repellents. In terms of duration, they were superior to diethylene glycol analogues; some were superior to deet. The n-heptyl monoether of triethylene glycol was twice as effective as deet under laboratory conditions (112).

These studies have shown that besides volatility, existence of subtle structure-specific factors may hold the key to the "intrinsic repellency" of a molecule. Recently, SRI has begun to apply a computerized method of structure-activity correlation employing multiple-regression analysis.

Another approach to the problem of developing repellents is to design compounds to anchor to the skin surface and act as a precursor-anchoring molecule slowly releasing the repellent moiety, the release would be effected by skin moisture, other emanations, or bacteria. Quintana and his associates synthesized several compounds including undecenoic acid esters, dihydroxyacetone monoesters, and amino analogues of ethyl hexanediol, but without much success (114)-(120). We synthesized some compounds with the same objective by forming low-volatility carbonyl addition derivatives of repellent alcohols and carbonyl compounds (120). Sufficiently rapid hydrolytic release of volatile, repellent moieties on the skin surface did not occur.

## D. Formulation of Repellents with Fixatives to Increase Protection Time

Numerous attempts have been made to increase protection time of repellents with special formulations. Several hundred were tested at the Orlando laboratory during World War II (121). None was effective longer than full-strength repellents (44). Efforts are being continued (105)(106). Smith, Kline, and French laboratories in 1951-53 tested 2000 formulations using different materials including powders, activators, antiperspirants, emulsifiers, stabilizers, and conditioners. The effectiveness of the formulation did not increase, although the cost did (44). Russian scientists used materials like phenylbenzoate, ethyl cinnamate, and Peruvian balsam without success (122)(123). Fixatives are commonly used in perfume formulations to diminish the evaporation of the scent. A fixative is relatively less volatile, and on addition to the perfume retains or fixes its odor. Such fixatives include musks, civet, and ambergris. These materials were never tried with repellents. We formulated four repellents with seven synthetic musks in three different formulations. The repellents used were deet, DMP, ethyl hexanediol, and *Indalone®*. The musks were *Tibetene® Moskene®*, musk ambrette, musk xylol, musk ketone, benzylcinnamate, and givambrol. The formulations used in repellent-to-musk ratios were 1:1, 1:2, and 1:3. Of the seven musks tested, four were effective in increasing protections time of deet significantly. These were the musks *Tibetene®* ambrette, givambrol, and xylol. Others were ineffective with deet; all seven were ineffective with the other three repellents. The increase in protection time with deet ranged from 12 to 88% depending on the musk and formulation (124).

We examined vanillin as a fixative for mosquito repellent using the four repellents mentioned above, plus triethylene glycol monohexyl ether, triethylene glycol monoheptyl ether, and triethylene glycol ethyl hexyl ether. The repellent-vanillan formulation ratios were 1:1, 1:2, and 1:3. Vanillin was effective with all the repellents. The different ratios in which vanillin was mixed with repellent were not significant except with deet. In most cases increase in protection time was more than 100%. The maximum increase was with deet (178%). With triethylene glycol ethyl hexyl ether applied at 0.16 mg/cm$^2$ mixed with an equal amount of vanillin, a protection of 21 hr was obtained against *A. aegypti* (125). These data suggest that if repellent loss through abrasion can be prevented, it is feasible to obtain protection for close to 24 hr with one application.

## E. Formulation of Repellents with Additives to Increase Abrasion and Water-Wash Resistance

Repellents are lost quickly with abrasion. In tropical areas, the problem is further compounded by sweating. Protection time under sweating conditions is considerably shorter (126)(127). A formulation designed to decrease this loss would provide longer protection. Christophers (128) attempted to increase the resistance of DMP to rubbing by formulating it with materials like china clay, shellac, and polystyrene, but without success. Kurtz et al. (129) tested several acrylic polymers, silicones, and polysaccharides combined with repellents. *Carboset* 526 (B.F. Goodrich, Co.) increased protection time of deet about 50% and also improved water-wash resistance, but cosmetically the polymer was unacceptable.

We studied a mixture of co-polymers of hydroxy-vinyl chloride-acetate and sebacic acid, maleic rosin ester, and glycolate plasticizer combined with deet. It withstood abrasion and was water-wash resistant (130). Deet applied at 1.6 mg/cm$^2$ and unprotected from abrasion during sleep at night does not last until morning (10-12 hr). The same quantity of deet mixed with one-third amount of polymer lasted close to 24 hr. Triethylene glycol monohexyl ether at 1.6 mg/cm$^2$ and protected from abrasion was effective for 16 to 18 hr without polymer; mixed with one-third amount of polymer, it protected for 24 to 49 hr in six trials. No precaution was taken to protect the repellent-treated surface from abrasion, and the subject slept normally. The water-wash resistance also improved several fold. The polymer mixture mixed with repellents, however, feels sticky on application for a few hours and the cosmetic acceptability problem requires solution.

## F. Repellents for Use on Clothing

Under tropical conditions when heavy clothing is intolerable and mosquito infestation is high, application of repellent on the skin alone may not be sufficient. Repellent-treated clothing in this case would be useful. The standard military repellent for treatment of clothing is a mixture containing 30% 2-butyl-2-ethyl-1-3-propanediol as a mosquito repellent, 30% *N*-butylacetanilide as a tick and flea repellent, 30% benzyl benzoate for protection against chiggers and fleas, and 10% *Triton X*-100 as emulsifier. The mixture is known as M-1960. Deet as a clothing repellent is even more effective than M-1960 against mosquitoes and fleas, but inferior against ticks (44). Repellents on clothing may last several days, and the cosmetic acceptability and toxicity requirements may not be as stringent.

## G. Use of Repellents on Wide-Mesh Netting

There has been an interest in repellents that, on application to wide-mesh netting, would prevent mosquitoes and other biting flies from access to the host without hampering his vision and air circulation. Medical entomologists in the U.S.S.R. have shown much interest in this area (131)-(135). Deet was effective on wide-mesh (18-mm) nets against sandflies (133) and black flies (132), while hexyl mendelate was most effective against *Aedes* spp. (134). In the United States, Gouck et al. (136) (137) and McGovern et al. (138) reported several compounds suitable for net treatment, some of them being carbanilates, benzamides, aliphatic amides, and immides (138). In other studies made in Thailand, deet or M-1960 when impregnated on wide-mesh net at a rate of 0.5 g of repellent on 1 g of net gave complete protection against *C. p. quinquefasciatus* and *A. aegypti* for 15 to 17 weeks (139). Grothaus et al. (140) reported 1-(o-ethoxybenzoyl) piperidine effective for more than 1 yr and o-ethoxy-N, N-dipropylbenzamide for more than 2 yr against mixed mosquito populations. The repellents were tested at 0.5 g/g net weight.

## H. Systemic Repellents

Two important drawbacks of topical repellents are the short duration of action due to removal from skin by absorption, evaporation, and abrasion and the sometimes incomplete coverage of vulnerable skin areas. A systemic repellent would be an ideal choice if available. The theoretical characteristics for such a repellent include effectiveness against a wide variety of hematophagous arthropods, oral route of administration, long duration of action (*ca.* 24 hr), low toxicity, and long shelf life (141) (142). Kingscote (143) administered several repellents orally to animals, and Bar-Zeev and Smith (144) injected deet intravenously to see if repellents worked systematically. The results were negative. Shannon (145) reported thiamine chloride (vitamin $B_1$) effective as a systemic repellent, but Wilson et al. (146) were unable to confirm its systemic repellency. It was also tested at the USDA laboratory, Gainesville, Florida, and was found ineffective (44). We found it ineffective topically and systematically (147). Strauss et al. (148) studied the effect of 113 drugs (vitamins included) for repellent effect on hospital patients, and found none promising.

## III. CONCLUSIONS

In view of the present crisis in the control of harmful insects due to insecticide resistance and the importance of a cleaner environment for our survival, the use of attractants and repellents as a method of control becomes of vital importance. There is need for potent mosquito attractants and repellents lasting a long time with one application. Our knowledge of attractants and repellents is too skimpy at the moment for us to realize these goals. We do not have chemicals that duplicate the attraction of a natural host. We have repellents, but we do not know why and how they work. Their effect on mosquito behavior has been studied (17) (18) (149) (150), but we only know what happens in the presence of a repellent—not why and how. Perhaps structure-activity correlation and multifactorial-regression analysis will provide a key to a more rational approach to synthesizing better repellents. Regarding how repellents work, Wright (151) noted that the far infrared spectra of repellents is similar to that of water, and suggested that repellents act by masquerading as water vapor. Since very high humidity repels mosquitoes, Wright claims that repellents would prove ineffective in a dry atmosphere. This hypothesis was tested by Hocking and Khan (17), and was found incorrect. We need more information not only regarding the mode of action of repellents but also regarding the structure and function of chemoreceptors of insects including mosquitoes. The literature was recently reviewed by Slifer (152). The recent work of Lacher (153) (154), Kellogg (155), Reddy (156), and Davis (157) (160) has contributed some understanding of the effect of repellents on mosquito receptors at the cellular level. Emerson, the American sage and philosopher, said that nature keeps much on her table but more in her closet. Regarding the sensory physiology of mosquitoes, the chemical nature of host doors, and the essential requirements for a compound to act as a repellent, the secrets are mostly in the closet. The need for a long-lasting repellent is obvious, especially now when chemical control of mosquitoes is becoming more and more difficult because of resistance to pesticides and to the demand for a cleaner environment, necessitating their decreasing use. The importance of research in this field cannot be overemphasized.

## REFERENCES

(1)  P. Karlson and A. Butenandt. 1959. Pheromones (ectohormones) in insects. *Ann. Rev. Entomol.* **4**:39-58.

(2)  P. Karlson and M. Luscher. 1959. "Pheromones": A new term for a class of biologically active substances. *Nature* **183**:55-56.

(3)  M. Jacobson. 1965. *Insect Sex Attractants.* Wiley, New York.

(4)  H.H. Shorey. 1973. Behavioral responses to insect pheromones. *Ann. Rev. Entomol.* **18**:349-380.

(5)  C.M. Gjullin, J.O. Johnsen, and F.H. Plapp. 1965. The effect of odors released by various waters on the oviposition sites selected by two species of *Culex*. *Mosquito News* **25**:268-271.

(6)  J.W. Kliever, T. Miura, R.C. Husbands, and C.H. Hurst. 1966. Sex pheromones and mating behavior of *Culiseta inornata* (Diptera: Culicidae). *Entomol. Soc. Am. Ann.* **59**:530-533.

(7)    T. Ikeshoji. 1966. Studies on mosquito attractants and stimulants. Part I. Chemical factors determining the choice of oviposition site by *Culex pipiens fatigans* and *pallens*. Part II. The presence in mosquito breeding waters of a factor which stimulates oviposition. *Jap. J. Expt. Med.* **36**:49-59; 67-72.

(8)    T. Ikeshoji, T. Umino, and S. Hirakoso. 1967. Studies on mosquito attractants and stimulants. Part IV. An agent producing stimulative effects for oviposition *Culex pipiens fatigans* in field water and the stimulative effects of various chemicals. *Jap. J. Expt. Med.* **37**:67-69.

(9)    T. Ikeshoji and M.S. Mulla. 1970. Oviposition attractants for four species of mosquitoes in natural breeding waters. *Entomol. Soc. Am. Ann.* **63**:1322-1327.

(10)   R.S. Soman and R. Reuben. 1970. Studies on the preference shown by ovipositing females of *Aedes aegypti* for water containing immature stages of the same species. *J. Med. Entomol.* **7**:485-489.

(11)   C.E. Osgood. 1971. An oviposition pheromone associated with the egg rafts of *Culex tarsalis. J. Econ. Entomol.* **64**:1038-1041.

(12)   E.I. Hazard, M.S. Mayer, and K.E. Savage. 1967. Attraction and oviposition stimulation of gravid female mosquitoes by bacteria isolated from hay infusions. *Mosquito News* **27**:133-136.

(13)   C.W. Thaggard and D.A. Eliason. 1969. Field evaluation of components for an *Aedes aegypti* (L.) oviposition trap. *Mosquito News* **29**:608-612.

(14)   J.D. Gillett, A.J. Haddow, and P.S. Corbet. 1962. The sugar feeding cycle in a cage population of mosquitoes. *Entomol. Exp. Appl.* **5**:223-232.

(15)   B. Hocking, W.R. Richards, and C.R. Twinn. 1950. Observations on the bionomics of some northern mosquito species (Culicidae: Diptera). *Can. J. Res.* D **28**:58-80.

(16)   M.W. Service. 1969. Observations on the ecology of some British mosquitoes. *Bull. Entomol. Res.* **59**:161-194.

(17)   B. Hocking and A.A. Khan. 1966. The mode of action of repellent chemicals against blood-sucking flies. *Can. Entomol.* **98**:821-831.

(18)   H. Kalmus and B. Hocking. 1960. Behavior of *Aedes* mosquitoes in relation to blood-feeding and repellents. *Entomol. Exp. Appl.* **3**:1-26.

(19)   B. Hocking. 1963. The use of attractants and repellents in vector control. *Bull. World Health Org.* **29** (Suppl.):121-126.

(20)   B. Hocking. 1971. Blood-sucking behavior of terrestial arthropods. *Ann. Rev. Entomol.* **16**:1-26.

(21)   W. Rudolfs. 1922. Chemotropism of mosquitoes. *N.J. Agric. Exp. Sta.* Bull. No. 367, pp. 1-23.

(22)   D.C.D. Happold. 1965. Mosquito ecology in Central Alberta. II. Adult population and activities. *Can. J. Zool.* **43**:821-846.

(23)   D.S. Bertram and I.A. McGregor. 1956. Catches in the Gambia, West Africa of *Anopheles gambiae* Giles and *A. gambiae* var. *melas* Theobald in entrance traps of a baited portable wooden hut, with special reference to the effect of wind direction. *Bull. Entomol. Res.* **47**:669-681.

(24)   A.A. Khan, H.I. Maibach, W.G. Strauss, and W.R. Fenley. 1966. Quantitation of effect of several stimuli on the approach of *Aedes aegypti. J. Econ. Entomol.* **59**:690-694.

(25)   M.T. Gillies, and T.J. Wilkies. 1969. A comparison of the range of attraction of animal baits and of carbon dioxide for some West African mosquitoes. *Bull. Entomol. Res.* **59**:441-456.

(26)   M.T. Gillies and T.J. Wilkies. 1970. The range of attraction of single baits for some West African mosquitoes. *Bull. Entomol. Res.* **60**:225-235.

(27)    M.T. Gillies and T.J. Wilkies. 1972. The range of attraction of animal baits and carbon dioxide for mosquitoes. Studies in a fresh water area of West Africa. *Bull. Entomol. Res.* **61**:389-404.

(28)    T.J. Headlee. 1934. Mosquito work in New Jersey for year 1933. *J. Proc. 21st Ann. Meeting N.J. Mosquito Ext. Assoc.* 8-37.

(29)    W.C. Reeves. 1951. Field studies on carbon dioxide as a possible host stimulant to mosquitoes. *Proc. Soc. Exp. Biol. Med.* **77**:64-66.

(30)    W.C. Reeves. 1953. Quantitative field studies on carbon dioxide chemotropism of mosquitoes. *Am. J. Trop. Med. Hyg.* **2**:325-331.

(31)    V.G. Newhouse, R.W. Chamberlain, J.G. Johnson, and W.G. Sudia. 1966. Use of dry ice to increase mosquito catches of the CDC miniature light trap. *Mosquito News* **26**:30-35.

(32)    R.R. Carestia and L.B. Savage. 1967. Effectiveness of carbon dioxide as a mosquito attractant in the CDC miniature trap. *Mosquito News* **27**:90-92.

(33)    J.R. Anderson and W. Olkowski. 1968. Carbon dioxide as an attractant for host-seeking Cephenemyia females. *Nature* **220**:190-191.

(34)    M.T. Gillies and W.F. Snow. 1967. A $CO_2$-baited sticky trap for mosquitoes. *Trans. Roy. Soc. Trop. Med. Hyg.* **61**:20.

(35)    E.R. Willis. 1947. The olfactory responses of female mosquitoes. *J. Econ. Entomol.* **40**:769-778.

(36)    A.W.A. Brown. 1951. Studies of the responses of the female *Aedes* mosquito. Part IV. Field experiments on Canadian species. *Bull. Entomol. Res.* **42**:575-582.

(37)    A.W.A. Brown, D.S. Sarkaria, and R.P. Thompson. 1951. Studies on the responses of the female *Aedes* mosquito. Part 1. The search for attractant vapors. *Bull. Entomol. Res.* **42**:105-114.

(38)    J.J. Laarman. 1955. The host-seeking behavior of the malaria mosquito *Anopheles maculipennis atroparvaus. Acta Leidensia* **25**:1-144.

(39)    R.H. Wright. 1962. The attraction and repulsion of mosquitoes. *World Rev. Pest Control* **1**:20-30.

(40)    M.S. Mayer and J.D. James. 1969. Attraction of *Aedes aegypti* to human arms, carbon dioxide, and air currents in a new type of olfactometer. *Bull. Entomol. Res.* **58**:629-642.

(41)    A.A. Khan and H.I. Maibach. 1966. Quantitation of effect of several stimuli on landing and probing by *Aedes aegypti. J. Econ. Entomol.* **59**:902-905.

(42)    A.A. Khan, W.G. Strauss, H.I. Maibach, and W.R. Fenley. 1967. Comparison of the attractiveness of the human palm and other stimuli to the yellow fever mosquito. *J. Econ. Entomol.* **60**:318-320.

(43)    A.A. Khan and H.I. Maibach. 1972 Effect of human breath on mosquito attraction to man. *Mosquito News* **32**:11-15.

(44)    C.N. Smith. 1970. Repellents for anopheline mosquitoes. *Misc. Publ. Entomol. Soc. Am.* **7**:99-115.

(45)    V.G. Dethier, L. Barton, Browne, and C.N. Smith. 1960. The designation of chemicals in terms of the response they elicit from insects. *J. Econ. Entomol.* **53**:134-136.

(46)    A.N. Clements. 1963. *The Physiology of Mosquitoes.* Macmillan, New York.

(47)    A.W.A. Brown. 1966. The attraction of mosquitoes to hosts. *JAMA* **196**:159-162.

(48)    F.M. Howlett. 1910. The influence of temperature upon the biting of mosquitoes. *Parasitology* **3**:479-484.

(49)    D.G. Peterson and A.W.A. Brown. 1951. Studies on the response of the female *Aedes* mosquito. Part III. The response of *Aedes aegypti* (L.) to a warm body

and its radiation. *Bull. Entomol. Res.* **42**:535-541.

(50)   A.A. Khan and H.I. Maibach. 1971. A study of the probing response of *Aedes aegypti*. 4. Effect of dry and moist heat on probing. *J. Econ. Entomol.* **64**:442-443.

(51)   M.K. Smart and A.W.A. Brown. 1956. Studies on the responses of the female *Aedes* mosquitoe. Part VII. The effect of skin temperature, hue and moisture on the attractiveness of the human hand. *Bull. Entomol. Res.* **47**:80-100.

(52)   H.K. Gouck and M.C. Bowman. 1959. Effect of repellents on the evolution of carbon dioxide and moisture from human arms. *J. Econ. Entomol.* **52**:1157-1159.

(53)   A.H. Parker. 1948. Stimuli involved in the attraction of *Aedes aegypti* to man. *Bull. Entomol. Res.* **39**:387-397.

(54)   L. Burgess. 1959. Probing behavior of *Aede aegypti* (L.) in response to heat and moisture. *Nature* **184**:1968-1969.

(55)   A.A. Khan and H.I. Maibach. 1970. A study of the probing response of *Aedes aegypti*. I. Effect of nutrition on probing. *J. Econ. Entomol.* **63**:974-976.

(56)   A.A. Khan and H.I. Maibach. 1971. A study of the probing response of *Aedes aegypti*. 2. Effect of dessication and blood feeding on probing to skin and an artificial target. *J. Econ. Entomol.* **64**:439-442.

(57)   A.A. Khan, H.I. Maibach, and W.G. Strauss. 1968. The role of convection currents in mosquito attraction to human skin. *Mosquito News* **28**:462-464.

(58)   J.J. Laarman. 1958. The host-seeking behavior of anopheline mosquito. *Trop. Geogr. Med.* **10**:293-305.

(59)   R. Brouwer. 1960. The attraction of carbon dioxide excreted by the skin of the arm for malaria mosquitoes. *Trop. Geogr. Med.* **12**:62-66.

(60)   R.P. Thompson and A.W.A. Brown. 1955. The attractiveness of human sweat to mosquitoes and the role of carbon dioxide. *Mosquito News* **15**:80-84.

(61)   U. Rahm. 1956. Zum Problem der Attraktion von Stechmucken durch den Menschen. *Acta Trop.* (Basel) **13**:319-344.

(62)   U. Rahm. 1957. Zur Bedentung des Duftes und des Schweisses bei der Attraktion von *Aedes aegypti* durch den Menschen. *Acta Trop.* **(Basel)** **14**:208-217.

(63)   A.A. Khan and H.I. Maibach. 1970. A study of the probing response of *Aedes aegypti* III. Effect of oviposition on probing. *J. Econ. Entomol.* **63**:2009.

(64)   V. Rahm. 1958. Die attraktive Wirkung der vom Menschen abgegebenen Duftstoffe auf *Aedes aegypti* L. *Z. Tropenmed. Parasit.* **9**:146-156.

(65)   R. Brouwer. 1960. Variations in human body odor as a cause of individual differences of attraction for malaria mosquitoes. *Trop. Geogr. Med.* **12**:186-192.

(66)   R.C. Muirhead-Thomson. 1951. The distribution of anopheline mosquito hosts among different age groups. A new factor in malaria epidemiology. *Brit. Med. J.* **1**:1114-1117.

(67)   T.A. Freyvogel. 1961. Ein Beitragzuden Problemen um die Blutmahlzeit von Stechmucken. *Acta Trop.* **18**:201-251.

(68)   H.I. Maibach, A.A. Khan, and W.G. Strauss, 1966. Attraction of humans of different age groups to mosquitoes. *J. Econ. Entomol.* **59**:1302-1303.

(69)   A.A. Khan, H.I. Maibach, W.G. Strauss, and W.R. Fenley. 1965. Screening humans for degrees of attractiveness to mosquitoes. *J. Econ. Entomol.* **58**:694-697.

(70)   A.A. Khan, H.I. Maibach, and W.G. Strauss. 1966. Mosquito probing as a measure of host attractiveness. *Mosquito News* **26**:210-212.

(71) W.G. Strauss, A.A. Khan, and H.I. Maibach. 1966. Pertinacity of host-seeking behavior of *Aedes aegypti. Nature* 210:759-760.

(72) E. Roubaud. 1920. Le conditions de nutrition des *Anopheles* en France (*Anopheles maculipennis*) et le role de betail dans la prophylaxie due paludisme. *Ann. Inst. Pasteur* (Paris) 34:181-228.

(73) L.J. Bruce-Chwatt and C.W. Goekel. 1960. A study of the blood-feeding patterns of *Anopheles* mosquitoes through precipitin test. *Bull. World Health Org.* 22:685-720.

(74) G.A.H. McClelland and B. Weitz. 1963. Serological identification of the natural hosts of *Aedes aegypti* (L.) and some other mosquitoes caught resting in vegetation in Kenya and Uganda. *Ann. Trop. Med. Parasitol.* 57:214-224.

(75) A.A. Khan, H.I. Maibach, W.G. Strauss, and J.L. Fisher. 1970. Differential attraction of the yellow fever mosquito to vertebrate hosts. *Mosquito News* 30:43-47.

(76) M. Bates and L.W. Hackett. 1939. The distinguishing characteristics of the populations of *Anopheles maculipennis* found in Southern Europe. Verh. VII. *Int. Kongr. Entomol.* (Berlin, 1938). 3:1555-1569.

(77) S.B. McIver. 1968. Host preferences and discrimination by the mosquitoes *Aedes aegypti* and *Culex tarsalis* (Diptera: Culicidae). *J. Med. Entomol.* 5:422-428.

(78) J. Reuter. 1936. Orienteerend onderzoek naar de orrzaak van het gedrag van *Anopheles maculipennis* Meigen bij de voedselkeuze. Thesis, Leiden.

(79) A.W.A. Brown. 1958. Factors which attract *Aedes* mosquitoes to humans. *Proc. 10 Int. Congr. Entomol.* 3:757-763.

(80) H.I. Maibach, A.A. Khan, W.G. Strauss, and T.R. Pearson. 1966. Attraction of anhidrotic subjects to mosquitoes. *Arch. Derm.* 94:215-217.

(81) A.A. Khan, H.I. Maibach, W.G. Strauss, and J.L. Fisher. 1969. Increased attractiveness of man to mosquitoes with induced eccrine sweating. *Nature* 223:859-860.

(82) W.A. Skinner, H. Tong, T. Pearson, W.G. Strauss, and H.I. Maibach. 1965. Human sweat components attractive to mosquitoes. *Nature* 207:661-662.

(83) W.A. Skinner, H. Tong, H. Johnson, H.I. Maibach, and D. Skidmore. 1968. Human sweat components attractancy and repellency to mosquitoes. *Experientia* 24:679-680.

(84) P.H. van Thiel and C. Weurman. 1947. L'attraction excercee sur *Anopheles maculipennis atroparvus* par l'acide carbonique dans l'appareil de cheix II. *Acta Trop.* (Basel) 4:1-9.

(85) L. Burgess and A.W.A. Brown. 1957. Studies on the responses of the female *Aedes* mosquito. Part VIII. The attractiveness of beef blood to *Aedes aegypti* (L.) *Bull. Entomol. Res.* 48:783-794.

(86) B. Schaerffenberg and E. Kupka. 1953. Orientierungsversuche an *Stomoxys calcitrans* und *Culex pipiens* mit einem Blutduftstof. *Trans. Int. Congr. Entomol.* IX., Amsterdam (1951). I:359-361.

(87) A.W.A. Brown and A.G. Carmichael. 1961. Lysine and alanine as mosquito attractants. *J. Econ. Entomol.* 54:317-324.

(88) T. Ikeshoji. 1967. Enhancement of the attractiveness of mice as mosquitoe bait by injection of methionine and its metabolites. *Jap. J. Sanit. Zool.* 18(2/8):101-107.

(89) T. Ikeshoji, T. Umino, and T. Sazuki. 1963. On the attractiveness of some amino acids and their decomposed products for the mosquito *Culex pipiens pallens. Jap. J. Sanit. Zool.* 14:152-156.

(90) H.P. Roessler. 1961. Versuche zur geruchlichen Anlockungweiblicher Stemucken (*Aedes aegypti* L., *Culicidae*). *Z. vgl. Physiol.* **44**:184-231.

(91) F. Acree Jr., R.B. Turner, H.K. Gouck, M. Beroza, and N. Smith. 1968. *L*-Lactic acid: A mosquito attractant isolated from humans. *Science* **161**:1346-1347.

(92) T.P. McGovern, H.K. Gouck, M. Beroza, and J.C. Ingangi. 1970. Esters of α-hydroxy-β-phenyl aliphatic acids that attract female yellow fever mosquitoes. *J. Econ. Entomol.* **63**:2002-2004.

(93) P. Granett. 1940. Studies of mosquito repellents. I. Test procedure and method of evaluating test data. II. Relative performance of certain chemicals and commercially available mixtures as mosquito repellents. *J. Econ. Entomol.* **33**:563-572.

(94) P. Granett. 1944. Paired product testing for the evaluation of mosquito repellents. *Proc. N.J. Mosquito Exterm. Assoc.* **31**:173-178.

(95) C.N. Smith, I.H. Gilbert, H.K. Gouck, M.C. Bowman, F. Acree Jr., and C.H. Schmidt. 1963. Factors affecting the protection period of mosquito repellents. *USDA Tech. Bull.* No. 1285.

(96) A.A. Khan, H.I. Maibach, and D.L. Skidmore. 1973. A study of insect repellents. 2. Effect of temperature on protection time. *J. Econ. Entomol.* **66**:437-438.

(97) B.V. Travis, F.A. Morton, H.A. Jones, and J.H. Robinson. 1949. The more effective mosquito repellents tested at the Orlando, Fla., laboratory, 1942-47. *J. Econ. Entomol.* **42**:686-94.

(98) B.V. Travis. 1950. Known factors causing variation in results of insect repellent tests. *Mosquito News* **10**:126-132.

(99) P. Granett and H. Haynes. 1945. Insect repellent properties of 2-ethyl-1, 3-hexanediol. *J. Econ. Entomol.* **38**:671-675.

(100) R.M. Altman. 1969. Repellent tests against *Anopheles albimanus* Wiedmann in the Panama Canal Zone. **Mosquito News 29**:110-112.

(101) I.H. Gilbert, J.E. Scanlon, and D.L. Baily. 1970. Repellents against mosquitoes in Thailand. *J. Econ. Entomol.* **63**:1207-1209.

(102) A.A. Khan, H.I. Maibach, and D.L. Skidmore. 1975. Insect repellents: Effect of mosquito and repellent related factors on protection time. *J. Econ. Entomol.* **68**:43-45.

(103) W.V. King. 1954. Chemicals evaluated as insecticides and repellents at Orlando, Fla. *USDA, Agric. Handb.* No. 69.

(104) Entomology Research Division, *Agric.* Res. Serv., USDA, 1967. Materials evaluated as insecticides, repellents and chemosterilants at Orlando and Gainesville, Florida 1952-64. *USDA Handb.* No. 340.

(105) I.H. Gilbert. 1971. Evaluation of repellent mixtures against three species of mosquitoes and stable flies. *USDA, ERD-ARS.* Special Rep. No. 71-01 G.

(106) I.H. Gilbert and N. Smith. 1972. Evaluation of individual repellents and repellent mixtures as skin applications against four species of mosquitoes, stable flies and deer flies. *USDA, ERD-ARS,* Special Rep. No. 72-02 G.

(107) H.L. Johnson, W.A. Skinner, H.I. Maibach, and T.R. Pearson. 1967. Repellent activity and physical properties of ring-substituted *N,N*-diethylbenzamides. **J.** Econ. Entomol. **60**:173-176.

(108) H.L. Johnson, W.A. Skinner, D. Skidmore, and H.I. Maibach. 1968. Topical mosquito repellents. II. Repellent potency and duration in ring-substituted *N,N*-dialkyl and amino alkylbenzamides. *J. Med. Chem.* **11**:1265-1268.

(109) G.S. Pervomaisky, V.T. Osunyan, V.B. Kazhdan, I.K. Mosly, A.K. Shustrov, B.C. Grawovsky, D.T. Zhodolev, and E.D. Kinaiva. 1967. A new insect repellent hexamethylenbutane sulfonamide. *Med. Parazitol.* **36**:730-733.

(110) F. Gualitieri, P. Tsakotellis, W. Skinner, H. Johnson, D. Skidmore, and H.

Maibach. 1973. Topical mosquito repellents VI: Sulfonamides and quinoline-4-carboxylic acid derivatives. *J. Pharmacol. Sci.* 62:849-851.

(111)  F. Gualtieri, H. Johnson, H. Tong, H. Maibach, D. Skidmore, and W. Skinner. 1973. Topical mosquito repellents V: Benzyl ethers. *J. Pharmacol. Sci.* 62:487-489.

(112)  H. Johnson, J. DeGraw, J. Engstrom, W.A. Skinner, V.H. Brown, D. Skidmore, and H.I. Maibach. 1975. Topical mosquito repellents VII: Alkyl triethylene glycol monoethers. *J. Pharm. Sci.* 64:693-695.

(113)  R.P. Quintana, L.R. Garson, and A. Lasslo. 1968. Synthesis of phenolic esters with potential long lasting insect repellent properties. *Can. J. Chem.* 46:2835-2842.

(114)  R.P. Quintana, L.R. Garson, and A. Lasslo. 1969. Monomolecular films of compounds with potential dermophilic and prophylactic properties. Ester of dihydroxyacetone. *Can. J. Chem.* 47:853-856.

(115)  L.R. Garson and R.P. Quintana. 1969. Esters of undecanoic acid with potential long lasting insect repellent activity. *J. Med. Chem.* 12:538-540.

(116)  R.P. Quintana, A. Lasslo, L.R. Garson, and C.N. Smith. 1970. Dermophilic insect repellents with perdurable efficacy. *J. Pharmacol. Sci.* 59:1503-1505.

(117) R.P. Quintana, L.R. Garson, A. Lasslo, S.I. Sanders, J.H. Buckner, H.K. Gouck, I.H. Gilbert, D.E. Weidhaas, and C.E. Schrek. 1970. Mosquito-repellent dihydroxy acetone monoesters. *J. Econ. Entomol.* 63:1128-1131.

(118)  R.P. Quintana, P.T. Mui, A. Lasslo, M.A. Boulware, C. Schreck, and H.K. Gouck. 1972. Synthesis of insect repellent amino analogs of 2-ethyl-1,3-hexanediol (Rutgers 612) *J. Med. Chem.* 15:1073-1074.

(119)  R.P. Quintana, P.T. Mui, R.G. Fisher, C. Schreck, and H.K Gouck. 1972. Dihydroxyacetone monoester mosquito repellents. Effect of branching cyclization and unsaturation upon repellent efficacy. *J. Econ. Entomol.* 65:66-69.

(120)  P. Tsakotellis, H.L. Johnson, W.A. Skinner, D. Skidmore, and H.I. Maibach. 1971. Topical mosquito repellents III: Carboxamide acetals and ketals and related carbonyl addition derivatives. *J. Pharmacol. Sci.* 60:84-89.

(121)  B.V. Travis, F.A. Morton, and W.W. Yates. 1956. Effectiveness of repellents against *Culicine* mosquitoes. *USDA ARS*-33-19.

(122)  S.I. Kharitonova and I.V. Koshkina. 1969. Poiski prolongatorov dlya repellentov. I. Laboratornye ispylaniya nekotorykh vestrchestv v Kachestve prolongatorov dimetilftalata. *Med. Parazitol.* 36:707-710.

(123)  I.V. Koshkina and S.I. Kharitonova. 1970. K voprosu ob etiltsellyuloze kak prolongatore repellentov. II *Med. Parazitol.* 39:224-227.

(124)  A.A. Khan, H.I. Maibach, and D.L. Skidmore. 1975. Addition of perfume fixatives to mosquito repellents to increase protection time. *Mosquito News* 35:23-26.

(125)  A.A. Khan, H.I. Maibach, and D.L. Skidmore. 1975. Addition of vanillin to mosquito repellents to increase protection time. *Mosquito News* 35:223-225.

(126)  I.H. Gilbert, H.K. Gouck, and C.N. Smith. 1957. New insect repellent. Part 1. *Soap Chem. Spec.* 33:115-117; 129-133.

(127)  I.H. Gilbert, H.K. Gouck, and C.N. Smith. 1957. New insect repellent. Part II. *Soap Chem. Spec.* 33:95-99; 109.

(128)  S.R. Christophers. 1947. Mosquito repellents. Being a report of the work of the mosquito repellent inquiry, Cambridge 1943-45. *J. Hyg.* 45:176:231.

(129)  A.P. Kurtz, J.A. Logan, and W.A. Akers. 1973. More effective topical repellents against malaria bearing mosquitoes. Rep. 13, Letterman Army Institute of Research, Presidio, San Francisco.

130)  A.A. Khan, H.I. Maibach, and D.L. Skidmore. 1976. Polymer to increase.

abrasion and water-wash resistance of mosquito repellents. Mosquito News (in press).

(131)   V.M. Saf'yanova. 1963. Results of testing individual bed nets impregnated with repellents for protection from sand flies. *Med. Parazitol.* 35:549-551.

(132)   M.L. Fedder, E.B. Kerbabaev, and A.N. Alekseyev. 1962. Effectiveness of some new repellents against black flies (Simuliidae). *Med. Parazitol.* 34:429-434.

(133)   V.M. Saf'yanova and V.M. Ktkov. 1964. Net impregnated with repellent for protection from sand flies. *Zdravookhr. Turkm.* 8:36-39.

(134)   D. Novak. 1965. Prispevek k orientacnim testum repelenta komaru. *Farmakoterapeuticke zpravy Spofa,* 11-1-1965, pp. 89-92.

(135)   D.T. Zhogolev. 1970. Use of nets and special wearing apparel impregnated with hexamethylene butane sulfonamide for protection against mosquito attacks in a hot climate. *Med. Parazitol.* 39:499.

(136)   H.K. Gouck, T.P. McGovern, and M. Beroza. 1967. Chemicals tested as space repellents against yellow fever mosquitoes. I. Esters. *J. Econ. Entomol.* 60:1587-1590.

(137)   D.R. Gouck, D.R. Godwin, C.E. Schreck, and N. Smith. 1967. Field tests with repellent-treated netting against black salt-marsh mosquitoes. *J. Econ. Entomol.* 60:1451-1452.

(138)   T.P. McGovern, M. Beroza, and H. Gouck. 1967. Chemicals tested as space repellents for yellow fever mosquitoes. II. Carbanilates, benzamides, aliphatic amides, and imides. *J. Econ. Entomol.* 60:1591-1594.

(139)   H.K. Gouck and M.A. Moussa. 1969. Field tests with bed nets treated with repellents to prevent mosquito bites. *Mosquito News* 29:263-264.

(140)   R.H. Grothaus, J.M. Hirst, H.K. Gouck, and D.E. Weidhaas. 1972. Field tests with repellent-treated wide-mesh netting against mixed mosquito populations. *J. Med. Entomol.* 9:149-152.

(141)   J.L. Sherman. 1966. Development of a systemic insect repellent. Symposium: Insects and disease. *JAMA* 196:166-168.

(142)   A.A. Kingscote. 1958. Orally administered insect repellents: Approaches and problems related to the search. *Proc. 10th Int. Congr. Entomol.,* Montreal 3:799-800.

(143)   A.A. Kingscote. 1952. Studies on the internal administration of known and potential insect repellents. *Can. Defense Res. Board, Environ. Protection Tech. Rep.* No. 2.

(144)   M. Bar-Zeev and C.N. Smith. 1959. Action of repellents feeding through treated membranes or on treated blood. *J. Econ. Entomol.* 52:263-267.

(145)   W. Shannon. 1943. Thiamine chloride—an aid in the solution of the mosquito problem. *Minn. Med.* 26:799-802.

(146)   C.W. Wilson, D.R. Mathieson, and L.A. Jachowski. 1944. Ingested thiamine chloride as a mosquito repellent. *Science* 100:147.

(147)   A.A. Khan, H.I. Maibach, W.G. Strauss, and W.R. Fenley. 1969. Vitamin $B_1$ is not a systemic mosquito repellent in man. *Trans. St. John's Hosp. Derm. Soc.* 55:99-102.

(148)   W.G. Strauss, H.I. Maibach, and A.A. Khan. 1968. Drugs and disease as mosquito repellents in man. *Am. J. Trop. Med. Hyg.* 17:461-464.

(149)   A.A. Khan. 1965. Effect of repellents on mosquito behavior. *Quaest. entomol.* 1:1-35.

(150)   A.A. Khan and H.I. Maibach. 1972. A study of insect repellents 1. Effect on the flight and approach by *Aedes aegypti. J. Econ. Entomol.* 65:1318-1321.

*(151)* R.H. Wright. *1957. A theory of olfaction and the action of mosquito repellents. Can. Entomol.* 89:518-528.

(152) E.H. Slifer. 1970. The structure of arthropod chemoreceptors. *Ann. Rev. Entomol.* **15**:121-142.

(153) V. Lacher. 1967. Elektrophyseologische Untersuchungen an einzelnen Geruchsrezeptoren auf den Antennen weiblicher Moskito (*Ades aegypti,* L.) *J. Insect Physiol.* **13**:1461-1470.

(154) V. Lacher. 1971. Electrophysiological equipment for measuring the activity of single olfactory nerve cells on the antennae of mosquitoes. *J. Econ. Entomol.* **64**:313-314.

(155) F.E. Kellogg. 1970. Water vapor and carbon dioxide receptors in *Aedes aegypti. J. Insect Physiol.* **16**:99-108.

(156) M.J. Reddy. 1970. The mode of action of insect repellents II: Electrophysiological studies. *Quaest. entomol.* **6**:353-363.

(157) E. Davis and C.S. Rebert. 1972. Elements of olfactory receptor coding in the yellow fever mosquito *Aedes aegypti. J. Econ. Entomol.* **65**:1058-1061.

(158) E. Davis. 1973. An electrophysiological approach to mosquito control. *R/D* **24**:32-34; 36.

(159) E. Davis. 1974. Identification of antennal chemoreceptors of *Aedes aegypti,* a correction. *Experientia* **30**:1282-1283.

(160) E. Davis and P.G. Sokolove. 1975. Temperature responses of antennal receptors of the mosquito *Aedes aegypti. J. Comp. Physiol.* **96**:223-236.

## CHAPTER 19

# Attractants for Fruit Fly Survey and Control

Derrell L. Chambers

*Insect Attractants, Behavior and Basic Biology Research Laboratory*
*Agricultural Research Service, USDA,*
*Gainesville, Florida*

To judge where we are now and where are we going in the use of fruit fly attractants we need an historical perspective on where we have been. Perhaps a description of the intensive, and largely mission-oriented work early researchers have done with attractants for tephritids may be perceived to have scope and implications not just within the international community of fruit fly specialists, but beyond that community as well.

Perhaps the history can best be put in perspective by quoting from Howlett (1), who in that charming style of 60 years ago wrote some things about the attraction of the oriental fruit fly, *Dacus dorsalis* (Hendel), and its kin, *D. zonatus* (Saunders), that are hauntingly familiar to readers of today's literature: "...I heard that a neighbor had been troubled by some kind of fly settling on him at a time when he was using oil of citronella as a mosquito deterrent." Howlett had been searching for a means of inducing *D. zonatus* females to lay eggs, and was experimenting in Pusa, India, with essential oils; he wetted a handkerchief with citronella, exposed it in the area of a peach orchard:

"...in less than half an hour the handkerchief, lying in a crumpled heap, was almost hidden by a crowd of *D. zonatus* and presented a very striking appearance. I jumped at once to the conclusion that the economic problem of how to destroy female fruit flies had found an easy solution, but on examination it was soon apparent that all the flies on the handkerchief were males. . . . Since the reaction was confined to the male sex and did not appear to be in any way connected with feeding habits, it seemed most reasonable to suppose that the smell might resemble some sexual odour of the female which in natural conditions served to guide the male to her."

Howlett overlooked the significance of application of a potent male attractant in manipulating reproduction (to which I will later return). However, Froggatt (2), who worked with Howlett, suggested the possibility of control by the male annihilation method. Also, Howlett subsequently (3) noted he was incorrect (insofar as we know) in equating citronella with the pheromone of the female, but he made a number of important observations relative to behavior, trapping, and olfactory morphology, and drew interesting conclusions that are, as I said, familiar:

"...the olfactory sense of flies may be highly developed in certain directions and within certain narrow limits, while outside these limits it is comparatively inoperative. We should expect to find on this hypothesis instances where the males were very sensitive to smell of the females or *vice versa,* the attractiveness, however, probably confined to one sex .... Regarding the matter as thus crudely put, we might look on each species as tuned to respond to three or four notes on the scale of smell, and we should find the most delicate adjustment and most accurate tuning in the direction of sexual smell, since errors of perception would be most disadvantageous to the species....If we accept for the moment some such view as this, then among those species in which the male finds the female by smell we must regard each one as an assemblage of individuals in which one sex is tuned to respond to a definite kind of molecular vibration corresponding to some compound or mixture of compounds emitted by the other sex, and these compounds would thus constitute definite specific characters. We might even perhaps go further and define some of the larger groups by those 'generic' smells which characterize certain kinds of chemical substances...and which depend on the presence of certain atoms or atomic groups of some particular configuration."

I do not know Howlett's background in chemistry or physics, so his propositions on mechanisms may be coincidentally similar to those being considered currently, but his theme is certainly apropos to most of the basic and applied theories now being considered and utilized.

By Howlett's time many papers had been published on the morphology, mechanism, and role of odor perception. According to Yearsley (4), more than 100 papers on the functions of the antenna were written between 1730 and 1883, so we are still splitting hairs, so to speak; see, for example, Pierret (5), who refers to pheromone tracking in 1841, and Ramsay (6), who postulated an infrared absorption theory of olfaction in 1882. However, I have a particular bias toward Howlett's observations and comments for their clarity, predictiveness, and eventual application, and find that the perspective I mentioned is akin to looking at ourselves through a rolled-up newspaper 60 years long; the image is somewhat reduced. If we are still image-conscious, let me quote from Lucretius, 65 B.C. [cited from Hocking (7)]:

"Now listen and I will deal with the question how the impact of odour affects the nose. First there must be a number of things from which rolls flowing a manifold stream of odours, and we must think these flow and are sped and scattered everywhere; but different odours are more fitted to different creatures, because of their differing forms. And therefore bees are drawn to any distance by the scent of honey. So smell is caused by streams of particles which enter the nostrils; and their different shapes affect creatures differently."

So much for the generalities, which leave us simply to work out the details. My particular detail is attractants for fruit fly survey and control. The development of attractants for tephritids, which are economically among the most dangerous of pests, has followed over the past 60 years the same trend currently seen in the more intensive research on a broad array of insect species. The initial step usually involves the use of behavior-modifying chemicals for development of an attractant to put in a trap for detection and monitoring. For tephritids, detection has special significance because every agriculturally developed area is anxious to avoid introduction of new species, and early detection is therefore imperative. Monitoring insects is important for the knowledge one gains and can apply to other methods of insect control in providing information on population delineation and appropriate timing and intensity of pest management. Along with or immediately following trap use of attractants comes incorporation of attractants into baits or in formulations for direct control. Manipulation of the reproduction of fruit flies with attractants came later than the aforementioned uses, but earlier than for most pest species. I believe, in fact, that the high degree of utility of the potent tephritid attractants that was developed in the 1950s helped give impetus to attractants research. As a result a high level of effort on behavior-modifying chemicals has now been reached, ranging from the study of processes of olfaction and behavior to development of technologies for manipulating insect populations. Such studies provide the most viable option to classical pesticides yet developed, as well as improved use patterns for insecticides where they are imperative. The following discussion of attractants for fruit flies provides some support for this opinion.

# I. HISTORY

## A. Development of Food Baits

By 1913 Howlett had obtained the separate principal components of citronella and bay oils (3). These were tested for attraction of the principal tephritid species of the Pusa region with the following results: methyl eugenol (4, allyl-1,2-dimethoxybenzene) attracted *D. dorsalis* and *D. zonatus* very strongly, and isoeugenol was attractive to *D. diversus* (Coquillett). Also, he states "The attraction of *Ceratitis capitata* (Wiedemann)" (the Mediterranean fruit fly) "by kerosene is now known to be similarly confined to the males and is apparently quite an analogous case."

By this time the classical poison-bait concept was being used for control of the Mediterranean fruit fly by Marsh in Hawaii (8), by Malley in South Africa, and by Berlese in Italy (9). They typically employed a stomach poison such as arsenic combined with a food attractant such as molasses.

Food attractants in traps and in baits received primary attention for the next 30 years, but with a few interesting exceptions. I suppose Howlett's observation on the *Dacus* species went unutilized because the species involved were not indigenous to agriculturally developed regions. Many references for this period are available. I will refer to those that seem to set trends and establish leads. As you note the materials mentioned, you will see a gradual building toward today's "state-of-the-art."

Wheat bran was used in baited containers by Newman (10) in Australia in the 1920s. Sugar, syrup, and molasses sprays containing arsenicals were employed in Florida in 1929 to combat the introduced Mediterranean fruit fly (11). Kerosene as well as fermenting solutions were employed in traps for the Mediterranean fruit fly in Florida in 1929 (11) and in Australia (12).

In 1931, Jarvis (13) reported a new lure for the Mediterranean fruit fly and the Queensland fruit fly, *Dacus tryoni* (Froggatt), composed of a mixture of imitation vanilla essence, household ammonia, and water. This lure was attractive to both sexes and was reported to be used to good effect in many orchards. A commercial product in fairly wide use at the time called Clensel (14) consisted of fatty acids, essential oils (probably citronella), ammonia, and glycerol. Terpinyl acetates were used in a male annihilation attempt against the natal fruit fly, *Ceratitis rosa* (Karsch), in South Africa (15). Consider the probable structure of these compounds (ring structures with *para* substituents) when I discuss later the development of synthetic lures for tephritids.

By 1937, McPhail (16) had developed in Mexico an invaginated glass trap, still commonly used as a standard for trap design, and a lure for the Mexican fruit fly, *Anastrepha ludens* , based on brown sugar and brewer's yeast which yield an attractive fermenting substance. Subsequently (17), he tested a variety of proteins and protein-containing substances which in solution were satisfactory field lures. He also considered products of protein hydrolysis, which, it will later be noted, became the standard lure for *Anastrepha* spp. McPhail (18) also conducted studies in Hawaii resulting in the discovery that linseed oil soap and corn oil soap (a fatty acid was probably the attractive component) were attractive to the melon fly, *Dacus cucurbitae* (Coquillett). Boyce and Bartlett (19) used these results to good effect in testing various mixtures of sugar, molasses, brewer's yeast, casein, *clensel,* and so on as lures for the walnut husk fly, *Rhagoletis completa* (Cresson). They found the most promising to be a mixture of 2% glycine and 3% sodium borate in water. Starr and Shaw (20) found that pyridine (a ring-structured compound) in alcohol solution enhanced the attractiveness of fermenting sugar lure to the Mexican fruit fly.

You will note that the various principal odoriferous products of the variety of brews cited are yeast and protein fermentation and decomposition products that yield alcohols, acetates, aldehydes, esters, and fatty acids; free fatty acids; and pyridine, essential oils, and other ring-structured compounds. Ripley and Hepburn (21) present an interesting analysis of food-bait fermentation processes and products.

Acid-hydrolyzed proteins derived from corn syrup were developed by Steiner (22) (23) and by Gow (24) in Hawaii; and a product of the A.E. Staley Manufacturing Company identified as *Staley's Protein Insecticide Bait No. 7*® (PIB-7) soon became the standard trapping and baiting lure for the *Anastrepha* species in Mexico and for the Mediterranean fruit fly in Hawaii.

## B. Development of Parapheromones

In 1952, Steiner (25) had confirmed Howlett's finding that methyl eugenol is an extremely potent attractant for the male oriental fruit fly. Steiner proposed the use of methyl eugenol in baits for suppression of this insect through removal of the males from the population (he coined the term *male annihilation*) resulting in female infertility (25) (26). Materials such as methyl eugenol, which have behavior-modifying characteristics but which are not known to be pheromones, shall be termed *parapheromones* herein.

The establishment in Hawaii of the oriental fruit fly in 1944 and concern that it, the Mediterranean fruit fly or melon fly (in Hawaii since about 1910 and 1895, respectively), or the Mexican fruit fly might be introduced to the continental U.S., gave impetus in the United States for the development of improved detection and control methods based on traps and lures. Thus, both fruit fly laboratories of the Agricultural Research Service, USDA (one in Mexico City, the other in Honolulu, Hawaii) intensified efforts to find more potent attractants and control treatments. The methodology was essentially empirical. Cages about 25 m³ in size were stocked with flies of mixed ages, sexes, and even species, and candidate attractants were exposed in them in solution or on wicks in miniature invaginated glass traps. In a period of about 5 yr, over 5000 materials were screened in Hawaii and over 8000 in Mexico City. The technique can be criticized as unsophisticated according to present standards for pheromone development. Yet the criticism can be easily refuted on the grounds that despite economic pressures and the state-of-the-art in 1950-1955, this olfactometer screening backed by field tests of promising candidates produced attractants in a relatively short time that have provided some of the greatest economic benefits yet demonstrated. These are: the development of inexpensive standardized commercial food baits for sprays and detection; the confirmation of methyl eugenol as an attractant for oriental fruit fly males; the discovery of potent parapheromones for the males of the Mediterranean fruit fly and the melon fly; and the reconsideration and development of male annihilation as a potent control technique. Unfortunately, no parapheromones have yet been reported for the *Anastrepha* or *Rhagoletis* species or for the important *Dacus olea* (Gmelin), the olive fly. Also, no attractants for females of any tephritid species are yet in use except for general food attractants.

In the case of the Mediterranean fruit fly, the oil of angelica seed (27) was the first of the attractants developed and was very timely in view of the introduction of the pest into Florida in 1956; it proved most useful in detection. Soon thereafter, the isopropyl and *sec*-butyl esters of cyclohexene-carboxylic acid were developed and substituted for the rarer and more expensive angelica seed oil (28). When siglure (6-methyl-3-cyclohexene-1-carboxylic acid) became available, it equalled or exceeded angelica seed oil when it was applied at a higher rate on a trap wick (29). Then when the *tert*-butyl 4 (or 5)-chloro-2-methyl-cyclohexanecarboxylic acid form of the molecule was developed [trimedlure, (30)], a truly potent standard commercially produced lure for the species was available. Guiotto et al. (31) propose that activity is related to presence of a methylcyclohexenic structure, based on studies of both natural and synthetic attractive chemicals.

The chemical 4-(*p*-hydroxyphenyl)-2-butanone acetate, known as *cue-lure* (for *D. cucurbitae),* was also developed and reported by Beroza and associates (32) following identification of benzylacetone and anisylacetone as leads for further screening in Hawaii.

Thus, we now have available methyl eugenol, which attracts male oriental fruit flies and *D. zonatus* (the mango fly of Asia); cue-lure, which attracts the male melon fly and Queensland fruit fly; trimedlure, which attracts the males of the Mediterranean fruit fly, and the natal fruit fly; and protein hydrolysates, which attract both sexes of virtually all tephritids.

Chemically, the similarity among the three synthetic attractants, methyl eugenol, cue-lure, and trimedlure, lies only in their basic ring structure with some of the substituent groups *para* to one another. The first two are the most similar since both have an aromatic ring and unsaturation in the major substituent group.

## II. IMPLEMENTATION

The utility that potent attractants and effective bait sprays can have in tephritid control is perhaps best visualized by the impact their development has had on the capability of county, state, and federal action agencies to manage introductions to the mainland of the U.S. Table 1 lists chronologically these introductions through 1971, the control methods used, and the approximate costs of eradication. Improved food-bait sprays had a considerable impact, but the major impact was that of the specific male parapheromones. Their specificity and potency has allowed much earlier detection and more precise delineation, resulting in more rapid response and treatment of less extensive areas in less expensive ways.

### A. Formulation of Parapheromones

In recent years parapheromones have become more acceptable—in both an economic and environmental sense—through formulations development. For

TABLE I.

Method and Cost of Controlling Tephritids Introduced to
the Mainland United States

| Species | Year | Location | Method | Cost ($)[a] |
|---------|------|----------|--------|-------------|
| *Ceratitis capitata* | 1929 | Florida | Clean Cultivation and molasses bait sprays | 20 million |
| *C. capitata* | 1956 | Florida | Protein bait sprays | 11 million |
| *C. capitata* | 1962-63 | Florida | Protein bait sprays | 1 million |
| *C. capitata* | 1966 | Texas | Protein bait sprays | 200 thousand |
| *Dacus cucurbitae* | 1956 | California | Protein bait sprays | 233 thousand |
| *D. dorsalis* | 1960 | Californis | Methyl eugenol traps | 77 thousand |
| *D. dorsalis* | 1964 | Florida | Methyl eugenol traps | 36 thousand |
| *D. dorsalis* | 1966 | California | Methyl eugenol traps and stations | 54 thousand |
| *D. dorsalis* | 1969 | Florida | Methyl eugenol traps | 45 thousand |
| *D. dorsalis* | 1969 | California | Methyl eugenol traps and liquid | 38 thousand |
| *D. dorsalis* | 1970 | California | Methyl eugenol traps and liquid | 42 thousand |
| *D. dorsalis* | 1971 | California | Methyl eugenol liquid | 12 thousand |

[a]Values are estimates of the combined costs of the three classes of participating agencies—county, state, and federal.

example, fluid male annihilation mixtures have been developed (33) (34) as substitutes for fibrous blocks used in oriental fruit fly eradication campaigns in the Mariana Islands in 1962-1963 (35) and in tests of suppression of the Queensland fruit fly in Australia (36). These mixtures were successful in the case of the recent oriental fruit fly introductions to California when thixotropic formulation was applied with hydraulic oil cans to poles, tree trunks, and so on (37). Also, Cunningham and associates (38) recently showed that aerial applications of as little as 9 g/ha of thickened liquid formulation containing 80% cue-lure, 5% technical naled, and 15% extender reduced populations of melon flies in 10 $km^2$ areas by 92 to 97%. However, environmental considerations may mean that applications in urban areas would be restricted to discrete ground- or air-applied deposits.

David Suda and associates (39) tested a variety of candidate solid carriers in Hawaii* for the three synthetic attractants. Solid carriers, although less efficient initially than spray applications, generally have a longer residual life and are in some respects more acceptable in that the lure and toxicant are held at a relatively few discrete sites and are released only slowly to the environment. Whether for trapping or for male annihilation, effort is needed in the area of formulation (this is as true for pheromones in general as it is for tephritid attractants). Moreover, a variety of application techniques should be developed in order that a formulation suited to each potential need will be available; whether the treatment area is limited or extensive in area; whether urban, cultivated (and what sort of cultivation), or wilderness; whether it is accessible readily or with difficulty; whether the control is for one species or several; whether manpower costs are high versus low; and so on.

## B. Male Annihilation

I would like to expand on the subject of male annihilation. Steiner's successes in using male annihilation in the Pacific Islands are quite well known (35); the oriental fruit fly was eradicated from Rota in 1962-1963 and subsequently from the rest of the Mariana Islands. As I commented earlier, I believe these successes provided considerable impetus for research on attractants. Conversely, these results are sometimes criticized as not being indicative of general usefulness because the fly populations were confined to the limits of the islands. However, additional examples are available. Bateman and associates (36) utilized cue-lure and malathion impregnated fibrous blocks to suppress a population of the Queensland fruit fly in a town some 16 km from Sydney and 6.4 km from a second town that was used as a control. They estimated that more than 100,000 males were killed and that

---

*Hawaiian Fruit Flies Investigations Laboratory, USDA, ARS, Hil, HI.

suppression early in the season was achieved but did not last. Subsequently (40) they treated more isolated townships with male annihilation separately and in combination with bait sprays; with reservations, they concluded that both treatments were effective.

The Queensland fruit fly was introduced into Easter Island in 1971. Bateman (41), working with the Chilean government, successfully eradicated the species, again using a combination treatment that consisted of 15 applications of food-bait sprays and one male-annihilation application (25-cm-long sections of thick string impregnated with cue-lure and technical malathion at the rate of 3 strings/1000 m$^2$). Treatment began in July and eradication was achieved by late August 1972.

The Japanese government is conducting an eradication campaign against the oriental fruit fly in the Ogasawara Islands. The status of the program is unknown to me. Iwahashi (42) commented on an apparent eradication from Kikai Island after 2 years of male-annihilation fiber blocks treated with methyl eugenol-naled. Then the fly was apparently reintroduced from nearby (27 km) Amami Oshima Island, possibly during a typhoon. Iwahashi also reported movement of flies from Chichi Jima to Haha Jima, a distance of 50 km, which suggests that reintroduction from nearby sources easily jeopardizes a suppression or eradication campaign.

However, the work of Arroyo (43) is more apropos to the question of isolated populations. He conducted a program against the Mediterranean fruit fly in Valencia in an area of 630 ha of citrus. The region was only semiisolated from other populations of the fly. Arroyo developed a very inexpensive paper cup trap in which he placed a small bottle containing 1 ml of trimedlure and 1 ml of dichlorvos; 60,000 traps were emplaced throughout the groves in the area. The rate of distribution ranged from about one trap per three trees to one per 12 trees. A control area treated with insecticide was established. In two seasons of comparison the insecticide-treated groves experienced fruit infestation rates of about 2 and 5%; in both seasons no infested fruit were found in the trapping area.

In Israel the Citrus Board conducts a citrus regionwide control program. In a series of tests in which various control methods were compared, the Cohens (44) included in 1965-1966 a test of male annihilation of Mediterranean fruit flies with plastic traps containing a plastic bottle with 2 ml of trimedlure with 1% dichlorvos; 17,000 such traps were emplaced in the Hadera coastal region, approximately one per 400 m$^2$ during the off-season summer months (compared with six per 100 m$^2$ in Arroyo's test). Food-bait sprays were applied as usual during the fruiting season. Comparison with areas receiving year-round bait sprays showed no difference: fruit in both areas were almost 100% free of infestation. Since trap catches during the off-season reached as high as 2000 per trap in groves soon after cessation of spraying and then dropped somewhat and leveled off, the male-annihilation test indicated high potential for inclusion in an integrated program.

The economy of trap emplacement versus mechanical application will obviously vary among locales. However, Arroyo's, and Cohen and Cohen's data give an indication of the distribution frequency possibly required by high-output point sources of trimedlure. At the Hawaiian Laboratory we developed information that high-output sources of cue-lure might be appropriate at a rate of one per 100 $m^2$ and of methyl eugenol at a rate of one per 5000 $m^2$. As Cunningham et al. (38) point out, small volumes of lure at a point source have a shorter residual life, but the number of insects killed per gram of lure is greater with numerous small point sources than with fewer large ones. I point again to the need for a variety of options; this is particularly true if one considers the use of behavior-modifying chemicals in integrated control programs. Integration of male-annihilation and sterile-insect releases is being investigated in Hawaii. Novel uses of behavior-modifying chemicals have been proposed, such as topical application to sterile insects (45).

## C. Traps

The subject of traps, at least for fruit fly workers, is one where imagination appears to play its greatest role. Each of us hopes to design a slightly more efficient trap (trap efficiency is an important consideration and deserves an entire separate paper). The "build-a-better-mousetrap" theory seems to come into play, each researcher hoping to develop a trap that bears his name. I recall that in Mexico City, we had a museum of traps that did not pass the survival-of-the-fittest test. It must have contained 50 or more designs, some approaching Klein's bottle in abstruseness. However, several designs have survived the test of time. McPhail's invaginated glass trap is still the standard for exposing liquid lures. Everybody calls the McPhail trap the McPhail trap except McPhail. He called it the invaginated glass trap, knowing full well it was design originating in lost Chinese history and that it had had long use there as a house fly trap for household use.

Steiner (46) developed a dry trap for use with the synthetic lures developed in Hawaii. It is much like a number of cylindrical cardboard or can traps—a horizontally suspended cylinder with end ports—but Steiner's is made of lucite. In detection programs in the continental U.S. it is typically baited with all three attractants, on wicks, with a small amount of lindane-chlordane powder in the bottom to deter ant predation. Addition of an insecticide to the lures kills the flies that enter the trap, and they are preserved in good condition. This is quite contrary to the wet traps (such as the McPhail trap) required for the *Anastrepha* species. Moreover, in wet traps that contain fermenting sugar or decomposing protein solutions, the captured flies become difficult to discern from the many decomposing blow flies, bottle flies, moths, and other insects that also come to the trap, and from which the tephritids must be sorted. The picture may be unappealing, but I recall one difficulty encountered in field tests in rural Mexico. The odor of fermenting brown sugar solution is quite similar to that of pulque, a popular beerlike fermentation product derived from agave plants. Many of our traps disappeared, or at best were emptied, for consumption by the local residents. I hope they filtered the liquid first.

Dr. David Nadel of the I.A.E.A., Vienna (47), devised a dry trap, now widely used, consisting of a vertically suspended polyethylene cylinder with ports under a cap through which flies enter. Bait is contained in a polyethylene bottle within, and evaporation rate may be controlled and standardized by varying the size of the bottle's cone-shaped tip. This is a distinct advantage over wicks, which have an uneven release rate and smaller reserve.

Numerous sticky traps have been designed. They have undoubtedly the highest retention efficiency, but their drawbacks so far have precluded general use in major campaigns. These include rapid decline in effectiveness of the adhesive as it becomes dirty, limited capacity, and messy servicing. Harris et al. (48) present useful sticky trap designs.

Formulations for liquid traps have also been improved. Lopez and Hernandez (49) added borax to PIB-7 and to enzyme-hydrolyzed cottonseed proteins, and thus discouraged putrefaction. Later, Lopez et al. (50) developed pellets of cottonseed hydrolysate and borax that are not only more attractive in solution but also greatly facilitate trap servicing (and they do not make an appealing drink). These pellets (5 g borax: 1 g hydrolysate; six per trap) are now in use in the U.S. detection program. Comparable alternatives based on locally available products have been developed in other countries.

## D. Spray Food-Bait Formulation

Food-bait spray techniques have also been improved. In the 1956 Mediterranean fruit fly campaign in Florida, a full-coverage bait spray was applied at a rate of 4.5 1/4000 m$^3$. It contained 900 g of malathion 25% wettable powder, and about 1 1 of PIB-7 (9). Metropolitan Miami was included in the nearly 2.5 million ha treated in this manner. During the course of development of the program, spraying of alternate strips was tested and found generally effective. Cohen and Cohen (44) used a similar bait and reported that spraying swaths separated by 10 to 12 m gave good results. They also employed spot sprays, applying bait to about 1000 cm$^2$ of the outer canopy of every other tree, and found this technique was a viable alternative to cover sprays. Lopez et al. (51) were able to protect a 17.5-ha portion of a large, heavily infested orange grove near Veracruz, Mexico, from Mexican fruit fly attack by applying about 180 ml/4000 m$^3$ of 1:4 mixture of 95% malathion and PIB-7 by squirting about 5 ml on every other tree, alternately, each week (about 4 spots/ 100 m$^2$). This amounted to only about 35 g/acre of technical malathion, compared with 114 g used in Florida. Finally, Cohen and Cohen (44) reported complete success with ultra-low-volume application of bait applied in a swath 40 to 50 m wide at the rate of 25 g technical malathion per acre. Also, trimedlure is now incorporated with protein bait sprays to good advantage in Israel (52).

## III. PHEROMONES

Although Howlett (1) had proposed that chemical attraction mediated mate-finding in fruit flies, it was not until 1959 that such behavior was conclusively described. Féron (53) described the mating behavior of the Mediterranean fruit fly in detail and reported the release by males of substances from their erectile anal ampoules that attract and excite female flies. Lhoste and Roche (54) described the glands that were probably active in pheromone production. Presumably the erect ampoule is coated with pheromone, which is dispersed by wing fanning.

Fletcher (55) (56) described rectal organs for production, storage, and release of a pheromone of the Queensland fruit fly and related the release to both audio calling and fanning, which is common among tephritids.

Nation (57) (58) demonstrated the presence and production of a male produced pheromone in *Anastrepha suspensa* (Loew), the Caribbean fruit fly, now a significant pest in Florida. The pleural regions of abdominal segments 3, 4, and 5 are distended, and wing fanning apparently disperses the pheromone, which is produced or transmitted by epidermally derived cells. Production too may be related to these cells and to the salivary glands, which extend into this region, as they do in *Rioxa pornia* (Walker), the "island fruit fly" (59). Recently, Nation (60) described the pheromone as having four components, two 9-carbon alcohols containing one and two double bonds, respectively, and two lactone esters. Perdomo (61), one of his students, conducted field studies that indicated potential field usefulness for the pheromone. In this case also, sound production is associated with fanning and pheromone release. Surprisingly, exposure of live males or of pheromone extract in the field in sticky traps resulted in capture of nearly equal numbers of *males* and of virgin females (61) (62), indicating a potential aggregating role.

Sex pheromones produced by male Mediterranean fruit flies that strongly attract female flies were identified by Jacobson and associates in 1973 (63) (64). The components were identified as (E)-6-nonenoate and (E)-6-nonen-1-ol. However, certain additional acidic components were required to achieve female attraction in large field cages. These components have now been identified as 15 fatty acids. When combined in their naturally occurring proportions with the alcohol and ester, they are attractive to virgin females in cage tests. As in the case of the Caribbean fruit fly pheromone, exposure in the field results in capture of both males and females, and its potency is equivalent to that of trimedlure (65). However, exposure of living males in special field traps results in better capture of females than does exposure of the synthetic pheromone, which may indicate the absence of necessary cues. These results, like those of Nation and Perdomo mentioned previously, indicate the complexity of fruit fly behavior in relation to pheromones and other cues.

I should mention that Dr. Haniotakis of the Nuclear Research Center, Demokritos, is working in Greece on identification of the pheromone of the olive fly. Dr. J. Tumlinson of the Insect Attractants and Basic Biology Research Laboratory, ARS, Gainsville, Florida, is providing consultative input now, and later will participate in chemical analyses.

The 9-carbon compounds found in both Mediterranean and Caribbean fruit flies are very similar; also (Z)-6-nonen-1-ol acetate was shown to stimulate ovipositional acitivity in the melon fly female (66).

Fletcher and Watson (67) noted ovipositional response by female Queensland fruit flies to 2-chloro-ethanol. Prokopy (68) noted that female apple maggot flies, *Rhagoletis pomonella* (Walsh), mark the fruit following oviposition with a pheromone that deters subsequent oviposition and mediates accommodation of larval density relative to food availability.

Probably all of us who have worked with fruit flies have noted that they fan and buzz and produce a characteristic odor at the time of mating (dusk in most species). In Mexico City in the 1950s we wished we had facilities to examine these phenomena. In Hawaii it had been noted that the melon fruit fly buzzed, fanned, *and* actually produced a *visible* as well as odorous material during this behavior. We examined this microscopically, by collection on filters, and with simple chemical tests (69). We found that the oriental fruit fly also produces this "smoke," and concluded that it was composed of inert tiny rodlike crystals; when they were solubilized in water and allowed to dry, they recrystallized. We now know that the smoke is composed of a combination of an oily substance coated on what are apparently inorganic phosphate crystals. The idea that these floating crystals serve as carriers of a pheromone is intriguing.

In our laboratory in Gainesville we have the facilities for examining the sounds associated with insect behavior. The sounds produced by the Caribbean fruit fly during flight, aggressive behavior, signaling, and precopulatory activity have been studied through use of an anechoic chamber and sound-analyzing equipment (70) (71). A few remarks are appropriate here on the nature of the signaling sound since it is apparently associated with the time of pheromone release and mate-finding, as such buzzing is in may tephritids. The sounds produced are not the result of true stridulation, as in the Queensland fruit fly which has tergal bristles stroked by the wings (72). Rather, they appear to result from characteristic patterns of wing movement during fanning, possibly through changes in both speed and articulation. Many analytic determinations of the sound have been made, such as waveform, sound-pressure levels, and fundamental and harmonic frequencies, but the simplest presentation is the waveform of the pulse (Fig. 1). Characteristically, the male produces a fairly regular sound over a period that may vary greatly in duration. However, at the termination of each fanning period, the fly produces a short, reproducible, and characteristic "chirp," which we believe carries the signaling code. Studies are underway to clarify the interrelationships between mate-finding, aggregation, pheromone action, and sound production.

## IV. CONCLUSION

It has been my intention to place the use of attractants for fruit flies in perspective, historically and currently, and I have cited only a few of the many possible references—those that seemed to me to have particular pertinence. The literature on fruit fly research is voluminous, and I apologize to those whose studies I have not had the space or insight to incorporate. I recommend the article by Bateman (73) for a comprehensive review of the literature.

I have noted that certain philosophies and hypotheses, discoveries and data, and, most importantly, applications, have appeared and reappeared. Each time improvements have been made in understanding and utility.

The comments made in Section I.A on the composition of attractive

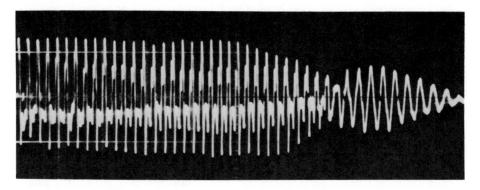

**FIGURE 1.** Signaling sound of the Caribbean fruit fly. A fairly uniform waveform of varied length is produced by wing fanning. Terminating these pulses is a short characteristic pulse believed to carry the signaling code.

substances emphasizes the "building-block" process in developing behavior-modifying chemicals. Comparison of those compounds with those discussed in Section III shows striking trends and similarities. We now have a few really potent tools, and there are more under development that can be integrated into fruit fly management programs. There need be no apology that some of these were developed rather empirically, and I am fully optimistic that an analytic approach to pheromone elucidation and behavioral analysis will complement these findings with even more useful management tools. In particular, we need techniques for manipulating and surveying the females of these species, and we seem to be approaching that capability. If, indeed, Fabré, Howlett, Lucretius, and the rest did leave to us only the details, I feel we are making up for lost time.

The development of food attractants and their incorporation into traps for survey and detection and into bait sprays for control has had great impact on the management of these pests. Similarly, specific male attractants have been discovered and utilized in trapping and control through male annihilation. Pheromones of fruit flies are under intensive investigation and hold promise of providing us with additional capabilities in detecting and managing these important pests.

Finally, I want to reemphasize that one critical area of research has to do with the formulation of behavior-modifying chemicals, and not only for tephritids. We must proceed to develop an arsenal of application methodologies in order to have techniques suitable for each contingency.

In proper context, the old adage may be revised—"An attractant in the bush may be worth two in the hand."

## REFERENCES

(1)  F.M. Howlett. 1912. The effect of oil of citronella on two species of *Dacus*. *Trans. Entomol. Soc. Lond.* **60**:412-418.

(2)  W.W. Froggatt. 1909. Official report on fruit fly and other pests in various countries, *Offic. Rep.*, N.S.W. Dept. Agric. 1907-1908.

(3)  F.M. Howlett. 1915. Chemical reactions of fruit flies. *Bull. Entomol. Res.* **6**:297-305.

(4)  M. Yearsley. 1905. A contribution to the study of the function of the antennae in insects. *Proc. Zool. Soc. Lond.* **77**:85-88.

(5)  A. Pierret. 1841. Remarques sur l'usage des antennes chez les insectes. *Bull. Soc. Entomol.* (Fr.) **8**:10-11.

(6)  W. Ramsay. 1882. On smell, *Nature* (Lond.) **26**:187-189.

(7)  B. Hocking. 1960. Smell in insects: A bibliography with abstracts (to December 1958). EP Tech. Rep. No. 8, Dept. Nat. Defense, Canada.

(8)  H.O. Marsh. 1910. Report of the assistant entomologist, Board Commissioners Agric. Forest Hawaii, pp. 152-159.

(9)  L.D. Christenson. 1958. Recent progress in the development of procedures for eradicating or controlling tropical fruit flies. *Proc. 10th Int. Congr. Entomol.* **3**:11-16.

(10)   L.J. Newman. 1926. Fruit fly *(Ceratitis capitata)*. Trapping or luring experiments. *J. Dept. Agriĉ. W. Austr.* Ser. 2, 3:513-515.

(11)   G.G. Rohwer. 1958. The Mediterranean fruit fly in Florida—past, present, and future. *Fla. Entomol.* 41:23-25.

(12)   P.C. Hely. 1931. Offic. Rep., N.S.W. Dept. Agric.

(13)   H. Jarvis. 1931. Experiments with a new fruit fly lure. *Queensland Agric. J.* 36: 485-491.

(14)   L.J. Newman and B.A. O'Connor. 1931. Fruit fly *(Ceratitis capitata)*. A further series of trapping or luring experiments. *J. Dept. Agric. W. Austr.,* Ser. 2, 8:316-318.

(15)   L.B. Ripley and G.A. Hepburn. 1935. Olfactory attractants for male fruit flies. *S. Afric. Dept. Agr. Entomol. Mem.* 9:3-17.

(16)   M. McPhail. 1937. Relation of time of day, temperature, and evaporation to attractiveness of fermenting sugar solution to Mexican fruit fly. *J. Econ. Entomol.* 30:793-799.

(17)   M. McPhail. 1939. Protein lures for fruit flies. *J. Econ. Entomol.* 32:758-761.

(18)   M. McPhail. 1943. Linseed oil soap—a new lure for the melon fly. *J. Econ. Entomol.* 36:426-429.

(19)   A.M. Boyce and B.R. Bartlett. 1941. Lures for the walnut husk fly. *J. Econ. Entomol.* 34:318.

(20)   D.F. Starr and J.G. Shaw. 1944. Pyridine as an attractant for the Mexican fruit fly. *J. Econ. Entomol.* 37:760-763.

(21)   L.B. Ripley and G.A. Hepburn. 1929. Studies on reactions of the natal fruit fly to fermenting baits. *S. Afric. Dept. Agric. Entomol. Mem.* 1:19-53.

(22)   L.F. Steiner. 1952. Fruit fly control in Hawaii with poison-bait sprays containing protein hydrolysates. *J. Econ. Entomol.* 45:838-843.

(23)   L.F. Steiner. 1955. Bait sprays for fruit fly control. *Agric. Chem.* 10:32-34; 113-115.

(24)   P.A. Gow. 1954. Proteinaceous bait for the oriental fruit fly. *J. Econ. Entomol.* 47:153-160.

(25)   L.F. Steiner. 1952. Methyl eugenol as an attractant for oriental fruit fly. *J. Econ. Entomol.* 45:241-248.

(26)   L.F. Steiner, and R.K.S. Lee. 1955. Large-area tests of a male-annihilation method for oriental fruit fly control. *J. Econ. Entomol.* 48:311-317.

(27)   L.F. Steiner, D.H. Miyashita, and L.D. Christentson. 1957. Angelica oils as Mediterranean fruit fly lures. *J. Econ. Entomol.* 50:505.

(28)   L.I. Gertler, L.F. Steiner, W.C. Mitchell, and W.C. Barthel. 1958. Esters of 6-methyl-3-cyclohexene-1-carboxylic acid as attractants for the Mediterranean fruit fly. *J. Agric. Food Chem.* 6:592-594.

(29)   L.F. Steiner, G.G. Rohwer, E.L. Ayers, and L.D. Christenson. 1961. The role of attractants in the recent Mediterranean fruit fly eradication in Florida. *J. Econ. Entomol.* 54:30-35.

(30)   M. Beroza, N. Green, S.I. Gertler, L.F. Steiner, and D.H. Miyashita. 1961. Insect attractants: New attractants for the Mediterranean fruit fly. *J. Agric. Food Chem.* 9:361-365.

(31)   A. Guiotto, U. Furnasiero, and F. Baccichetti. 1972. Investigations on attractants for males of *Ceratitis capitata*. *Farmaco. ed. Scientifica* 27:663-669.

(32)   M. Beroza, B.H. Alexander, L.F. Steiner, W.C. Mitchell, and D.H. Miyashita. 1960. New synthetic lures for the male melon fly. *Science* 131:1044-1045.

(33)   W.G. Hart, L.F. Steiner, R.T. Cunningham, S. Nakagawa, and G.J. Farias. 1966.

Glycerides of lard as an extender for cue-lure, medlure, and methyl eugenol in formulations for programs of male annihilation. *J. Econ. Entomol.* 59:1395-1400.

(34) K. Ohinata, L.F. Steiner, and R.T. Cunningham. 1971. Thixcin® E as an extender of poisoned male lures used to control fruit flies in Hawaii. *J. Econ. Entomol.* 64:1250-1252.

(35) L.F. Steiner, W.G. Hart, E.J. Harris, R.T. Cunningham, K. Ohinata, and D.C. Kamakahi. 1970. Eradication of the oriental fruit fly from the Mariana Islands by the methods of male annihilation and sterile insect release. *J. Econ. Entomol.* 63:131-135.

(36) M.A. Bateman, A.H. Friend, and F. Hampshire. 1966. Population suppression in the Queensland fruit fly, *Dacus (Strumeta) tryoni.* I. The effects of male depletion in a semi-isolated population. *Austr. J. Agric. Res.* 17:687-697.

(37) D.L. Chambers, R.T. Cunningham, R.W. Lichty, and R.B. Thrailkill. 1974. Pest control by attractants: A case study demonstrating economy, specificity, and environmental acceptability. *BioScience* 24:150-152.

(38) R.T. Cunningham, D.L. Chambers, L.F. Steiner, and K. Ohinata. Thixcin-thick-ened sprays of cue-lure plus maled: Investigation of application rates for use in programs of melon fly male annihilation. *J. Econ. Entomol.* (in press).

(39) D. Suda. Personal communication.

(40) M.A. Bateman, A.H. Friend, and F. Hampshire. 1966. Population suppression in the Queensland fruit fly, *Dacus (Strumeta) tryoni.* II. Experiments on isolated populations in western New South Wales. *Austr. J. Agric. Res.* 17:699-718.

(41) M.A. Bateman. 1974. The eradication of the Queensland fruit fly from Easter Island. *FAO Plant Protect. Bull.* 21:114.

(42) O. Iwahashi. 1972. Movement of the oriental fruit fly among islets of the Ogasawara Islands. *Environ. Entomol.* 1:176-179.

(43) M. Arroyo. 1972. La lucha contra las plagas, pasado, presente y . . . futuro? *Rev. Univ. Madrid* 21:249-280.

(44) I. Cohen and J. Cohen. 1967. Centrally organized control of the Mediterranean fruit fly *(Ceratitis capitata* Weid.) in citrus groves in Israel. Agrotech. Div., Citrus Board of Israel.

(45) D.L. Chambers, K. Ohinata, M. Fujimoto, and S. Kashiwai. 1972. Treating tephritids with attractants to enhance their effectiveness in sterile-release programs. *J. Econ. Entomol.* 65:279-282.

(46) L.F. Steiner. 1957. Low-cost plastic fruit fly trap. *J. Econ. Entomol.* 50:508-509.

(47) D. Nadel. Unpublished.

(48) E.J. Harris, S. Nakagawa, and T. Urago. 1971. Sticky traps for detection and survey of three tephritids. *J. Econ. Entomol.* 64:62-65.

(49) F. Lopez-D and O. Hernandez-B. 1967. Sodium borate inhibits decomposition of two protein hydrolysates attractive to the Mexican fruit fly. *J. Econ. Entomol.* 60:137-140.

(50) F. Lopez-D, L.F. Steiner, and F.R. Holbrook. 1971. A new yeast hydrolysate-borax bait for trapping the Mediterranean fruit fly. *J. Econ. Entomol.* 64:1541-1543.

(51) F. Lopez-D, D.L. Chambers, M. Sanchez-R., and H. Kamasaki. 1969. Control of the Mexican fruit fly by bait sprays concentrated at discrete locations. *J. Econ. Entomol.* 62:1255-1257.

(52) R. Galun. Personal communication.

(53) M. Féron. 1959. Attraction chimique due male de *Ceratitis capitata* Wied. (Dipt. Trypetidae) pour la femelle. *C.R. Seances Acad. Sci.,* Ser. D., Sci. Nat. (Paris) *248:* 2403-2404.

(54) J. Lhoste and A. Roche. 1960. Organes odoriferantes des males de *Ceratitis capitata* Wied. (Dipt. Trypetidae). *Bull. Soc. Entomol. Fr.* 65:206-210.

(55) B.S. Fletcher. 1968. Storage and release of a sex pheromone by the Queensland fruit fly, *Dacus tryoni* (Diptera: Trypetidae) *Nature* 219:631-632.

(56) B.S. Fletcher. 1969. The structure and function of the sex pheromone glands of the male Queensland fruit fly, *Dacus tryoni. J. Insect Physiol.* 15:1309-1322.

(57) J.L. Nation. 1972. Courtship behavior and evidence for a sex attractant in the male Caribbean fruit fly, *Anastrepha suspensa. Ann. Entomol. Soc. Am.* 65:1364-1367.

(58) J.L. Nation. 1974. The structure and development of two sex specific glands in male Caribbean fruit flies. *Ann. Entomol. Soc. Am.* 67:731-734.

(59) G. Pritchard. 1967. Laboratory observations on the mating behavior of the island fruit fly, *Rioxa pornia* (Diptera: Tephritidae). *J. Austr. Entomol. Soc.* 6:127-132.

(60) J.L. Nation. 1975. The sex pheromone blend of Caribbean fruit fly males: Isolation, biological activity and partial chemical characterization. *Environ. Entomol.* 4:27-30.

(61) A.J. Perdomo. 1974. Sex and aggregation pheromone bioassays and mating observations of the Caribbean fruit fly, *Anastrepha suspensa* (Loew), under field conditions. Ph.D. dissertation, Univ. Florida, Gainesville.

(62) A.J. Perdomo, R.M. Baranowski, and A.J. Nation. Recapture of virgin female Caribbean fruit flies with traps baited with males. *Fla. Agric. Exp. Sta. J.,* Ser. No. 4992 (in press).

(63) M. Jacobson, K. Ohinata, D.L. Chambers, W.A. Jones, and M.S. Fujimoto. 1973. Insect sex attractants: 13. Isolation, identification, and synthesis of sex pheromones of the male Mediterranean fruit fly. *J. Med. Chem.* 16:248-251.

(64) K. Ohinata, M.S. Fujimoto, D.L. Chambers, M. Jacobson, and D.C. Kamakahi. 1973. Mediterranean fruit fly: Bioassay techniques for investigating sex pheromones. *J. Econ. Entomol.* 66:812-814.

(65) K. Ohinata. Personal communication.

(66) I. Keiser, R.H. Kobayashi, D.H. Miyashita, M. Jacobson, E.J. Harris, and D.L. Chambers. 1973. *Trans*-6-nonen-1-ol acetate: An ovipositional attractant and stimulant of the melon fly. *J. Econ. Entomol.* 66:1355-1356.

(67) B.S. Fletcher and C.A. Watson. 1974. The ovipositional response of the tephritid fruit fly, *Dacus tryoni,* to 2-chloro-ethanol in laboratory bioassays. *Ann. Entomol. Soc. Am.* 67:21-23.

(68) R.J. Prokopy. 1972. Evidence for a marking pheromone deterring repeated oviposition in apple maggot flies. *Environ. Entomol.* 1:326-332.

(69) K. Ohinata and D.L. Chambers. Unpublished.

(70) J.C. Webb, J.L. Sharp, D.L. Chambers, and J.C. Benner. Acoustical properties of the flight activities of the Caribbean fruit fly. *J. Exp. Biol.* (in press).

(71) J.C. Webb, J.L. Sharp, D.L. Chambers, and J.C. Benner. The analysis and identification of sounds produced by the male Caribbean fruit fly, *Anastrepha suspensa* (Loew). *J. Exp. Biol.* (in press).

(72) J. Monro. 1953. Stridulation in the Queensland fruit fly, *Dacus (Strumeta) tryoni* Frogg. *Austr. J. Sci.* 16:60.

(73) M.A. Bateman. 1972. The ecology of fruit flies. *Ann. Rev. Entomol.* 17:493-518.

## CHAPTER 20

# Manipulation of Insect Pests of Stored Products

W. E. Burkholder

*Stored Product and Household*
*Insects Laboratory*
*Agricultural Research Service, USDA*
*Department of Entomology*
*University of Wisconsin*
*Madison, Wisconsin*

More effective control of storage pests in large granaries, small farm storages, ships, warehouses, stores, houses, and apartments could mean an immediate increase in the world's edible grain and food without any change in agricultural productivity. This is particularly significant when the money and the use of energy necessary to increase food production are considered. Indeed, in view of the hunger in the world today, loss of food and food products in storage and in transit due to pests is tragic.

Several new techniques for controlling stored-product insects have been devised in recent years. These include the use of inert gases, radiation, pathogens, growth regulators, and pheromones, all of which can be combined with the older methods of control—such as sanitation, inspection, good packaging, providing good storage facilities, and using pesticides—to achieve an integrated approach to the control of stored-product insects.

In working toward the prevention (1)(2) of infestations rather than control, biologically active agents such as pheromones might provide a new way of manipulating the pests. However, before pheromones can be used either for the surveillance or the control of stored-product insects, they must provide effective, predictable, and reproducible results. The greatest success may be obtained with insect species that do not feed in the imaginal stage and with those that usually have well-developed mating adaptations such as sex pheromones that ensure reproduction early in the adult stage. Also, pheromones are especially promising against low—but not-to-be-tolerated— populations of stored-product insects. In such situations habitats are usually restricted, and the insects usually populate small foci that expand only gradually. The development of simple traps baited with insect pheromones or attractants to lure the pests from their hiding places may enable us to determine the proper time for efficient control of stored-product pests, to minimize the number of applications of a pesticide, to estimate population levels, and to aid in the identification of problem species. Food-processing plants, warehouse managers, and pest-control personnel and their pest-management advisors could utilize such pheromone traps. The resulting records of infestations would be valuable for inspectors and an important reference of the species involved (in the event of claims of insect damage against a company). Traps could also be an effective tool for inspectors looking for previously undetected insect infestations.

Indeed, early detection is the key to effective control of insect damage and infestation of stored products. Often those responsible for insect control are not aware of the existence of an incipient infestation until the product itself is infested. Surely, then, prevention is of greater importance than control of insect infestations of stored products.

A system of certification of food products such as grain might be established in which the status of infestation is designated (3). Domenichini (4) suggested that a document should accompany a commodity in transit. It could contain all relevant data on origin, transportation, transit, and control measures. These would be especially important when stored-product insects are found in export cargoes of United States grain and grain products upon delivery to foreign ports, even though the commodities were inspected before shipment and found to be apparently free of insects (5).   ·

Products intended for human consumption, such as dry milk, flour, nuts, spices, dried fruits, and processed cereals, must be entirely free of living or dead insects or evidence of their presence. Infestation of such commodities as animal feeds and certain raw foods may be tolerated, though they can present serious problems when the commodities are held in the same facilities as food products that must be entirely free of insects.

With this background, I will discuss a system of surveillance, detection and control of stored-product insects that uses pheromones to manipulate some aspects of the behavior of these insects.

## I. SURVEILLANCE AND DETECTION OF INFESTATION

Sex pheromones often elicit extremely high levels of response by insects and provide a powerful means of drawing them into trapping devices. Food attractants are likewise useful in trapping, but the competition of the stored or residual food lessens their effectiveness. Light traps too are useful, and their effectiveness could undoubtedly be enhanced by the use of pheromones, but they have some disadvantages, such as cost, maintenance, electrical demand, and, perhaps most important, dependence on flying rather than crawling adults. Also, the light traps are ineffective in locating the source of the trapped adults. This requirement is so important that it weighs in favor of the use of large numbers of pheromone traps placed in, on, or under a commodity so that the insects can be caught as they emerge.

Accounts of pheromone research concerning stored-product pests in the families Dermestidae, Anobiidae, and Pyralidae, were presented by Burkholder (2)(6), Nakajima (7), and Levinson (8), who recently provided an excellent review. The matter of prime importance—determining the presence of insects and locating them within a warehouse or building—can be accomplished by distributing pheromone-baited traps throughout the site. Suspected infestations may then be confirmed by inspection of the nearby commodities and storage area. At the same time it is important to make species determinations of infesting insects and to keep accurate records for future reference. Finally, control procedures may be modified once the species of insect and population densities are determined.

The results of a study of seasonal patterns of emergence of *Attagenus* and *Trogoderma* in warehouses in Milwaukee provide an example. Corrugated paper traps treated with the appropriate pheromones and an insecticide were placed in the areas. Larger numbers of males of the target species were recovered in the pheromone traps than in control traps not treated with pheromones; a previously unknown infestation of *Trogoderma* was detected, the site of infestation was located, and the species of the trapped insects was determined (9). Determination of population levels based on trap catches may also be possible, though many ecological and physical factors will need to be considered and evaluated to ensure accurate estimates.

## II. CONTROL OF INFESTATION

The use of pheromone traps to maintain an infestation of stored-product insects at or below the economic threshold is theoretically feasible, but this approach will most likely need to be combined with other control measures.

Two methods of using the traps for control that appear most promising are the disruption of mating communication and the removal of enough males from a population so as to lower reproduction by females. In tests with *Attagenus megatoma* (F.), a single treatment with high levels of pheromone in sealed 208-liter fiber drums was sufficient to reduce significantly mating activity for up to one year (10). However, the relatively large amounts of pheromone necessary for such inundation could cause several problems. For example, the stored products and containers might absorb enough pheromone to attract insects to the products after they leave the warehouse and enter marketing channels. Also the pheromone-treated warehouse might draw insects to it from neighboring untreated facilities. Finally, the insects might adopt alternative methods of communication. Brady et al. (11) exposed *Plodia interpunctella* (Hubner) and *Cadra cautella* (Walker), normally infesting three commercial storage facilities, to continuously high levels of synthetic sex pheromone for up to 99 days. The mating activity of females sampled at intervals was not reduced (based on spermatophore counts) compared with that of control females.

Nevertheless, in preliminary field trials, Wheatley (12) reduced the fertility of female *C. cautella* collected in a warehouse from 80% to about 30% by using a combination of the three components of the synthetic pheromone (natural ratio) as a confusion agent. He also obtained a similar reduction by using the minor component only, and suggested that it had an inhibitory effect on copulation. In addition, he showed that the synthetic pheromones, when used as a bait in traps, greatly increased the number of adult males caught in traps of various designs, and so may be of interest in sampling low-level populations of the pest. However, the direct removal of insects from a population would be most efficient if we could remove the virgin females. Such a system may be practical with the pheromone produced by the male bruchid, *Acanthoscelides obtectus* (Say) (13). The removal of mated females may also be useful; however, information concerning the reproductive biology of the insect would be needed before we could judge the value of the system.

## III. PHEROMONES AS LURES FOR INOCULATION DEVICES AND PATHOGEN DISSEMINATION

The habitats of storage insects provide ideal conditions for the dissemination of protozoan diseases, particularly those caused by dessication-resistant spore-formers. Although the areas of infestation are initially small and localized, they contain highly concentrated populations of insects that lend themselves to disease epizootics.

With an effective pheromone and luring device, it should be just as easy to expose the attracted insects to a pathogen as to an insecticide. Of course with a pathogen, the hoped-for result would not be immediate kill; rather the infected or spore-laden insect would return to its natural habitat and infect others of its kind. This idea appears especially promising with many stored-product insect pests. We know their specific pheromones, and the pathogens that are highly virulent to them; we have effective traps; and we are concerned with commodities held in a given, usually confined area. Also, in such sites, baits containing pesticides would be of considerable concern.

Protozoan infections have long been recognized as regulators of insect populations (14). However, McLaughlin (15) thinks the incidence of disease due to protozoans in insect populations may be underestimated. There are several well-documented examples of naturally occurring protozoan infections that have had considerable impact on pest insect populations.

McLaughlin (16) demonstrated the effective use of a bait containing a feeding stimulant and the protozoan, *Mattesia grandis* McLaughlin, in spreading the pathogen throughout cotton fields for the control of the boll weevil, *Anthonomus grandis* Boheman.

At the present time we are evaluating in the laboratory and field the concept of utilizing pheromones for luring pest insects to treatment foci for dispersal of agents deleterious to the stored-product pest populations. We have recently completed a preliminary laboratory study to determine how rates of infection of the *Trogoderma* protozoan pathogen, *Mattesia trogodermae* Canning, are influenced by mode of transmission, dosage, and host species (17). Male *Trogoderma glabrum* (Herbst) beetles were drawn to a corrugated cardboard innoculation device treated with a component of the pheromone $(Z)$-14-methyl-8-hexadecen-1-ol that attracts *Trogoderma* when they were dusted with *Mattesia* spores diluted with cellulose powder. The result was contamination of 96% of the test insects placed in the arena; only 56% picked up spores from innoculation devices containing only *Mattesia* spores (no pheromone).

The disease is transferred as follows: the lured males are attracted by the female sex pheromone, where they are contaminated with disease organism in the inoculation device. They must then leave the inoculation device, locate a female, and contaminate the female by contact or copulation. Thereafter the disease may spread to the eggs during oviposition or to the larvae that ingest dead disease-carrying adults. Also, transfer may take place by using larval attractants, female attractants, or aggregating compounds. If successful, this novel approach should be of economic benefit against *Trogoderma* infestations and would also serve as a model system for developing other such control programs.

## IV. SUMMARY AND PROSPECTS FOR THE FUTURE

During our pheromone-pathogen project and related studies, we have spent considerable time studying all aspects of the behavioral biology of stored-product insects, particularly those aspects that relate to courtship and mating.

For example, our recent studies (18) have revealed that adult male *Trogoderma glabrum* show distinct diel behavior patterns: they remain concealed for most of the light and dark periods, but have an 8-hr active period centered near the mid-point of the photophase. During this period, they move to unsheltered areas near their places of concealment and remain passive but alert for the presence of female-released sex pheromone. If they are stimulated by pheromone, males begin a random searching effort; but searching becomes directional at low air velocities, which implies anemotaxis. This male active period has been shown to coincide with the period of maximum sensitivity of the males to the sex pheromone.

Also, other recent studies (19) have clarified the multistep sequence in the response of *T. glabrum* to females. The first step (arousal/searching) involves airborne pheromones and is mediated using the antennae; the second (attempted mating) involves contact chemoreception and is mediated using receptors on the mouthparts. Thus, the function of the multicomponent pheromone system in *Trogoderma* spp., particularly *T. glabrum*, is becoming clearer. Some components [(*E*)-14-methyl-8-hexadecen-1-a1] may be primarily involved with attraction; others may serve to stimulate copulation.

Now in cooperation with Dr. R.M. Silverstein, Syracuse University, we are clarifying the complete multicomponent pheromone system for several species of *Trogoderma*. A glass chamber suitable for holding females that allows maximum calling and pheromone release is used. The volatiles pass into a column packed with Porapak Q. Thus, the volatile pheromones can be studied by extracting the Porapak, as can the less volatile components by extracting and washing of the females and substrate.

In addition, the role of nonpheromone chemical inhibition of response of *Trogoderma* is being studied as a means of control. This concept is basically the same as that used with odor-control systems for man. A compound is released that will render ineffective the unwanted stimulus. However, as with other pheromone control systems, this method would be effective probably only with low population levels.

We hope that the information obtained from these studies will enable us to design a trap and a pheromone control procedure that will meet the requirements set forth earlier for effective surveillance and insect manipulation by pheromones, and will be a predictable and reproducible system.

## REFERENCES

(1)   L.S. Henderson. 1967. Preventing insect infestations. *Proc. Grain Cereal Prod. Sanit. Conf.* Feb. 16-17, pp. 35-42.

(2)   W.E. Burkholder. 1974. Programs utilizing pheromones in survey or control: Stored product pests. In *Pheromones.* M.C. Birch, Ed. American Elsevier, New York, pp. 449-452.

(3)   W. Klassen. 1973. Personal communication.

(4)   G. Domenichini. 1974. International and national problems arising in storage. *EPPO Bull.* 4:305-307.

(5)   R.R. Cogburn. 1973. Stored product insect populations in port warehouse of the gulf coast. *Environ. Entomol.* 2:401-407.

(6)   W.E. Burkholder. 1970. Pheromone research with stored product Coleoptera. In *Control of Insect Behavior by Natural Products.* D.L. Wood, R.M. Silverstein, and M. Nakajima, Eds. Academic, New York, pp. 1-20.

(7)  M. Nakajima. 1970. Studies on sex pheromones of the stored grain moths. *In Control of Insect Behavior by Natural Products.* D.L. Wood, R.M. Silverstein, and M. Nakajima, Eds. Academic, New York, pp. 209-211.

(8)  H.Z. Levinson. 1974. Possibilities of using insectistatics and pheromones in the control of stored product pests. *EPPO Bull.* **4**:391-416.

(9)  A.V. Barak and W.E. Burkholder. 1976. Trapping studies with dermestid sex pheromones. *Environ. Entomol.* **5**:111-114.

(10) W.E. Burkholder. 1973. Black carpet beetle: Reduction of mating by megatomoic acid the sex pheromone. *J. Econ. Entomol.* **66**:1327.

(11) U.E. Brady, E.J. Jay, L.M. Redlinger, and G. Pearman. 1975. Mating activity of *Plodia interpunctella* and *Cadra cautella* during exposure to synthetic sex pheromone in the field. *Environ. Entomol.* **4**:441-444.

(12) P.E. Wheatly. 1975. Research being undertaken by the Tropical Stored Products Center in the United Kingdom and Overseas. *EPPO Bull.* **4**:495-500.

(13) D.F. Horler. 1970. An allenic ester produced by the male dried bean beetle, *Acanthoscelides obtectus* (Say). *J. Chem. Soc.* (c):859-862.

(14) W.M. Brooks. 1971. Protozoan infections of insects with emphasis on inflammation. *Proc. IVth Int. Colloq. Insect Pathol. & Soc. Invert. Pathol.* College Park, Md., pp. 11-27.

(15) R.E. McLaughlin. 1971. Use of protozoans for microbial control of insects. *In Microbial Control of Insects and Mites.* H.D. Burges and N.W. Hussey, Eds. Academic, New York, pp. 151-172.

(16) R.E. McLaughlin. 1966. Infection of the boll weevil with *Mattesia grandis* induced by a feeding stimulant. *J. Econ. Entomol.* **59**:909-911.

(17) C.P. Schwalbe, W.E. Burkholder, and G.M. Boush. 1974. *Mattesia trogodermae* infection rates as influenced by mode of transmission, dosage and host species. *J. Stored Prod. Res.* **10**:161-166.

(18) T.J. Shapas and W.E. Burkholder. Unpublished.

(19) R.E. Greenblatt, W.E. Burkholder, J.C. Cross, R.C. Byler, and R.M. Silverstein. 1976. Chemical communication in the mating behavior of *Trogoderma glabrum* (Herbst) (Coleoptera: Dermestidae). J. Chem. Ecol. **2**:285-297.

## CHAPTER 21

# Manipulation of Insect Pests of Agricultural Crops

H.H. Shorey

*Department of Entomology*
*University of California*
*Riverside, California*

Behavior-modifying chemicals have had a small role in the management of insect pests of agricultural crops for a number of years. They have been principally employed as bait in survey traps to indicate the distribution or abundance of certain pests that are stimulated to approach the chemical source. For instance, traps baited with eugenol and related chemicals have been useful in surveys to detect the spread of introduced pest species, such as the Japanese beetle and European chafer. The biologic roles of the chemicals in the normal life of these species are often not clear; most of the chemicals have been developed for survey purposes by empirical field-screening of known synthetic attractants and their analogues. However, we might expect that the chemicals are usually identical or similar to naturally occurring feeding attractants.

Behavior-modifying chemicals have also been incorporated with insecticides, forming toxic baits. These chemicals might act as attractants, causing approach of pest insects from a distance, or as arrestants or feeding stimulants, causing the insects to remain in contact with or to ingest the toxicants. Materials such as molasses have been used in toxic baits broadcast on the ground for cricket control or sprayed on plants for control of moths such as the pink bollworm and bollworm.

With the recent advent of synthetic pheromones, man may have much more powerful behavior-manipulating tools than he had when only nonpheromone chemicals were available. This is because species survival may often depend on the insects making appropriate responses when they perceive pheromones released from conspecifics. On the other hand, plant-produced chemicals that are feeding or ovipositional attractants or stimulants may often be less essential to species survival; this is particularly the case if the pest insects have a variety of types of natural host plants on which they may feed or lay their eggs. Also, we may speculate that pheromones are more powerful in their action than chemicals involved in feeding or oviposition; very minute quantities of a pheromone may be as biologically active as much greater quantities of a chemical indicative of a feeding or oviposition site.

The pheromones that appear to have the greatest utility in pest-management programs on agricultural crops are those that stimulate the approach of conspecifics from a distance. A large class of such chemicals consists of the sex pheromones. They are released by one sex and cause premating behavioral reactions, often including approach from a distance, by the opposite sex. Another class of chemicals constitutes the aggregation pheromones. These are often released by one or the other sex and cause the approach of conspecifics of both sexes. The biologic roles of aggregation pheromones are usually obscure, although many of them may serve to bring other individuals to a suitable food source that has been located by the pheromone-releasing individual.

With the notable exception of the aggregation pheromone of the boll weevil, almost all research directed toward the use of pheromones in agricultural pest management has been conducted on the sex pheromones released by female moths prior to mating.

It can not be stressed too strongly that the key to devising effective systems for managing pest insects through manipulation of their responses to chemicals is the acquisition of an intimate knowledge of the insects' use of the chemicals in their normal day-to-day lives. The slow pace at which we are approaching truly practical pheromonal pest-management systems probably reflects in large part our tendency to bypass these important studies of normal behavior.

Two main strategies are emerging for the use of pheromones in pest management. These strategies are almost direct opposites of each other. One involves the use of a pheromone for stimulation of the normal approach response of the responding insects, except that the response is manipulated in such a way that the insects end up in a location that is disadvantageous for them and advantageous for man. For instance, by following chemical cues and approaching the source of a synthetic sex pheromone, male moths might become ensnared in a trap instead of arriving near a pheromone-releasing female. The other strategy involves the disruption of the normal chemical communication behavior of the insects. In this case, male moths might be rendered incapable of responding to or locating the source of a natural pheromone, and the females would remain unmated. These strategies are considered in more detail in the following sections.

## I. STIMULATION OF THE APPROACH RESPONSE OF PEST INSECTS

Pheromones that stimulate approach responses can be used as bait in traps. Great care must be exercised in the design of an effective trap. It must be appropriate so that the insects, in displaying their normal behavior, will freely enter it and be captured. Variables that should be considered include the size, color, and shape of the trap, the type of orifice through which the insects will enter, the height above the ground, the vegetative habitat in which the trap is deployed, and the concentration of pheromone issuing from the trap (1) (2).

Height above the ground is a very important variable, and responding insects may be caught in maximal numbers only when the trap is at the appropriate elevation. Among a number of moth species, best responses of males occur when pheromone-baited traps are suspended near the top of the foliage canopy, regardless of whether the vegetation is cabbages or forest trees (3)-(5). This generality does not always apply, however, and optimal heights should be determined from behavioral analyses of each species. Also, the optimal height for a given species may differ when certain environmental factors such as wind velocity vary from low to high (4).

Some insects are highly restricted to their vegetative habitat. Males of the pink bollworm and the sunflower moth are rarely caught in pheromone-baited traps located away from the type of crop in which they developed as larvae (6) (7). On the other hand, some insects are relatively independent of vegetative hosts in their pheromone behavior. Males of the cabbage looper moth are attracted equally to pheromone-baited traps placed in cabbage fields or over bare ground (8).

The concentration of pheromone emitted from a trap may be especially critical. Often, the catches of male moths increase as the concentration of pheromone is increased, up to a certain maximum level. After that, further increases in pheromone concentration are related to corresponding decreases in numbers of males trapped (9) (10). Furthermore, the concentration that captures the greatest number of males may vary from one trap to another, depending on the details of trap design (10). We suspect that the narrow range of the most effective concentrations is related to the behavior of the responding moths. If the concentration near the orifice of the trap is higher than that likely to occur near a normal, pheromone-releasing female, then the males might sense that they are exposed to a concentration indicative that a female is nearby when they are still some distance downwind from the trap. The males might then be stimulated to stop upwind orientation and to initiate short-range, visually oriented searching behavior to locate the nonexistent female.

Pheromone-baited traps have two potential uses in agricultural pest management: survey for the distribution or abundance of the pests and direct control by removal of the trapped insects from the population.

## A. Pheromone-Baited Traps For Survey of Pest Insects

The only practical usage of pheromones to date in pest-management programs has been as a survey tool. Pheromone-baited survey traps have been used either to monitor the spread of a potential pest into previously noninfested areas or to monitor the population level of an established pest so that conventional pest-control techniques can be applied at the most appropriate time.

Pheromone-baited traps are presently used in large areas of the United States to monitor the spread of such introduced insects as the pink bollworm. The trapping technique may be infinitely more efficient than the alternative—visual searching for signs of damage by the insects (11) (12). The potential range of insect species that can be surveyed in this way is limited only by the range of species that use pheromone signals for inducing aggregation of conspecifics.

In principle, the use of pheromone-baited traps to monitor the build-up of an established pest population so as to predict when insecticidal control methods are necessary should also be highly effective. Often the male and female adults must utilize pheromone communication systems and mate before the damaging larval stage can be produced. Therefore, if the number of males captured in traps can be correlated with subsequent larval populations, spray schedules can be efficiently timed. Such correlations have increased the efficiency of insecticidal control measures directed against the codling moth and the pink bollworm (13)-(17). However, the correlations may often be poor and thus lead to misjudgments. For example, differing weather patterns may cause trapping results not to correlate well with subsequent egg-laying by female codling moths. Therefore, codling moth pheromone trap surveys have had to be supplemented by determination of seasonal oviposition patterns of laboratory-reared females or by the inspection of fruit for larval entries (13) (15).

## B. Pheromone-Baited Traps for Direct Control of Pest Insects

When used for direct control, the trapping method must remove sufficient insects to cause a reduction in the pest population of the next generation. The proportion of insects that must be removed depends on the sex that is trapped and the biology of the species involved. In those species in which virgin females are trapped, the reduction in the subsequent generation may be directly related to the proportion removed from the population. If females are trapped after mating and laying of some of their eggs, the efficiency of the method decreases. If males are trapped, a very large proportion probably would have to be captured before any noticeable diminution of the next generation would occur. This is especially true for the many insect species in which both males and females are capable of mating many times. If 90% of the males were captured, the remaining 10% might be sufficient to inseminate most of the females.

The trap may be a conventional device in which the attracted insects are killed or otherwise neutralized; it also could be a source of chemical sterilant that only neutralizes the reproductive capacity of a male plus, perhaps, a female with which it mates; or it may be an area into which the attracted insects are "herded." The latter approach has proved effective with the boll weevil, in which both sexes approach a source of grandlure, the male-produced aggregation pheromone (18)-(21). One control strategy calls for the weevils to be attracted into specific areas of cotton fields baited with pheromone, whereupon insecticides sprayed over those areas kill the insects.

The trapping approach, using female sex-pheromone-baited traps, is often called the *male annihilation technique.* Considerable effort has gone into the evaluation of this technique against a number of moth species. In general, preliminary results have shown that the method may have some usefulness in keeping pest populations from increasing if they are already at a very low level; however, little success has been shown against infestations that are already at a damaging level (22)-(27). This restriction to low-density populations seems reasonable if we consider that the field traps must compete directly with pheromone-releasing females for attracting males. If over 90% of the males must be removed, and if the concentration of pheromone leaving a trap can not be much higher than the concentration near a pheromone-releasing female, then many more traps than natural females would probably be necessary. Also, the proportion of male moths that a pheromone-baited trap removes from the environment may decrease as the absolute density of the pest population in the field increases (16) (24) (28). This factor is perhaps directly related to the numbers of competing wild females in the field.

There may be more latitude in the critical concentration of pheromone leaving a trap than that indicated above. For example, the most effective trap evaluated by Sharma et al. (10) for capturing males of the cabbage looper moth released pheromone at an optimum rate of 1 $\mu$g/min, whereas the average female release rate is near 0.01 $\mu$g/min (29). Also, for some species, *synergists* may be used to potentiate the activity of a pheromone; however, recent work indicates that these synergists may often be components of the natural pheromone blend that is released also by the living females (30) (31).

A pheromone-baited trap, because it releases the chemical continuously, may have a timing advantage over pheromone-releasing females. Females of many insect species have well-defined periods during the day or night when they release pheromone. Males of these species tend to be maximally responsive to the pheromone at the same time, although their period of responsiveness is often broader than the females' pheromone-release period (32). Thus, males may be attracted to and be ensnared in traps before most of the females start to release pheromone.

On the other hand, other stimuli released from females, in addition to a pheromone, may be needed to guide the short-range approach of males efficiently. Male moths of many species visually orient to females when they perceive high pheromone concentrations (32). In some cases, then, the presence of synthetic pheromone in the field might even be a disadvantage, stimulating males to activity wherein they might find females by reacting to other short-range orientation stimuli (22) (33).

The interpretation of results obtained during an attempt to control a pest insect through male trapping may be confounded with communication disruption (see Section II). If sufficient pheromone is released from the traps into the air, some males may be captured and some may be rendered incapable of orienting to either natural or artificial pheromone sources. This dual effect—trapping some males and disrupting communication of the rest—may not be a disadvantage; it may only make our interpretation of why most females remained virgin difficult (24).

## II. DISRUPTION OF PHEROMONE COMMUNICATION

Even before the sex pheromones of any pest insects had been identified, a number of writers proposed that if sufficient synthetic pheromone were distributed in the air, normal pheromone-communication systems of certain species would be disrupted and the sexes might be incapable of locating each other for mating (34)-(41). This approach has a great deal of appeal and has been pursued extensively in recent years, with very promising results.

The behavioral mechanisms that cause communication disruption are poorly known. Some evidence based on laboratory experiments, plus considerable speculation, indicates that three factors are involved: sensory adaptation, central nervous system (CNS) habituation, and "confusion."

Sensory adaptation to odors is a common phenomenon in both vertebrates and invertebrates. If an animal is exposed to a constant level of odorant, its olfactory sensory receptors soon stop reporting to the CNS that the odor is present (42). The sensory receptors again become responsive to the odor within a few seconds or minutes after the stimulus is removed.

Habituation is similar in principle to sensory adaptation, except the phenomenon occurs within the CNS. If an animal is exposed to a stimulus and if its subsequent responses do not lead to a suitable end result (mating in the case of a male insect stimulated by female sex pheromone), then it tends to be less responsive when it perceives the stimulus again. Habituation of male insects to sex pheromones may persist for many minutes, or even hours (43) (44).

"Confusion" is a direct result of competition between the pheromone released from synthetic sources and the identical pheromone normally released from the insects themselves in the field. If the number of synthetic sources greatly exceeds the number of natural pheromone-releasing insects, or if the sources release much more pheromone than the natural insects, then the responding insects, if they are not adapted or habituated, will be more likely to approach the synthetic sources.

Disruption of communication has been studied mostly with synthetic pheromones that are identical to the natural pheromones of a particular species. However, disruption might also be accomplished by the use of parapheromones [nonpheromone chemicals that cause behaviors identical to those caused by natural pheromones (45)] or by antipheromones (nonpheromone chemicals that directly block or inhibit responsiveness of insects to their natural pheromones).

## A. Disruption by Pheromones

The first demonstration of the feasibility of the disruption approach was accomplished by Gaston et al. (46). They released looplure, the synthetic pheromone of the cabbage looper moth, from an array of 100 evaporators spaced 3 m apart in 0.1-ha plots and found that male moths were rendered completely incapable of locating pheromone-releasing females used as bait in traps in the centers of the plot. Since that time, disruption of sex-pheromone communication through atmospheric permeation with synthetic pheromones has been demonstrated in a wide variety of moth species [see references in (47)].

Especially in the case of the gypsy moth (a pest of forests, and thus not considered further here) and the pink bollworm moth, programs using synthetic pheromones for disruption of premating communication seem to be coming close to practical realization. For both species, it appears likely that the release of pheromone in concentrations of about 15 gm/ha over the entire pest season may cause most females to remain infertile and thus may cause substantial reduction in the larval pest population of the next generation (48)-(51). The disruption systems for both species appear to be nearing the level at which their use is competitive with standard insecticide-control techniques. Additional fine-tuning of the systems, through additional studies of the normal communication behavior of the insects and through the engineering of techniques for evaporating the pheromone into the atmosphere at the correct locations and the correct times, should lead to further improvements in the degree of pest control afforded.

Increased knowledge of the normal behavior of the insects is particularly important. As one example of this, pink bollworm moths normally aggregate near the top of the cotton foliage canopy before pheromone communication occurs (4), and synthetic pheromone dispensers placed at that elevation are normally most effective in disrupting communication (52). However, on windy nights the moths move down to near the soil surface prior to pheromone communication, and this microenvironment can then be permeated with synthetic pheromone. In the end, synthetic pheromone must be maintained at sufficient concentrations in all environments where premating communication is likely to occur. This is a formidable challenge, considering that our present knowledge is so inadequate that even the normal location of mating is unknown for most pest species.

Two differing strategies for deploying synthetic pheromones for communication disruption have been studied. One strategy involves widely separated evaporative substrates, spaced as far as 0.1 to 1 km apart, with each substrate releasing relatively massive quantities of pheromone into the air. The other strategy is based on a large number of small substrates placed close together in the field. The small substrates could be dispersed by aircraft or other conventional insecticide-application techniques, and each substrate would release a relatively small quantity of pheromone.

Working with the cabbage looper moth, Shorey and Gaston (53) proposed that neither the separation between substrates nor the release rate of pheromone from each substrate was the critical factor. Rather, the important factor was the absolute quantity of pheromone released into the atmosphere over each unit area of land in a given time. Farkas et al. (54) have obtained over 90% disruption of premating communication among cabbage looper moths by using synthetic pheromone dispensers placed 400 m apart.

At the other end of the spectrum—and probably the most practical disruption method for many pest species—is the use of microcapsules containing pheromone (48) (49) (51). The microcapsules are often approximately 100 $\mu$ in diameter; to be effective many capsules must be distributed over a given area.

### B. Disruption by Parapheromones

In some cases, chemicals other than pheromones may cause behavioral activity similar to that caused by the natural pheromone of a species. These parapheromones are often very similar in chemical structure to natural pheromones, and they might be biologically active because they stimulate the natural pheromone receptors. However, because of the extreme selectivity of the insect receptor mechanism for the correct pheromone structure, most parapheromones would be expected to have considerably lower biologic activity than the natural pheromones (47).

If parapheromones do indeed stimulate natural pheromone receptor sites, then the insect not only could be stimulated to activity by them, but also its normal pheromone communication system could be disrupted by them. This has been found to be true in the case of the pink bollworm. Hexalure, a parapheromone of the pink bollworm moth, is approximately 100-fold less stimulatory to male pink bollworms than the natural pheromone, gossyplure (55). Hexalure can be used to disrupt premating communication between males and females of this species, although approximately 100-fold more hexalure is needed to bring about an effect equivalent to that provided by synthetic gossyplure (50). When 330 gm of hexalure per hectare were volatilized into the atmosphere of cotton fields during a 16-week growing season, the resulting disruption of moth sex pheromone communication led to a 75% reduction in the numbers of damaging larvae attacking the cotton bolls (56).

Some chemicals may disrupt moth sex pheromone communication even though they do not stimulate the approach of males of the species involved. Thus, Kaae et al. (57) found that looplure not only disrupts communication between males and females of the cabbage looper, alfalfa looper, and soybean looper—all species that apparently use this chemical as a major component of their sex pheromone—but it also disrupts communication of the cotton bollworm, tobacco budworm, pink bollworm, and yellow-striped armyworm—species whose males are not attracted to looplure. Such a finding gives some hope that a few key chemicals might be found that could be employed together to disrupt premating communication of a number of pest species that attack a given crop. On the other hand, because quantitatively much larger amounts of a parapheromone might have to be used to cause a disruptive effect equivalent to that caused by the natural pheromone (58), nonpheromone chemicals might have to be used in such large amounts that they are not economically feasible.

## C. Disruption by Antipheromones

Certain nonpheromone chemicals, when released into the air immediately adjacent to pheromone-releasing female moths, prevent males from orienting to those females (26) (31) (33) (59)-(67). However, later research has shown that the same chemicals have little or no effect in preventing males from approaching females if the chemicals are used to permeate a large mass of air surrounding the insects (65) (68)-(72). Apparently, then, these antipheromones only operate to inhibit male approach when they constitute a locus in the vicinity of the females.

It is known that a number of related moth species "share" certain pheromone components—that is, they have one or more attractive chemicals in common. Some of the species also produce compounds that act as inhibitors to males of the related species, preventing their approach for mating (31). These inhibitors may constitute the class of chemicals called *antipheromones.*

It seems likely that antipheromones may have little practical value in disruption of communication. If they are used to permeate an air mass, if the insects become adapted or habituated to them, and if the receptor sites for the antipheromones are different from the sites for the natural pheromones, then the insects might remain fully responsive to the natural pheromones and be capable of communicating in their normal manner (65) (69) (71).

## III. CONCLUSIONS

Chemical manipulation of the behavior of insect pests of agricultural crops has enormous potential for use in pest-management systems. As extensive studies of the normal behavior of the insects are conducted, many more avenues for manipulation in addition to those surveyed in this chapter may become apparent. Two examples are the dispersion-inducing pheromones of aphids and of apple maggot flies. Aphids of a number of species, when disturbed by possible predators, release "alarm pheromones" that cause nearby aphids to cease feeding and to move away from the site of the volatile chemicals (73)-(75). The aphids may even drop from the plants. Possibly, schemes could be developed for the use of these chemicals to control pest aphid populations. A female apple maggot fly, after laying an egg in a fruit, marks the surface of the fruit with a pheromone that deters other females from ovipositing therein (76) (77). The other females are thus caused to disperse and to locate unmarked fruit for oviposition. Possibly the pheromone could be sprayed in orchards so that female flies would be inhibited from laying their eggs in any of the fruit. Only further research will reveal the practical potentialities for the manipulation of these and other behavioral characteristics as means of pest control.

An agricultural crop is often threatened by a complex of pest insect species. Pheromonal control of one of the species may be of little value if normal insecticide application schedules must be followed for control of a number of other species. On the other hand, often one or two key species are the only primary pests of a crop and require rigorous suppression techniques. If insecticide treatments could be minimized and if populations of beneficial insects increased, many secondary pests might present little hazard to the crop. These secondary pests often are abundant because insecticides upset the normal biotic balance within the agricultural ecosystem. The extreme selectivity of pheromonal, behavioral control of a few key pests might cause a number of potential secondary pests to pose less of a problem.

Behavior-modifying chemicals should be considered as potentially powerful pest-management tools, and should be intelligently integrated with other tools to make the environment less suitable for survival or reproduction of agricultural pests.

# REFERENCES

(1)    A.K. Minks and S. Voerman. 1973. Sex pheromones of the summerfruit tortrix moth, *Adoxophyes orana:* Trapping performance in the field. *Entomol. exp. appl.* **16**:541-549.

(2)    A.K. Minks, J. Ph. W. Noordink, and C.A. Van den Anker. 1971. Recapture by sex traps of *Adoxophyes orana,* released from one point in an apple orchard. *Mededelingen Fakulteit Landbouw-Wetenschappen Gent* **36**:274-282.

(3)    M.T. AliNiazee and E.M. Stafford. 1972. Sex pheromone studies with the omnivorous leafroller, *Platynota stultana* (Lepidoptera: Tortricidae): Effect of various environmental factors on attraction of males to the traps baited with virgin females. *Ann. Entomol. Soc. Am.* **65**:958-961.

(4)    R.S. Kaae and H.H. Shorey. 1973. Sex pheromones of Lepidoptera. 44. Influence of environmental conditions on the location of pheromone communication and mating in *Pectinophora gossypiella. Environ. Entomol.* **2**:1081-1084.

(5)    C.A. Miller and G.A. McDougall. 1973. Spruce budworm moth trapping using virgin females. *Can. J. Zool.* **51**:853-858.

(6)    R.K. Sharma, R.E. Rice, H.T. Reynolds, and H.H. Shorey. 1971. Seasonal influence and effect of trap location on catches of pink bollworm males in sticky traps baited with hexalure. *Ann. Entomol. Soc. Am.* **64**:102-105.

(7)    G.L. Teetes and N.M. Randolph. 1970. Color preference and sex attraction among sunflower moths. *J. Econ. Entomol.* **63**: 1358-1359.

(8)    C.A. Saario, H.H. Shorey, and L.K. Gaston. 1970. Sex pheromones of noctuid moths. XIX. Effect of environmental and seasonal factors on captures of males of *Trichoplusia ni* in pheromone-baited traps. *Ann. Entomol. Soc. Am.* **63**:667-672.

(9)    L.K. Gaston, H.H. Shorey, and C.A. Saario. 1971. Sex pheromones of noctuid moths. XVIII. Rate of evaporation of a model compound of *Trichoplusia ni* sex pheromone from different substrates at various temperatures and its application to insect orientation. *Ann. Entomol. Soc. Am.* **64**:381-384.

(10)    R.K. Sharma, H.H. Shorey, and L.K. Gaston. 1971. Sex pheromones of noctuid moths. XXIV. Evaluation of pheromone traps for males of *Trichoplusia ni. J. Econ. Entomol.* **64**:361-364.

(11)    D.S. Moreno, J. Fargerlund, and J.G. Shaw. 1973. California red scale: Captures of males in modified pheromone traps. *J. Econ. Entomol.* **66**:1333.

(12)    J.G. Shaw, D.S. Moreno, and J. Fargerlund. 1971. Virgin female California red scales used to detect infestations. *J. Econ. Entomol.* **64**:1305-1306.

(13)    W.C. Batiste, A. Berlowitz, W.H. Olson, J.E. DeTar, and J.L. Joos. 1973. Codling moth: Estimating time of first egg hatch in the field—A supplement to sex attractant traps in integrated control. *Environ. Entomol.* **2**:387-391.

(14)    H.F. Madsen and J.M. Vakenti. 1972. Codling moths: Female-baited and synthetic pheromone traps as population indicators. *Environ. Entomol.* **1**:554-557.

(15)    H.F. Madsen and J.M. Vakenti. 1973. Codling moth: Use of Codlemone®-baited traps and visual detection of entries to determine need of sprays. *Environ. Entomol.* **2**:677-679.

(16)    H. Riedl and B.A. Croft. 1974. A study of pheromone trap catches in relation to codling moth (Lepidoptera: Olethreutidae) damage. *Can. Entomol.* **106**:525-537.

(17)    N.C. Toscano, A.J. Mueller, V. Sevacherian, R.K. Sharma, T. Niilus, and H.T. Reynolds. 1974. Insecticide applications based on hexalure trap catches versus

automatic schedule treatments for pink bollworm moth control. *J. Econ. Entomol.* 67:522-524.

(18)   T.B. Davich, D.D. Hardee, and M.J. Alcála. 1970. Long-range dispersal of boll weevils determined with wing traps baited with males. *J. Econ. Entomol.* 63:1706-1708.

(19)   D.D. Hardee, O.H. Lindig, and T.B. Davich. 1971. Suppression of populations of boll weevils over a large area in West Texas with pheromone traps in 1969. *J. Econ. Entomol.* 64:928-933.

(20)   E.P. Lloyd, M.E. Merkl, F.C. Tingle, W.P. Scott, D.D. Hardee, and T.B. Davich. 1972. Evaluation of male-baited traps for control of boll weevils following a reproduction-diapause program in Monroe County, Mississippi. *J. Econ. Entomol.* 65:552-555.

(21)   S.H. Roach, L. Ray, A.R. Hopkins, and J.M. Taft. 1971. Comparison of attraction of wing traps and cotton trap plots baited with male boll weevils for overwintered weevils. *Ann. Entomol. Soc. Am.* 64:530-531.

(22)   E.A. Cameron. 1973. Disparlure: A potential tool for gypsy moth population manipulation. *Bull. Entomol. Soc. Am.* 19:15-19.

(23)   E.H. Glass, W.L. Roelofs, H. Arn, and A. Comeau. 1970. Sex pheromone trapping red-banded leaf roller moths and development of a long-lasting polyethylene wick. *J. Econ. Entomol.* 63:370-373.

(24)   E.F. Taschenberg, R.T. Cardé, and W.L. Roelofs. 1974. Sex pheromone mass trapping and mating disruption for control of redbanded leafroller and grape berry moths in vineyards. *Environ. Entomol.* 3:239-242.

(25)   K. Trammel, W.L. Roelofs, and E.H. Glass. 1974. Sex-pheromone trapping of males for control of redbanded leafroller in apple orchards. *J. Econ. Entomol.* 67:159-164.

(26)   W.L. Roelofs and J.P. Tette. 1970. Sex pheromone of the oblique-banded leaf roller moth. *Nature* 226:1172.

(27)   W.L. Roelofs, E.H. Glass, J. Tette, and A. Comeau. 1970. Sex pheromone trapping for red-banded leaf roller control: Theoretical and actual. *J. Econ. Entomol.* 63:1162-1167.

(28)   J.F. Howell. 1974. The competitive effect of field populations of codling moth on sex attractant trap efficiency. *Environ. Entomol.* 3:803-807.

(29)   L.L. Sower, L.K. Gaston, and H.H. Shorey. 1971. Sex pheromones of noctuid moths. XXVI. Female release rate, male response threshold, and communication distance for *Trichoplusia ni. Ann. Entomol. Soc. Am.* 64:1448-1456.

(30)   A. Hill, R. Cardé, A. Comeau, W. Bode, and W. Roelofs. 1974. Sex pheromones of the tufted apple bud moth *(Platynota idaeusalis). Environ. Entomol.* 3:249-252.

(31)   W.L. Roelofs and A. Comeau. 1971. Sex attractants in Lepidoptera. In: Proc. 2nd Int. Congr. Pesticide Chem., Vol. 3. *Chemical Releasers in Insects.* Gordon and Breach, New York, pp. 91-114.

(32)   H.H. Shorey. 1973. Behavioral responses to insect pheromones. *Ann. Rev. Entomol.* 18:349-380.

(33)   J.A. Klun and J.F. Robinson. 1971. European corn borer moth: Sex attractant and sex attraction inhibitors. *Ann. Entomol. Soc. Am.* 64:1083-1086.

(34)   A.L. Babson. 1963. Eradicating the gypsy moth. *Science* 142:447-448.

(35)   E.D. Burgess. 1964. Gypsy moth control. *Science* 143:526-527.

(36)   M. Jacobson and M. Beroza. 1963. Chemical insect attractants. *Science* 140:1367-1373.

(37)   R.H. Wright. 1964. After pesticides—what?*Nature* 204:121-125.

(38)  R.H. Wright. 1964. 'Metarchon': a new term for a class of non-toxic pest control agents. *Nature* **204**:603-604.

(39)  R.H. Wright. 1964. Insect control by nontoxic means. *Science* **144**:487.

(40)  R.H. Wright. 1965. Metarchons: Insect control through recognition signals. *Bull. Atomic Sci.* **21**:28-30.

(41)  R.H. Wright. 1965. Finding metarchons for pest control. *Nature* **207**:103-104.

(42)  T.L. Payne. 1974. Pheromone perception. In *Pheromones*. M.C. Birch, Ed. American Elsevier, New York, pp. 35-61.

(43)  R.J. Bartell and L.A. Lawrence. 1973. Reduction in responsiveness of males of *Epiphyas postvittana* (Lepidoptera) to sex pheromone following previous brief pheromonal exposure. *J. Insect. Physiol.* **19**:845-855.

(44)  R.M.M. Traynier. 1970. Habituation of the response to sex pheromone in two species of Lepidoptera, with reference to a method of control. *Entomol. exp. appl.* **13**:179-187.

(45)  R.S. Kaae, H.H. Shorey, S.U. McFarland, and L.K. Gaston. 1973. Sex pheromones of Lepidoptera. XXXVII. Role of sex pheromones and other factors in reproductive isolation among ten species of Noctuidae. *Ann. Entomol. Soc. Am.* **66**:444-448.

(46)  L.K. Gaston, H.H. Shorey, and S.A. Saario. 1967. Insect population control by the use of sex pheromones to inhibit orientation between the sexes. *Nature* **213**:1155.

(47)  H.H. Shorey. 1975. Concepts and methodology involved in pheromonal control of Lepidoptera by disruption of premating communication. *In Crop Protection Agents—Their Biological Evaluation*. N.R. McFarlane, (Ed.) Academic, New York (in press).

(48)  M. Beroza, C.S. Hood, D. Trefrey, D.E. Leonard, E.F. Knipling, W. Klassen, and L.J. Stevens. 1974. Large field trial with microencapsulated sex pheromone to prevent mating of the gypsy moth. *J. Econ. Entomol.* **67**:659-664.

(49)  E.A. Cameron and C.P. Schwalbe. 1974. Disruption of gypsy moth mating with microencapsulated disparlure. *Science* **183**:972-973.

(50)  R.S. Kaae, L.K. Gaston, and H.H. Shorey. Sex pheromones of Lepidoptera. Disruption of *Pectinophora gossypiella* populations in cotton by permeation of the atmosphere with gossyplure. Manuscript in Preparation.

(51)  C.P. Schwalbe, E.A. Cameron, D.J. Hall, J.V. Richerson, M. Beroza, and L.J. Stevens. 1974. Field tests of microencapsulated disparlure for suppression of mating among wild and laboratory-reared gypsy moths. *Environ. Entomol.* **3**:589-592.

(52)  J.R. McLaughlin, H.H. Shorey, L.K. Gaston, R.S. Kaae, and F.D. Stewart. 1972. Sex pheromones of Lepidoptera. XXXI. Disruption of sex pheromone communication in *Pectinophora gossypiella* with hexalure. *Environ. Entomol.* **1**:645-650.

(53)  H.H. Shorey and L.K. Gaston. 1974. Programs utilizing pheromones in survey or control: The cabbage looper. In *Pheromones*. M.C. Birch, Ed. American Elsevier, New York, pp. 421-425.

(54)  S.R. Farkas, H.H. Shorey, and L.K. Gaston. 1974. Sex pheromones of Lepidoptera. The use of widely separated evaporators of looplure for the disruption of pheromone communication in *Trichoplusia ni*. *Environ. Entomol.* **3**:876-887.

(55)  H.E. Hummel, L.K. Gaston, H.H. Shorey, R.S. Kaae, K.J. Byrne, and R.M. Silverstein. 1973. Clarification of the chemical status of the pink bollworm sex pheromone. *Science* **181**:873-875.

(56)   H.H. Shorey, R.S. Kaae, and L.K. Gaston. 1974. Sex pheromones of Lepidoptera. Development of a method for pheromonal control of *Pectinophora gossypiella* in cotton. *J. Econ. Entomol.* 67:347-350.

(57)   R.S. Kaae, J.R. McLaughlin, H.H. Shorey, and L.K. Gaston. 1972. Sex pheromones of Lepidoptera. XXXII. Disruption of intraspecific pheromone communication in various species of Lepidoptera by permeation of the air with looplure or hexalure. *Environ. Entomol.* 1:651-653.

(58)   R.S. Kaae, H.H. Shorey, L.K. Gaston, and H.E. Hummel. 1974. Sex pheromones of Lepidoptera: Disruption of pheromone communication in *Trichoplusia ni* and *Pectinophora gossypiella* by permeation of the air with nonpheromone chemicals. *Environ. Entomol.* 3:87-89.

(59)   M. Beroza. 1967. Nonpersistent inhibitor of the gypsy moth sex attractant in extracts of the insect. *J. Econ. Entomol.* 60:875-876.

(60)   M. Beroza, R.T. Staten, and B.A. Bierl. 1971. Tetradecyl acetate and related compounds as inhibitors of attraction of the pink bollworm moth to the sex lure hexalure. *J. Econ. Entomol.* 64:580-582.

(61)   D.O. Hathaway, T.P. McGovern, M. Beroza, H.R. Moffitt, L.M. McDonough, and B.A. Butt. 1974. An inhibitor of sexual attraction of male codling moths to a synthetic sex pheromone and virgin females in traps. *Environ. Entomol.* 3:522-524.

(62)   W.L. Roelofs, A. Comeau, and R. Selle. 1969. Sex pheromone of the oriental fruit moth. *Nature* 224:723.

(63)   W.L. Roelofs and R.T. Cardé. 1974. Sex pheromones in the reproductive isolation of lepidopterous species. In *Pheromones*. M.C. Birch, Ed. American Elsevier, New York, pp. 96-114.

(64)   W.L. Roelofs and A. Comeau. 1971. Sex pheromone perception: Synergists and inhibitors for the red-banded leaf roller attractant. *J. Insect Physiol.* 17:435-448.

(65)   L.L. Sower, K.W. Vick, and K.A. Bull. 1974. Perception of olfactory stimuli that inhibit the responses of male phycitid moths to sex pheromones. *Environ. Entomol.* 3:277-279.

(66)   J.H. Tumlinson, E.R. Mitchell, S.M. Browner, M.S. Mayer, N. Green, R. Hines, and D.A. Lindquist. 1972. *cis*-7-Dodecen-l-ol, a potent inhibitor of the cabbage looper sex pheromone. *Environ. Entomol.* 1:354-358.

(67)   K.W. Vick and L.L. Sower. 1973. Z-9, Z-12-Tetradecadien-l-ol acetate: An inhibitor of the response to the sex pheromone of *Plodia interpunctella. J. Econ. Entomol.* 66:1258-1260.

(68)   J.R. McLaughlin, L.K. Gaston, H.H. Shorey, H.E. Hummel, and F.D. Stewart. 1972. Sex pheromones of Lepidoptera. XXXIII. Evaluation of the disruptive effect of tetradecyl acetate on sex pheromone communication in *Pectinophora gossypiella. J. Econ. Entomol.* 65:1592-1593.

(69)   J.R. McLaughlin, E.R. Mitchell, D.L. Chambers, and J.H. Tumlinson. 1974. Perception of Z-7-dodecen-l-ol and modification of the sex pheromone response of male loopers. *Environ. Entomol.* 3:677-680.

(70)   E.R. Mitchell, W.W. Copeland, A.N. Sparks, and A.A. Sekul. 1974. Fall armyworm: Disruption of pheromone communication with synthetic acetates. *Environ. Entomol.* 3:778-780.

(71)   G.H.L. Rothschild. 1974. Problems in defining synergists and inhibitors of the oriental fruit moth pheromone by field experimentation. *Ent. exp. appl.* 17:294-302.

(72)    L.L. Sower, K.W. Vick, and J.H. Tumlinson. 1974. (*Z*,*E*)-9,12-Tetradecadien-l-ol: A chemical released by female *Plodia interpunctella* that inhibits the sex pheromone response of male *Cadra cautella*. *Environ. Entomol.* 3:120-122.

(73)    W.S. Bowers, L.R. Nault, R.E. Webb, and S.R. Dutky. 1972. Aphid alarm pheromone: Isolation, identification, synthesis. *Science* 177:1121-1122.

(74)    C.J. Kislow and L.J. Edwards. 1972. Repellent odour in aphids. *Nature* 235:108-109.

(75)    L.R. Nault, L.J. Edwards, and W.E. Styer. 1973. Aphid alarm pheromones: Secretion and reception. *Environ. Entomol.* 2:101-105.

(76)    R.J. Prokopy. 1972. Evidence for a marking pheromone deterring repeated oviposition in apple maggot flies. *Environ. Entomol.* 1:326-332.

(77)    R.J. Prokopy and G.L. Bush. 1972. Mating behavior in *Rhagoletis pomonella* (Diptera: Tephritidae). III. Male aggregation in response to an arrestant. *Can. Entomol.* 104:275-283.

CHAPTER 22

# Manipulation of Forest Insect Pests

David L. Wood

*Department of Entomological Sciences*
*University of California*
*Berkeley, California*

Research on insect behavior-modifying chemicals or insect behavior regulators (IBRs) has grown prolifically during the past decade (1)-(6). Over 100 active compounds have been isolated and identified, most of these being either pheromones or allomones. The potential usefulness of these chemicals for the survey and control of pest species, concern for contamination of the environment with pesticides, and the increase sophistication of microchemical techniques (7) account, in large part, for this period of very active research.

Only a few IBRs have been utilized in pest-management systems, and most of these have been in agricultural crops and stored food products (Chapters 19-21) (8). Two such compounds are in *operational* (i.e., meet certain standards of safety and cost-effectiveness) use in forest ecosystems today. The attractant sex pheromones, *cis*-7,8-epoxy-2-methyloctadecane (9) and *trans*-9-dodecenyl acetate (10), are presently being used for the detection of spread of the gypsy moth, *Porthetria dispar* (L.) (11), and the European pine shoot moth, *Rhyacionia buoliana* (Schiff.) (12), respectively. Both of these insects are forest pests that have been introduced into North America from Europe. The operational use of IBRs in any population context—for example, prediction and supression, and for prophylactic treatments—is, as far as I am aware, nonexistent. The potential of pheromones for survey and control of bark beetles (Scolytidae) has been reviewed recently by Bakke (13), Bedard and Wood (14), and Vité (15).

Decisions as to when, where, and how to treat a pest population should be an outcome of some economic analysis, no matter how primitive. This analysis simply is the ratio of the benefits to be gained by the treatment to the costs of treatment. In the case of forest pests, this analysis traditionally has been based upon an estimate of the volume of trees that would die and/or the volume of growth increment that would be lost without the treatment. In most cases the cost of the treatment has been only an estimate of the cost of the toxicant and the application, and has not included environmental costs.

The evolution to pesticides safer than those currently used in forestry has been painfully slow as evidenced by the fact that DDT was sprayed in 1974 over nearly one-half million acres of forest and brush lands in Idaho, Oregon, and Washington in an attempt to lessen the impact of defoliation by the Douglas-fir tussock moth, *Orygia pseudotsugata* (McD.) (16). Aerial applications of nonspecific insecticides continue to be made against other pests such as the gypsy moth and the eastern and western spruce budworm, *Choristoneura fumiferana* (Clemens) and *C. occidentalis* (Freeman), respectively.

The modes of action and properties of IBRs offer significant advantages over insecticides and insect growth regulators (IGRs) (17). IBRs are highly selective for a single and occasionally a few closely related species, are active in extremely small quantities, and are biodegradable. Although very few data are available on toxicity, IBRs [except for some of the defensive substances (allomones)] appear to be nontoxic or only slightly toxic to insects and warm-blooded animals. Further, these compounds elicit specific behavior patterns, and thus the behavior of pest species can be manipulated to increase the rate of mortality. They provide alternatives to and also extend the use of existing physicochemical control methods.

# I. FOREST PEST-MANAGEMENT SYSTEMS CONCEPT

## A. Prediction and Suppression

The use of IBRs in forest pest management for prediction and direct suppression of populations should be visualized in a systems context (Fig. 1) (18) (19). A treatment aimed at reducing the pest population density may cause changes in forest stand parameters such as mortality rate, growth rate, density, age, and species distributions. Such changes must produce sufficient protection of existing values or a gain in values to justify the cost of treatment (20). After a particular effect on forest stand parameters has been demonstrated, a preliminary analysis of the potential benefits and costs should be presented to funding agencies prior to undertaking the developmental research necessary to demonstrate the efficacy and safety of a treatment utilizing IBRs. It is possible that because forest crops grow more slowly and yield a lower per acre value then many agricultural crops, the proposed tactic would not be economically feasible even if it were biologically effective.

This describes the basic dilemma for forest pest management regardless of the pest- (insect or pathogen) suppression treatment proposed—that is, insecticides, microbials, IGRs, IBRs, and so on. Efficacy and safety data must be obtained for any proposed treatment tactic that utilizes an IBR so that registration can be sought from regulatory agencies (e.g., the Environmental Protection Agency). The cost of the treatment is integrated with the benefits to be saved or gained, which in turn is determined from the impact assessment

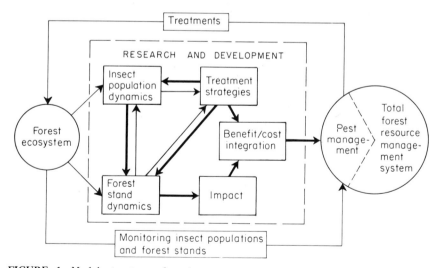

**FIGURE 1.** Model structure of an insect pest-management system, with research and development components and the action sequence (18).

(Fig. 1). Because the benefits and costs must be distributed over the length of the rotation (i.e., seedling to harvest), which may vary between 20 and 100 years for sawtimber, the monetary values may change considerably with time (21). This analysis is greatly complicated when products other than timber are considered. Thus we must determine the effects of the treatment on such parameters of the pest population dynamics as density, trend in density, degree of aggregation, and effects of natural enemies. In the interests of safety, data on the effects of the treatment on man and other nontarget organisms are essential. Such information is available for very few, if any, treatment tactic. It appears that there are only two forest pests for which this information is being sought to foster the development of an IBR treatment. These are the gypsy moth (22) and the western pine beetle, *Dendroctonus brevicomis* (LeC.) (14). In addition, data are being obtained on the attractant pheromones of a devastating pest of shade and forest trees, the smaller European elm bark beetle, *Scolytus multistriatus* (Marsham) (23) (24).

## B. Prophylactic Treatments

The use of IBRs in prophylactic treatments on single or small groups of trees probably would not require the extensive and detailed population assessment that has been described for the "trap-out" or "interruption" methods (14).* It is likely that the protection of foliage and/or bark would vary according to the size of the population, and therefore estimates of efficacy should be related to some estimate of population size (e.g., number and location of trees infested with bark beetles in the vicinity of the treatment). However, the systems diagram still applies for this type of treatment whether it is used for individual or small groups of trees, or over large areas. Cost and safety data are still required.

## II. CURRENT RESEARCH

The following discussion is focused on those forest pest insects for which IBRs are currently being developed for survey and/or control of populations and for prophylactic treatments.

------------

*Interruption* (5) is used throughout this chapter to describe the following statistical effects of an IBR: lowered catch at traps baited with attractants or attractant-producing substrates (e.g., calling female moths); lowered mating success; and lowered attack densities by bark beetles on trees. This statistical definition of *interruption* is more appropriate than terms such as *disruption, antiattractant, inhibition, confusion*, and so on, because the exact behavioral and physiologic mechanisms are not understood.

## A. Defoliators

### 1. Douglas-Fir Tussock Moth (O. pseudotsugata)

The sex pheromone of this major forest pest has recently been identified as cis-6-heneicosen-11-one (25). Studies are currently underway throughout western North America to determine how this compound can be used to survey populations of this insect and to interrupt mating in order to reduce populations.

### 2. European Pine Shoot Moth (R. buoliana)

No developmental research program is reported to be underway to utilize the identified sex pheromone for the prediction and suppression of this plantation and ornamental tree pest. The attraction of males to live females has been interrupted by several 12-carbon chain acetates. (26) (27).

### 3. Gypsy Moth (P. dispar)

An intensive research effort is currently underway to interrupt mating by the use of microencapsulated disparlure (cis-7,8-epoxy-2-methyloctadecane) applied by aircraft. (22) (28)-(31). In several tests in Pennsylvania and Massachusetts between 1972 and 1974 the treatment reduced the proportion of female moths fertilized. This is the first aerial application of an IBR where a definitive biologic effect has been recorded since the successful oriental fruit fly (Dacus dorsalis Hendel) eradication program in 1962-1963 on the Mariana Islands (1). However, as Cameron (28) states,

"It is generally agreed that disparlure alone, whether in traps or in broadcast applications, is incapable of having much effect on populations except when they are at the very low [as yet undefined] level ... this chemical is useful and operational [Cameron's italics] for survey and detection, but is not ready for operational—and I stress operational—use in the other contexts: population monitoring and/or prediction, eradication of isolated infestations, the establishment of a barrier zone, or use in a chronic infestation." The problems encountered in the development of disparlure for operational use have centered around formulations, aerial application methodology, evaluation procedures such as plot size and population sampling methods, and poorly understood behavioral mechanisms.

Field tests of the olefin precursor of disparlure, 2-methyl-cis-7-octadecene (32), failed to interrupt mate-finding or reduce trap catches to levels theorized to be necessary for population control (30).

## 4. Larch Budmoth [Zeiraphera diniana (Guenée)]

The sex pheromone, trans-11-tetradecenyl acetate, is being used to monitor population levels of this insect (33). This is the only forest insect outside of North America for which an IBR may be close to operational use. Experiments are underway with the cis-isomer of this compound to interrupt the attractant response.

## 5. Spruce Budworm (C. fumiferana)

Virgin females were used as a pheromone source in survey traps from 1966 to 1971, and the catch of males was compared to larval densities, estimated from branch samples and by beating the foliage, in the same areas (34). Although no analyses were presented, the highest trap catch was correlated with the highest larval counts. Presently, the sex pheromone, trans-11-tetra-decenal (35) is being used to monitor population fluctuations of this serious forest pest at endemic levels (34). However, the catch of males in green Sectar* traps baited with the synthetic Sprudamone† was poorly correlated with larval counts obtained by beating foliage. This may have been a result of an ineffective attractant delivery system.

The sex pheromone and two compounds that interrupt the attractant response [trans-11-tetradecenyl acetate and trans-11-tetradecenol (36)] are presently being investigated for their promise in population suppression by interrupting male attractant behavior and thus lowering the number of females that are mated (35). This treatment would be timed to coincide with the initial population increase from endemic levels. Because of the tremendous numbers of insects produced, treatments aimed at populations in the outbreak phase probably cannot sufficiently interrupt attraction to prevent a high level of mating (34). Male trapping is considered to be impractical because of the large number of traps that would be required to treat the extensive, inaccessible infestation areas.

## B. Bark Beetles

### 1. Ambrosia beetle [Gnathotrichus sulcatus (LeConte)]

#### a. Trap-Out

Insects are attracted to sticky traps baited with attractant compounds, and die in the sticky substrate. The attractant, sulcatol [6-methyl-5-hepten-2-ol (37)], has been shown to have potential as a survey method for this

---

*3-M Company, St. Paul, Minn.
†Trade name for trans-11-tetradecenal Zoëcon, Palo Alto, Calif.

beetle (38). A trap-out experiment was planned for 1975 in and around a sawmill in British Columbia (39).

## 2. *California Five-spined Ips* [Ips paraconfusus (Lanier)]

In 1966 ipsenol [(-)-2-methyl-6-methylene-7-octen-4-ol], (+)-*cis*-verbenol and ipsdienol [(+)-2-methyl-6-methylene-2,7-octadien-4-ol] were identified as the principal components of the attractant pheromone produced by this bark beetle species (40). In 1968 the ternary mixture was found to be attractive under field conditions (41). Subsequently, Vité and co-workers [summarized in (42)] showed that these compounds in a variety of combinations were attractants for three other species of *Ips* that occur in the southern United States. Recently ispenol has been shown to interrupt the attractant response of *I. pini* (Say), to its own pheromone (43). However, as yet no research has been undertaken to develop the above compounds for survey and suppression of these pest species.

## 3. *Douglas-Fir Beetle* [Dendroctonus pseudotsugae Hopkins]

### a. Baited Tree

Living Douglas-firs [*Pseudotsuga menziesii* (Mirb.) Franco] in areas that had been scheduled for logging were baited with frontalin [1,5-dimethyl-6,8-dioxabicylo [3.2.1] octane (44)]. Following mass aggregation, the beetles were removed from the forest in the infested logs (45). This method may have promise for univoltine species such as this one, but further tests are required to determine the effects on population trends over larger areas. The disadvantages are that each tree must be visited to place the attractant chemical; the baits must be in place prior to emergence of the overwintering generation; the newly infested trees must be removed from the forest prior to emergence; and natural enemies are removed along with the pest species. Furthermore, roads may not exist or be passable in areas to be logged in the spring.

### b. Prophylaxis

Compounds are placed on a substrate to prevent landing, boring, and/or feeding. Brood production was greatly reduced in felled Douglas-firs treated with 3-methyl-2-cyclohexen-l-one (46). This compound was identified as an attractant pheromone from female beetles (47), but it was later shown also to interrupt the response of this beetle to other attractant mixtures (48) (49). The goal of this research is to prevent an increase in the population by lowering the number of beetles aggregating on and boring into cut and blown-down trees.

## 4. *Mountain Pine Beetle* [D. ponderosae Hopkins]

### a. Baited Tree

Living western white pines [*Pinus monticola* Dougl.]that had been treated with the persistent insecticide, lindane, were baited with the attractant mixture of *trans*-verbenol and α-pinene (50). Results of this experiment were inconclusive partly because of problems in the application of the insecticide. The attractants were effective in inducing aggregation on the treated trees, but the high rate of tree morality indicated that the insecticide was not effective (50). This method has several disadvantages compared to trapping on sticky substrates: the application of insecticides to the boles of large trees is costly and difficult; natural enemies that are attracted to trees under attack are killed along with the pest insect; and a persistent, chlorinated hydrocarbon insecticide used over large areas creates problems of environmental contamination.

## 5. *Smaller European Elm Bark Beetle* (S. multistriatus)

### a. Trap-Out

Small-scale experiments ($<$ 1 sq. mile) were performed in Detroit in 1974 to assess the potential of multilure in a trap-out strategy. Multilure is a mixture of two female-produced compounds (multistriatin: 2,4-dimethyl-5-ethyl-6,8-dioxabicyclo [3.2.1] octane and 4-methyl-3-heptanone) and one compound from the host plant (α-cubebene) (51). Although many beetles were trapped, the effect on population and damage trend could not be determined. Trap-out experiments with small isolated populations were to be conducted in Hamilton, New York (23) and Ft. Collins, Colorado (24) during 1975 in an attempt to slow the spread of Dutch elm disease, of which this bark beetle is the vector. Also, experiments with large populations were to be repeated in Detroit during 1975 (24).

### b. Prophylaxis

Juglone (5-hydroxy-1,4-naphthoquinone), a natural constituent of the nonhost plant, shagbark hickory [*Carya ovata* (Mill.) K. Koch], deterred feeding by *S. multistriatus* adults on elderberry pith assay disks treated with benzene extracts of the bark of the host species, American elm [*Ulmus americana* (L.)] (52). Quinols (*p*-hydroquinone) have been found to stimulate this beetle to feed on similar assay disks. Quinones (*p*-benzoquinone and several 1,4-naphthoquinones) have been shown to inhibit the feeding response to *p*-hydroquinone when presented on the same assay discs (53). Two feeding stimulants, (+)-catechin-5-β-D-xylopyranoside and lupeyl cerotate, have been isolated from the bark of American elm (54). The application of feeding deterrents to the host tree has not been reported, but its promise for

prevention of feeding in the crotches of young twigs and the subsequent introduction of the Dutch elm disease organism would appear to justify limited field trials. Also, if the feeding response can be prevented on the larger limbs and on the bole, the aggregation pheromone would not be produced and thus infestations would be decreased or possibly prevented.
or possibly prevented.

## 6. Southern Pine Beetle [D. frontalis Zimm.]

### a. Baited Tree

Loblolly pines (*Pinus taeda* L.) have been killed with the herbicide, cacodylic acid, and then baited with frontalure, a mixture of α-pinene and frontalin (55). The trees are subsequently attacked, and the adults and their progeny succumb in the poisoned tissues. This method requires that individual trees be visited and that some estimate be made of the number of trees that will be attacked during the treatment period. Because our present predictive capabilities are not adequate, either too many or not enough trees will be poisoned with the herbicide. In either case, living trees must be sacrificed in order to kill the pest. Further, it has been demonstrated that brood survival can be high in poisoned trees (56).

## 7. Spruce Beetle [D. rufipennis Kirby]

### a. Baited Tree

As with lodgepole pine, spruce trees (*Picea* spp.) have been treated with lindane and baited with frontalin (57). The arriving beetles are killed after landing on the trees, and additional beetles die when nearby trees are attacked but not killed. This method is apparently not in use today in either Canada or the United States.

## 8. Western Pine Beetle (D. brevicomis)

### a. Trap-Out

In 1970 a field experiment was initiated at Bass Lake, California, in an attempt to suppress the western pine beetle over a 65-km$^2$ area using synthetic pheromones (14). Sticky traps were baited with *exo*-brevicomin (*exo*-7-ethyl-5-methyl-6,8-dioxabicyclo [3.2.1] octane) (58), frontalin, and myrcene, and deployed on 161-m centers in four 1.3 km$^2$ suppression plots and on a 0.8-km grid over the entire area. A reduction in ponderosa pine [*P. ponderosa* Laws.]mortality for the entire season was recorded—that is, from 283 ± 94 (1969-70 overwintering infested trees) to 91 ± 28 trees after the suppression period. Although many predators [*Temnochila chlorodia* (Mann.)] were also trapped during the suppression period, it could not be determined what, if any, effects this had on populations of *D. brevicomis* or other bark beetle species in the area.

### b. Interruption

Mass aggregation is reduced to below the threshold number required to kill trees (14) (59). Studies of this suppression method have been underway in California since 1972 by Bedard and co-workers (60). *Exo*-brevicomin, frontalin, and myrcene were evaporated alone and in the ternary mixture from devices in grids of varying dimensions. A range of concentrations was tested by varying the amount released at each point and the number of points per unit area. Effectiveness of the compounds to interrupt aggregation behavior was determined by measuring the number of beetles responding to a single trap located in the center of the grid and baited either with bolts from trees under mass attack or the ternary mixture. The ternary mixture was the most effective in lowering the catch at the center trap. Further tests are required to determine how long this level of effectiveness can be maintained and over how large an area. The above three compounds have been encapsulated* to allow 60-day release periods, and monitored in both the field and laboratory for two seasons to assess the variability in release rates under varying environmental conditions. Only large-scale field experiments, which are designed to account for immigration and emmigration, can establish the efficacy of this suppression method. Planning for these experiments is currently underway (60).

### c. Prophylaxis

Verbenone (61) (62) has been used successfully to prevent attack by the western pine beetle on small ponderosa pines (60). In three replications, only the untreated tree was killed. Further tests were planned for 1975.

### III. CONCLUSIONS

Many of the problems (63) identified at an earlier symposium on this subject (i.e., inadequate knowledge of behavior; mass rearing; trapping; concentration, isolation and identification of compounds that occur in minute amounts in complex mixtures; synergism; masking; and synthesis) have been solved, at least in part, and as a result, new active compounds have been identified from many of the most destructive forest pest species in North America. The identified compounds have been synthesized in quantities sufficient for large-scale field tests. However, very few IBRs have been

---

*Concerco, Seattle, Wash.

identified from the forest pest species occurring in other parts of the world. One should ask at this point whether the publicity given to the promise of these chemicals in pest suppression has been justified. It is important to note that only very limited attempts have been made to develop the use of any of the available compounds in pest-management programs. In the few cases where attempts have been made, the reasons for failure or limited success have not been clear. Undoubtedly, the high cost of the research following the identification phases and the inherent complexities of research on the dynamics of highly mobile, widely distributed pest populations, are the underlying causes for the current underdeveloped status of IBRs in forest pest management.

The isolation and identification of IBRs from particular pest species appear to have been a result of a variety of largely unrelated decisions by research administrators, funding agencies, and the scientists themselves. Now that IBRs are available for a number of pest species, some priority system needs to be established in order to allocate the scarce resources available to pursue the more difficult and expensive phases of the research. Priorities should be based upon the economic significance of the pest and upon some estimate of feasibility and promise of success. Resources are limited because private industry has not generally undertaken the research necessary to develop IBRs to the operational stage. This is probably due, in part, to high development costs, low market potential because of high specificity, proprietary problems, and the complexity of demonstrating efficacy and environmental safety for registration of the compounds by regulatory agencies. Because these problems are so diverse and complex, a much closer collaboration between private industry and public institutions will be required to develop these compounds for application in pest-management systems.

There is little doubt that research on IBRs, especially the pheromones, is being pursued vigorously in the three expanded research and development programs recently funded by the United States Department of Agriculture on the gypsy moth, Douglas-fir tussock moth, and southern pine beetle. Research on the western pine beetle (14) and the gypsy moth (22) has advanced to the large-scale field experiment stage. However, there is a continuing need for an improved understanding of pest population biology. The development of new population modeling techniques will accelerate this understanding so that the most promising management strategies can be evaluated using computer simulations (18) (19). The research needed to test these compounds adequately in most cases has not even been initiated. In those few instances where it has, the strategies selected have been largely based on intuition as a result of the lack of adequate information on population size and movement, and the relationship between population and trend in damage.

A cornucopia of compounds exists in both the insect and its host plant that elicit the basic behavior described by Dethier et al. (64)—that is, arrestment, attraction, deterrency repellency, and stimulation (locomotion feeding and oviposition). However, the area of chemical ecology that concerns the host plant has been largely overshadowed by the work on pheromones, allomones, and hormones. The behavioral and chemical methodology developed during the past decade has direct application in the study of insect chemosensory behavior in relation to the host plant. Although not demonstrated as yet, these compounds would appear to have great promise in pest-management systems. The challenge rests especially with the biologist, who must recognize and describe this great variety of chemosensory behavior so that the highly developed microchemical techniques can be applied in a meaningful way to the identification of these chemical stimuli. The full potential of IBRs in pest-management systems cannot be realized until their effects on individuals can be extended to the dynamics of pest populations over large areas.

One can imagine multiple-compound treatments where various combinations of plant- and insect-derived compounds will be used to manipulate insect behavior at one to several times during the life cycle of the pest. Further, the development of a population monitoring and predictive capability for important forest insect pests using these chemicals would be of tremendous value to the land manager who must consider whether or not to mobilize a treatment effort. This exploitation of the insect's natural chemical communications system is the most promising alternative to the use of broad-spectrum insecticides for the chemical manipulation of pest species.

*Note added in proof:* Although several attractant pheromones were indicated to be present in the hindguts of several European species of *Ips* (p. 375) (Vité, J.P., A. Bakke and J.A.A. Renwick. 1972. Pheromones in *Ips* (Coleoptera: Scolytidae): Occurrence and production. Can. Entomol. 104:1967-1975), Bakke was the first to show that *Ips duplicatus* was attracted to one of these compounds (ipsdienol) in the field (Bakke, A. 1975. Aggregation pheromone in the bark bettle *Ips duplicatus*. Norw. J. Entomol. 22:67-69).

## ACKNOWLEDGEMENTS

I very much appreciate the review of this manuscript by W.D. Bedard, L.E. Browne, P.A. Rauch, and R.M. Silverstein. Browne suggested *insect behavior regulators* (IBRs) to designate insect behavior-modifying chemicals. The work reported herein on forest pest-management systems was supported in part by grants from the National Science Foundation/Environmental Protection Agency (NSF-6B-34718/BMS75-04223), U.S. Forest Service, and the Rockefeller Foundation. The findings, opinions, and recommendations are those of the author, and not necessarily those of the University of California, the National Science Foundation, or the Environmental Protection Agency.

# REFERENCES

(1)  M. Beroza, Ed. 1970. *Chemicals Controlling Insect Behavior.* Academic, New York.

(2)  E. Sondheimer and J.B. Simeone, Eds. 1970. *Chemical Ecology.* Academic, New York.

(3)  D.L. Wood, R.M. Silverstein, and M. Nakajima, Eds. 1970. *Control of Insect Behavior by Natural Products.* Academic, New York.

(4)  M. Jacobson. 1972. *Insect Sex Pheromones.* Academic, New York.

(5)  M.C. Birch, Ed. 1974. *Pheromones.* North-Holland, Amsterdam.

(6)  U.E. Brady, Ed. 1974. *Pheromones, II:* Current Research. MMS Information, New York.

(7)  J.C. Young and R.M. Silverstein. 1975. Biological and chemical methodology in the study of insect communication systems. In *Methods in Olfactory Research.* D.G. Moulton, A. Turk, and J.W. Johnson, Jr., Eds. Academic, New York,

(8)  pp. 75-161.

(9)  M.C. Birch. 1974. *Pheromones.* North-Holland, Amsterdam.
     moth: Its isolation, identification, and synthesis. *Science* (Wash.) **170**:87-89.

(10) R.G. Smith, G.E. Daterman, G.D. Daves, Jr., K.D. McMurtrey, and W.L. Roelofs. 1974. Sex pheromones of the European pine shoot moth: Chemical identification and field tests. *J. Insect Physiol.* **20**:661-668.

(11) M. Beroza, B.A. Bierl, J.G.R. Tardif, D.A. Cook, and E.C. Paszek. 1971. Activity and persistence of synthetic and natural sex attractants of the gypsy moth in laboratory and field trials. *J. Econ. Entomol.* **64**:1499-1508.

(12) G.E. Daterman. 1974. Synthetic sex pheromone for detection survey of European pine shoot moth. U.S.D.A. Forest Ser. Res. Paper PNW-180.

(13) A. Bakke. 1973. Bark beetle pheromones and their potential use in forestry. *IEPP/EPPO Bull.* **9**:5-15.

(14) W.D. Bedard and D.L. Wood. 1974. Programs utilizing pheromones in survey and control: Bark beetles—the western pine beetle. In *Pheromones.* M.C. Birch, Ed. North-Holland, Amsterdam, pp. 441-449.

(15) J.P. Vité. 1975. Moglichkeiten und Grenzen der Pheromonauwendung in der Borkenkaferbekampfung. *Z. Angew. Entomol.* **77**:325-329.

(16) R.F. Harwood. 1975. Economics, esthetics, environment, and entomologists: the tussock moth dilemma. *Environ. Entomol.* **4**:171-174.

(17) J.P. Tette. 1974. Pheromones in insect population management. In *Pheromones.* M.C. Birch, Ed. North-Holland, Amsterdam, p. 339.

(18) D.L. Wood. 1974. Progress Report, Pine Bark Beetle Subproject. In The Principles, Strategies and Tactics of Pest Population Regulation and Control in Major Crop Ecosystems, Vol. 2. Conceptualized by the Integrated Pest Management Coordinat-

ing Committee for the western pine beetle subprogram. The contributors were: F.W. Cobb, Jr., D.L. Dahlsten, B. Ewing, J.R. Parmeter, Jr., P.A. Rauch, W.E. Waters, and D.L. Wood, University of California, Berkeley; W.D. Bedard, C.J. DeMars, and B.H. Roettgering, U.S.D.A., Forest Service, Berkeley and San Francisco; and R.M. Bramson, Pacific Organization Development Associates, Oakland.

(19)    R.W. Stark. 1975. Forest insect pest management. In *Introduction to Insect Pest Management*. W.H. Luckmann and R.H. Metcalf, Eds. Wiley, New York, pp. 509-528.

(20)    W.E. Waters. 1974. Systems approach to managing pine bark beetles. In *Proc. Southern Pine Beetle Symp*. T.L. Payne, R.L. Coulson, and R.C. Thatcher, Eds. Texas A & M University and U.S.D.A. Forest Service, pp. 12-14.

(21)    D.E. Teeguarden. 1975. Benefit-cost analysis in forest pest management. In *Proc. Western Forest Insect & Disease Work Conf.*, Monterey, Calif., February.

(22)    A.E. Cameron. 1974. Programs utilizing pheromones in survey and control: The gypsy moth. In *Pheromones*. M.C. Birch, Ed. North-Holland, Amsterdam, pp. 431-435.

(23)    G.N. Lanier. 1975. College of Environmental Science and Forestry, State University of New York, Syracuse. Personal communication.

(24)    G.T. Peacock and R.A. Cuthbert. 1975. Northeastern Forest Experiment Station, Forest Service, U.S.D.A., Delaware, Ohio. Personal communication.

(25)    R.G. Smith, G.E. Daterman, and G.D. Daves, Jr. 1975. Douglas-fir tussock moth: Sex pheromone identification and synthesis. *Science* (Wash.) **188**:63-64.

(26)    G.E. Daterman, G.D. Daves, Jr., and M. Jacobson. 1972. Inhibition of pheromone perception in European pine shoot moth by synthetic acetates. *Environ. Entomol.* **1**:382-383.

(27)    R. Lange and D. Hoffman. 1972. Essigsaure-*cis*-dodecen-(9)-yl-ester, ein inhibitor des Sexualpheromons des Kiefernknospertriebwicklers. *Naturwissenschaften* **59**:217.

(28)    E.A. Cameron. 1974. Integrated control and the role of chemical ecology: the gypsy moth (Lepidoptera). Paper presented to 46th Annual Meeting of Eastern Branch, Entomol. Soc. Am., Hershey, Pa. Sept. 25-27.

(29)    E.A. Cameron, C.P. Schwalbe, M. Beroza, and E.F. Knipling. 1974. Disruption of gypsy moth mating with microencapsulated disparlure. *Science* (Wash.) **183**:972-973.

(30)    E.A. Cameron, C.P. Schwalbe, L.J. Stevens, and M. Beroza. 1975. Field tests of the olefin precursor of disparlure for suppression of mating in the gypsy moth. *J. Econ. Entomol.* **68**:158-160.

(31)    M. Beroza, C.S. Hood, D. Trefrey, D.E. Leonard, E.F. Knipling, W. Klassen, and L.J. Stevens. 1974. Large field trial with microencapsulated sex pheromone to prevent mating of the gypsy moth. *J. Econ. Entomol.* **67**:659-664.

(32)    R.T. Cardé, W.L. Roelofs, and C.C. Doane. 1973. Natural inhibitor of the gypsy moth sex attractant. *Nature* **241**:474-475.

(33)    G. Benz and G. von Salis. 1973. Use of synthetic sex attractant of larch bud moth, *Zeiraphera diniana* (Gn.), in monitoring traps under different conditions, and antagonistic action of *cis*-isomer. *Experientia* **29**:729-730.

(34)    C.J. Saunders. 1974. Programs utilizing pheromones in survey and control: Forest Lepidoptera—the spruce budworm. In *Pheromones*. M.C. Birch, Ed. North-Holland, Amsterdam, pp. 435-441.

(35)  J. Weatherston, W.H. Roelofs, A. Comeau, and C.J. Saunders. 1971. Studies of physiologically active arthropod secretions. X. Sex pheromone of the eastern spruce budworm, *Choristoneura fumiferana* (Lepdioptera: Tortricidae). *Can. Entomol.* 103:1741-1747.

(36)  C.J. Saunders, R.J. Bartell, and W.L. Roelofs. 1972. Field trials for synergism and inhibition of *trans*-11-tetradecenal, sex pheromone of the eastern spruce budworm. *Can. Dept. Environ.. Bi-mon. Res. Notes* 28:9-10.

(37)  K.J. Byrne, A.A. Swigar, R.M. Silverstein, J.H. Borden, and E. Stokkink. 1974. Sulcatol: Population aggregation pheromone in the scolytid beetle, *Gnathotrichus sulcatus. J. Insect Physiol.* 20:1895-1900.

(38)  J.A. McLean and J.H. Borden. Survey of a sawmill population of *Gnathotrichus sulcatus* (Coleoptera: Scolytidae) using the pheromone, Sulcatol. *Can. J. For. Res.* (In Press).

(39)  J.H. Borden. 1975. Department of Biological Sciences, Simon Fraser University, Burnaby, British Columbia. Personal communication.

(40)  R.M. Silverstein, J.O. Rodin, and D.L. Wood. 1966. Sex attractants in frass produced by male *Ips confusus* in ponderosa pine. *Science* (Wash.) 154:509-510.

(41)  D.L. Wood, L.E. Browne, W.D. Bedard, P.E. Tilden, R.M. Silverstein, and J.O. Rodin. 1968. Response of *Ips confusus* to synthetic sex pheromones in nature. *Science* (Wash.) 159:1373-1374.

(42)  J.H. Borden. 1974. Aggregation pheromones in the Scolytidae. In *Pheromones*. M.C. Birch, Ed. North-Holland, Amsterdam. pp. 135-160.

(43)  M.C. Birch and D.L. Wood. 1975. Mutual inhibition of the attractant pheromone response by two species of *Ips* (Coleoptera: Scolytidae). *J. Chem. Ecol.* 1:101-113.

(44)  G.W. Kinzer, A.F. Fentiman, Jr., T.F. Page, Jr., R.L. Foltz, J.P. Vité, and G.B. Pitman. 1969. Bark beetle attractants: Identification, synthesis and field bioassay of a new compound isolated from *Dendroctonus. Nature* (Lond.) 221:477-478.

(45)  J.A.E. Knopf and G.B. Pitman. 1972. Aggregation pheromone for manipulation of the Douglas-fir beetle. *J. Econ. Entomol.* 65:723-726.

(46)  M.M. Furniss, G.E. Daterman, L.N. Kline, M.D. McGregor, G.C. Trostle, L.F. Pettinger, and J.A. Rudinsky. 1974. Effectiveness of the Douglas-fir beetle anti-aggregative pheromone, methylcyclohexeonone at three concentrations and spacings around felled trees. *Can Entomol.* 106:381-392.

(47)  G.W. Kinzer, A.F. Fentiman, Jr., R.L. Foltz, and J.A. Rudinsky. 1971. Bark beetle attractant: 3-methyl-2-cyclohexen-1-one isolated from *Dendroctonus pseudotsugae. J. Econ. Entomol.* 64:970-971.

(48)  M.M. Furniss, L.N. Kline, R.F. Schmitz, and J.A. Rudinsky. 1972. Tests of three pheromones to induce or disrupt aggregation of Douglas-fir beetles (Coleoptera: Scolytidae) on live trees. *Ann. Entomol. Soc. Am.* 65:1227-1232.

(49)  J.A. Rudinsky, M.M. Furniss, L.N. Kline, and R.F. Schmitz. 1972. Attraction and repression of *Dendroctonus pseudotsugae* (Coleoptera: Scolytidae) by three synthetic pheromones in traps in Oregon and Idaho. *Can. Entomol.* 104:815-822.

(50)  G.B. Pitman. 1971. *Trans*-verbenol and alpha-pinene: Their utility in manipulation of the mountain pine beetle. *J. Econ. Entomol.* 64:426-430.

(51)  G.T. Pearce, W.E. Gore, R.M. Silverstein, J.W. Peacock, R.A. Cuthbert, G.N. Lanier, and J.B. Simeone. 1975. Chemical attractants for the smaller European elm bark beetle, *Scolytus multistriatus* (Coleoptera: Scolytidae). *J. Chem. Ecol.* 1:115-124.

(52)  B.L. Gilbert, J.E. Baker, and D.M. Norris. 1967. Juglone (5-hydroxy-1,4-naphthoquinone) from *Carya ovata,* a deterrent to feeding by *Scolytus multistriatus. J. Insect Physiol.* **13**:1453-1459.

(53)  D.N. Norris. 1970. Quinol stimulation and quinone deterrency of gustation by *Scolytus multistriatus* (Coleoptera, Scolytidae). *Ann. Entomol. Soc. Am.* **63**:476-478.

(54)  R.W. Doskotch, D.K. Chatteric, and J.W. Peacock. 1970. Elm bark derived feeding stimulants for the smaller European elm bark beetle. *Science* (Wash.) **167**:380-382.

(55)  J.P. Vité. 1970. Pest management systems using synthetic pheromones. *Contrib. Boyce Thompson Inst. Plant Res.,* **24**:343-350.

(56)  R.N. Coulson, F.L. Oliveria, T.L. Payne, and M.W. Houseweart. 1973. Variables associated with use of frontalure and cacodylic acid in suppression of the southern pine beetle. I. brood reduction in trees treated with cacodylic acid. *J. Econ. Entomol.* **66**:897-899.

(57)  E.D.A. Dyer. 1973. Spruce beetle aggregated by the synthetic pheromone frontalin. *Can. J. Forest Res.* **3**:486-494.

(58)  R.M. Silverstein, R.G. Brownlee, T.E. Bellas, D.L. Wood, and L.E. Browne. 1968. Brevicomin: principal sex attractant in the frass of the female western pine beetle. *Science* (Wash.) **159**:889-890.

(59)  D.L. Wood. 1972. Selection and colonization of ponderosa pine by bark beetles. In *Insect/Plant Relationships.* H.F. van Emden, Ed. Sym. Roy. Entomol. Soc. (Lond.), No. 6. Blackwell Oxford pp. 101-117.

(60)  W.D. Bedard, P.E. Tilden, D.L. Wood, and L.E. Browne. 1975. Personal communication.

(61)  J.A.A. Renwick. 1967. Identification of two oxygenated terpenes from the bark beetles *Dendroctonus frontalis* and *Dendroctonus brevicomis. Contrib. Boyce Thompson Inst. Plant Res.* **23**:355-360.

(62)  J.A.A. Renwick and J.P. Vité. 1970. Systems of chemical communication in *Dendroctonus. Contrib. Boyce Thompson Inst. Plant Res.* **24**:283-292.

(63)  D.L. Wood, R.M. Silverstein, and M. Nakajima. 1969. Pest control. *Science* (Wash.) **164**:203-210.

(64)  V.G. Dethier, L.B. Brown, and C.N. Smith. 1960. The designation of chemicals in terms of the responses they elicit from insects. *J. Econ. Entomol.* **53**:134-136.

## CHAPTER 23

# Trapping with Behavior-Modifying Chemicals: Feasibility and Limitations

A.K. Minks

*Laboratory for Research on Insecticides*
*Wageningen, The Netherlands*

Traps baited with behavior-modifying chemicals are being used on a large scale for monitoring insect pests. Attractants placed in traps have been applied for several decades for monitoring some harmful tephritid flies. The use of sex pheromones in traps for monitoring lepidopteran or coleopteran pests is more recent, mainly because our knowledge of the chemistry of behavior-modifying chemicals, particularly of the sex pheromones, has only recently grown in a spectacular way. During the past 10 years the chemical structures of the pheromones of more than 100 moths and about 20 coleopterans have been elucidated. The majority of these species are economically important.

Why does trapping with attractants and pheromones receive so much interest? One important reason is the easy handling of such traps and the simple materials needed. Another reason is that the species-specific action of the traps containing behavior-modifying chemicals have an advantage over other sampling methods, such as light traps, suction traps, or lure pots, where time-consuming sorting of the catches is often necessary. Both reasons appeal to scientists, as well as to agricultural advisors and growers, who can now carry out part of the observations themselves. Four ways of using attractant traps will be discussed in more detail:

1. Detection of the presence or spread of an insect pest
2. Estimation of the population of an insect pest

3. Mass-trapping or male-annihilation technique
4. Method of assessment for the disruption technique


## I. TRAPPING FOR DETECTION OF AN INSECT PEST

For practical detection purposes, attractant/pheromone traps are being used on a large scale. The requirements that the traps must meet are limited. For example, it is important that specimens of the target insect are caught, but not necessarily large numbers of them. Easy handling and species-specific action constitute two characteristics of pheromone traps that give them an advantage over other types of traps. Attractant/pheromone traps are extremely suitable for survey of large areas. Examples are the monitoring of the gypsy moth, the spruce budworm moth, and of various species of pine beetles over large areas of impenetrable pine forests (Chapter 22). The use of pheromone traps to detect tephritids is of special significance. Although there are some tephritid pests almost everywhere, every agriculturally developed area is anxious to avoid introduction of those species not already present (Chapter 19). Also, the traps have great advantages in complex pest systems, such as those prevalent in many apple orchards where many similar leafroller species can cause different degrees of damage to the crop and where the identity of the species can easily be mistaken.

For pheromone traps, identification and subsequent synthesis of the pheromone are not always necessary; just the use of living virgin female insects or crude extracts from female insects as bait is sufficient to obtain useful information for detecting pheromones in limited areas (1)-(3). Female insects have to be collected from the field or must be reared in the laboratory, and the replacement of living virgin females (every 4-7 days) is more laborious than replacement of dispensers with synthetic pheromone (every 6-8 weeks). Finally, the females are only attractive during their active period, while synthetic pheromone dispensers work continuously.

Detection can aid in recording the spread or dispersal of a target insect to places where it has not been observed before and where it might cause new infestations. Early detection of exotic introductions is often imperative (Chapter 19), and traps may be an essential part of quarantine measures. Situations do exist where the mere presence of specimens of a certain insect is intolerable—for instance, in the case of certain stored-product pests (Chapter 20). If the target insects had already been recorded in a given place for several years, then one would want to know exactly when the first specimens appeared. Early warnings can obviate the need for preventive sprayings or other measures until they are really needed at a later date.

The trapping technique may be a more effective method of detection than visual checks (e.g., egg counts, larval counts, or determination of the degree of crop damage), which are often difficult to observe, particularly if one wishes information on the status of an infestation as early in the season as possible. Still, one must make egg or larval counts for reference purposes when the attractant/pheromone traps are to be used in order to estimate the abundance of a target insect.

## II. TRAPPING FOR ESTIMATING THE POPULATION OF AN INSECT PEST

If trap catches are used for estimating a population of the target insect, complex problems arise. It is difficult to relate the numbers of a certain insect species caught in traps to the numbers of that insect in the field, to the numbers of larvae of the following generation, or to plant damage levels (4). In general, the numbers of a certain insect caught in a trap are determined by the following factors:

1. Population density
2. Site of the population
3. Behavior
4. Numbers of traps in an area
5. Trap design
6. Concentration of attractant/pheromone per trap

### A. Population Density

A low population density may yield a low number of insects in the traps; a high density may be reflected by a large number of insects captured. But all of the above-mentioned factors clearly indicate that merely the numbers of insect catches from traps do not give sufficient information about population density. The general experience is that sex-pheromone traps easily detect low populations. Under these conditions, they are highly sensitive. However, when the population rises, the traps often attract fewer moths than one would expect from other observations. With the summer fruit tortrix moth, *Adoxophyes orana,* for instance, we have compared light-trap and sex-trap catches for several years: sex traps are much better during the first flight in June with few moths present, whereas in the August-September flight with much larger numbers of moths, light traps record these large numbers of moths more accurately (1) (4). Howell (5) described similar experiences during his investigations on the codling moth, *Laspeyresia pomonella,* and Miller and McDougall (6) also found that the attractive capacity of females of the spruce budworm moth, *Choristoneura fumiferana,* decreased sharply with increasing population density.

Several authors have explained this phenomenon as *competition effect.* The amount of pheromone emitted by the many wild females competes with pheromone evaporating from the traps. This explanation is, in my opinion, not satisfactory (Chapter 22), if we assume a 1:1 sex ratio (Chapter 19), because the number of males is also correspondingly higher in increasing populations. It is also possible that the decreased attractancy of the traps is due to modification of the behavior of the insects in high populations (see Section II.C). Competition, however, may not always be operating. Traps baited with muscalure, the sex pheromone of the housefly, *Musca domestica*, perform very well at high population densities, but do not catch any flies at low densities (7) The short-range attraction of the pheromone may explain the difference in the catches in this case.

## B. Site of the Population

Traps are installed at places where the presence of the target insect is suspected. However, the distribution of most insects in the field is often very patchy. For this reason trap catches show a great variation, and it is often difficult to collect statistically significant data and to reach firm conclusions from the data. Several times insects accumulated around the traps they were attracted to, but were not (yet) caught. Therefore, traps should be moved around from time to time.

## C. Behavior

Trap catches are related to flight activity, which in turn is influenced by environmental factors such as temperature, light, wind, and humidity. The influence of temperature has been described for *Adoxophyes orana* (8) and for *Choristoneura fumiferana* (6). Both species exhibit a positive relationship between trap catches and temperature. The effects of light on the oriental fruit moth, *Grapholitha molesta* (9), and wind effects on the cotton leafworm, *Spodoptera littoralis* (10), have been investigated. Also, it was found that the position of the traps in the crop is important because the form and height of the crop plant influence flight behavior and mating activity of the insect (11) (12). The optimum elevation of the traps for an insect species can vary because the site of mating can be influenced by wind velocity (13).

Vegetative habitat can also have an effect on the flight behavior of insects. Rahn (14) suggested that for the leek moth, *Acrolepia assectella*, the presence of its plant host favors a meeting of the sexes. Pink bollworm moths are rarely caught in traps outside the crop on which their larvae are feeding (15). For more polyphagous moths like *Adoxophyes orana* and *Trichoplusia ni*, the crop/larvae connection does not seem to be so strong.

Finally, physiologic differences in insects under different conditions may lead to different behaviors: the proportionate numbers of pink bollworm moths attracted to their natural pheromone, gossyplure, and to an empirically found attractant, hexalure, vary during the course of the cotton-growing season (16).

## D. Numbers of Traps in an Area

The spacing of traps can have a strong effect on the catches. Our own experience tells us that an overlapping radius of action may exist when traps baited with 100 µg of *Adoxophyes orana* pheromone are placed in a grid at intervals of 3 m. Traps along the outside of the grid catch more moths than traps in the center (17). Widening the grid reduces the overlapping attractancy of the traps. The distance at which the difference in catches disappears is the maximum range of attraction. There are as yet no good quantitative data on the influence of several traps on one another in a grid. Releasing and recapturing marked insects at different distances from a trap is a good technique for testing attractancy. If these tests are performed in the laboratory, however, the insects should be checked for possible differences—for instance, in flight behavior—from wild insects (18).

## E. Trap Design

Many trap designs have been devised; generally moths have been caught in cylindrical traps like the "classical" ice-cream cups, in pieces of plastic tube, in triangular-shaped celluloid, and also in the commercially available, ready-for-use traps such as SECTAR® and PHEROCON® traps. The attracted moths are caught on the sticky substance inside the traps. Beetles are, for the most part, trapped in winged traps; they fly against the sticky wings of the traps. Sometimes pieces of sticky iron mesh are used to trap the smaller elm bark beetle, *Scolytus multistriatus* (19). For tephritid flies, the invaginated glass-trap of McPhail is the best-known type for use with liquid attractants, but for these flies various types of sticky traps have also been used (Chapter 19).

For several insect species, investigators have attempted to design traps that will capture a great number of insects. The resulting traps, thus far, have large capacity, but are inefficient. Many authors have used the term *trap efficiency* in a narrow sense; however, efficiency is not related specifically to trap design, but is influenced by the whole complex of factors previously enumerated. A trap is considered "efficient" if it catches a high proportion of a sample of released marked insects. However, environmental conditions and behavioral factors—not the trap itself—determine the trap's efficiency. Traps that perform well do not have to be improved upon ad infinitum when they are used for detection and survey purposes. It is much more important to use the same type of trap all the time—than it is to introduce different traps—in order to facilitate accurate assessments of trapping data.

## F. Concentration of Attractant/Pheromone per Trap

Generally, when the concentration of attractive bait is increased, the number of captured insects also rises to an optimum point, beyond which an equilibrium may be established. Often, the number of captures even decreases. (20).

For monitoring purposes, a gradual evaporation of the attractive substance is desirable. Otherwise, the traps require too much maintenance. To regulate the rate of evaporation, many "keepers" have been developed, the polyethylene caps for moth traps and the fibrous blocks for tephritid fly traps are the best-known examples.

Clearly, in light of the many factors influencing the efficiency of traps and the difficulties in relating trap catches to population density, attractant and pheromone traps are in most pest situations not the right means of estimating population densities. One cannot use the number of caught insects as economic thresholds of population density. Egg counting, larval counting, and harvest checking, although more laborious and time-consuming, constitute better methods. Also, the relation between levels of damage and population density is often uncertain. Perhaps the only trapping method that can give reliable correlations between trap captures and population density is the release-recapture technique. From the proportions of marked and wild insects in the trap catches, one can determine the proportion of the wild population taken by the traps.

A number of studies have described the problems of estimating an insect population by sex-pheromone traps. Studies on tortricid orchard pests like the codling moth are cases in point. But they did not indicate any firm thresholds or suggest how spraying should be timed (3) (21) (22). Riedl and Croft (23) made a thorough analysis of damage levels in relation to trap catches; they tried to develop some computer models of this relationship. De Jong and van Dieren (24) analyzed different sampling methods, including sex-pheromone trapping, and, by computer, related these to the population dynamics of *Adoxophyes orana.*

In spite of these excellent studies, I believe that the whole matter is too complicated to allow the design of such a simple formula dictate an appropriate reaction on our part to all situations occurring during the growing season. For practical reasons, I prefer the heat-summation method of estimating populations; the emergence of the larvae can be predicted by using both information from trap catches of the preceding adult generation and the average temperatures at the time of egg incubation. As far as I know, this method has been developed only for some moths: for the codling moth (25) (26) and for *Adoxophyes orana* (27). This method allows one to apply the sprays at the most effective times so that the number of sprayings can be decreased. This combination of trap-catch and temperature data can probably be adjusted quite easily, and applied to other insects.

## III. MASS-TRAPPING OR MALE-ANNIHILATION TECHNIQUE

Mass-trapping techniques are designed to attract insects to traps and then to kill or to neutralize them before they have an opportunity to reproduce. These traps compete with the attractive females or (sometimes) males of the wild population, and all the factors described for detection and survey also apply to this direct-control method. Here the need is even greater to optimize trapping conditions because it is imperative to remove the majority ($> 90\%$) of the population. Also, extensive studies on all the factors influencing trap catches are necessary for mass trapping to have any chance of success. Here, too, I have doubts about the reliability of quantitative trapping.

Low populations of the red-banded leafroller can be kept reasonably under control, but there has been little headway made against infestations that have already reached a damaging level (28) (29). However, experiments with the red-banded leafroller, *Argyrotaenia velutinana*, and the grape berry moth, *Paralobesia viteana,* indicated substantial reduction in damage in mass-trapped fields (30). Extensive mass-trapping experiments with the western pine beetle, *Dendroctonus brevicomis,* at Bass Lake, California, justify more research; there were encouraging results during the first years (31). There are also the well-known examples of the fruit fly eradication campaigns in the Pacific islands, where tephritid attractants have been combined with a killing or sterilizing agent. Yet these attractants worked well only in isolated areas like islands and valleys, and this constitutes a serious restriction to the use of mass trapping: mass trapping can be adopted only for direct control, and only if immigration of new insects into the treated area can be prevented. Thus, trapping for purposes of detection remains an essential part of any post-treatment operation. To assess the level of infestation other methods of estimation than trapping are needed.

## IV. METHOD OF ASSESSMENT FOR THE DISRUPTION TECHNIQUE

Mating-disruption experiments have as yet been undertaken only with the sex pheromones of such moth species as the gypsy moth, *Porthetria dispar* (32) (33), the cabbage looper *Trichoplusia ni* (34), the pink bollworm, *Pectinophora gossypiella* (Chapter 21), the red-banded leafroller (35), and the grape berry moth (30).

It is no simple matter to assess the effect of a disruption experiment. In one study the reduced catches of sex-pheromone traps installed in the treated area were used as the only criterion for a successful experiment (36).

However, a reduced attractancy to the monitoring traps does not necessarily mean a reduced mating frequency, which is the principle aim of a disruption experiment. Mating frequency can easily be determined by spermatophore counts in females. Caged or tethered females can be used (34), or the females must be trapped. To trap females of some species, like the oriental fruit moth, one can use lure pots filled with wine, brown sugar, and terpinyl acetate (37). Other methods of estimation, like egg counts, larval counts, or harvest checks, can give additional information, essential for the assessment of the disruption technique.

I consider the use of monitoring sex traps in disruption experiments useful only as an indicator that the amount of pheromone used for the disruption is sufficiently high, and not as an overall measure of success or failure of the technique.

## V. SUMMARY

Traps baited with behavior-modifying chemicals can be used in four different ways—that is, for detecting the presence of an insect pest, for estimating its population density, for removing it *en masse* from an area, and for assessing the mating disruption technique. Trapping for detecting the presence of an insect pest is widely used and it is the only successful method in practice. Quantitative estimation of a population by attractant/pheromone traps is almost impossible because the trapping is influenced by too diverse a set of factors. The outlook for pest control by the use of mass trapping does not look very promising because it is difficult to remove more than 90% of the population, the amount necessary in many pest situations. Sex-pheromone traps have only a limited use for assessing the effect of mating-disruption experiments.

## REFERENCES

(1)   A.K. Minks. 1969. The use of sex traps for phenological observations on the summerfruit tortrix moth, *Adoxophyes orana* (F.v.R). *Meded. Rijksfak. Landbouwwetensch. Gent* 34:628-636.

(2)   M.T. Aliniazee and E.M. Stafford. 1971. Evidence of a sex pheromone in the omnivorous leafroller, *Platynota stultana* (Lepidoptera: Tortricidae): Laboratory and field testing of male attraction to virgin females. *Ann. Entomol. Soc. Am.* 64:1330-1335.

(3)   H.F. Madsen and J.M. Vakenti. 1972. Codling moths: Female-baited and synthetic pheromone traps as population indicators. *Environ. Entomol.* 1:554-557.

(4)   A.K. Minks. 1975. Die mögliche Anwendung von Sexualpheromonen für die Bekämpfung des Apfelschalenwicklers, *Adoxophyes orana*, im holländischen Obstbau. *Z. ang. Entomol.* 77:330-336.

(5)  F.J. Howell. 1974. The competitive effect of field populations of codling moth on sex attractant trap efficiency. *Environ. Entomol.* 3:803-807.

(6)  C.A. Miller and G.A. McDougall. 1973. Spruce budworm moth trapping using virgin females. *Can. J. Zool.* 51:853-858.

(7)  D.A. Carlson and M. Beroza. 1973. Field evaluation of (Z)-9-tricosene, a sex attractant pheromone of the housefly. *Environ. Entomol.* 2:555-559.

(8)  A.K. Minks and J.Ph.W. Noordink. 1971. Sex attraction of the summerfruit tortrix moth, *Adoxophyes orana:* Evaluation in the field. *Entomol. exp. appl.* 14:57-72.

(9)  G.H.L. Rothschild and A.K. Minks. 1974. Time of activity of male oriental fruit moths at pheromone sources in the field. *Environ. Entomol.* 3:1003-1007.

(10)  D.G. Campion. 1974. The use of sex pheromones in the control of *Spodoptera littoralis* Boisd. *EPPO Bull.* 4:357-362.

(11)  M.T. Aliniazee and E.M. Stafford. 1972. Sex pheromone studies with the omnivorous leafroller, *Platynota stultana* (Lepidoptera: Tortricidae): Effect of various environmental factors on attraction of males to the traps baited with virgin females. *Ann. Entomol. Soc. Am.* 65:958-961.

(12)  C.A. Saario, H.H. Shorey, and L.K. Gaston. 1970. Sex pheromones of noctuid moths. XIX. Effect of environmental and seasonal factors on captures of males of *Trichoplusia ni* in pheromone-baited traps. *Ann. Entomol. Soc. Am.* 63:667-672.

(13)  R.S. Kaae and H.H. Shorey. 1973. Sex pheromones of Lepidoptera. 44. Influence of environmental conditions on the location of pheromone communication and mating in *Pectinophora gossypiella*. *Environ. Entomol.* 2:1081-1084.

(14)  R. Rahn. 1968. Rôle de la plante-hôte sur l'attractivité sexuelle chez *Acrolepia assectella* Zeller (Lep. Plutellidae). *C.R. Acad. Sci. Paris* 266:2004-2006.

(15)  R.K. Sharma, R.E. Rice, H.T. Reynolds, and H.H. Shorey. 1971. Seasonal influence and effect of trap location on catches of pink bollworm males in sticky traps baited with hexalure. *Ann. Entomol. Soc. Am.* 64:102-105.

(16)  H.H. Shorey. Personal Communication.

(17)  A.K. Minks. 1975. Biological aspects of the use of pheromones in integrated control with particular reference to the summerfruit tortrix moth, *Adoxophyes orana. C. R. 5e Symp. Lutte integrée en vergers. OILB/SROP* 1975, pp. 295-302.

(18)  A.K. Minks. 1971. Decreased sex pheromone production in an in-bred stock of the summerfruit tortrix moth, *Adoxophyes orana. Entomol. Exp. Appl.* 14:361-364.

(19)  J.W. Peacock, R.A. Cuthbert, W.E. Gore, G.N. Lanier, G.T. Pearce, and R.M. Silverstein. 1975. Collection on Porapak Q of the aggregation pheromone of *Scolytus multistriatus* (Coleoptera: Scolytidae). *J. Chem. Ecol.* 1:149-160.

(20)  L.K. Gaston, H.H. Shorey, and C.A. Saario. 1971. Sex pheromones of noctuid moths. XVIII. Rate of evaporation of a model compound of *Trichoplusia ni* sex pheromone from different substrates at various temperatures and its application to insect orientation. *Ann. Entomol. Soc. Am.* 64:381-384.

(21)  E. Mani, Th. Wildbolz, and W. Riggenbach. 1972. Die Männchenfalle, eine neue Prognosenmethode für den Apfelwickler; Resultate 1969-1971. *Schweiz. Z. Obst. Weinbau* 81:337-344.

(22)  G.C. Rock and D.R. Yeargan. 1974. Flight activity and population estimates of four apple insect species as determined by pheromone traps. *Environ. Entomol.* 3:508-510.

'(23)  H. Riedl and B.A. Croft. 1974. A study of pheromone trap catches in relation to codling moth (Lepidoptera: Olethreutidae) damage. *Can. Entomol.* **106**:525-537.

(24)  D.J. de Jong and J.P.A. van Dieren. 1974. Population dynamics of the summer fruit tortricid *Adoxophyes orana* F.v.R. in relation to economic threshold levels. *Meded. Fak. Landbouwwetensch. Gent.* **39**:777-788.

(25)  W.C. Batiste, A. Berlowitz, W.H. Olson, J.E. DeTAR, and J.L. Joos. 1973. Codling moth: Estimating time of first egg hatch in the field—A supplement to sex-attractant traps in integrated control. *Environ. Entomol.* **2**:387-391.

(26)  P.J. Charmillot, M. Baggiolini, and G. Fiaux. 1975. Les phéromones en lutte intégrée, cas du carpocapse. *C. R. 5e Symp. Lutte intégrée en vergers. OILB/SROP,* 1975, pp. 303-313.

(27)  A.K. Minks and D.J. de Jong. 1975. Determination of spraying dates for *Adoxophyes orana* by sex pheromone traps and temperature recordings. *J. Econ. Entomol.* **68**:729-732.

(28)  W.L. Roelofs, E.H. Glass, J. Tette, and A. Comeau. 1970. Sex pheromone trapping for red-banded leaf roller control: Theoretical and actual. *J. Econ. Entomol.* **63**:1162-1167.

(29)  K. Trammel, W.L. Roelofs, and E.H. Glass. 1974. Sex-pheromone trapping of males for control of redbanded leafroller in apple orchards. *J. Econ. Entomol.* **67**:159-164.

(30)  E.F. Taschenberg, R.T. Cardé, and W. L. Roelofs. 1974. Sex pheromone mass trapping and mating disruption for control of redbanded leafroller and grape berry moths in vineyards. *Environ. Entomol.* **3**:239-242.

(31)  W.D. Bedard and D.L. Wood. 1974. Programs utilizing pheromones in survey and control: Bark beetles—the western pine beetle. In *Pheromones.* M.C. Birch, Ed. North-Holland, Amsterdam, pp. 441-449.

(32)  M. Beroza, C.S. Hood, D. Trefrey, D.E. Leonard, E.F. Knipling, W. Klassen, and L.J. Stevens. 1974. Large field trial with microencapsulated sex pheromone to prevent mating of the gypsy moth. *J. Econ. Entomol.* **67**:659-664.

(33)  E.A. Cameron, C.P. Schwalbe, M. Beroza, and E.F. Knipling. 1974. Disruption of gypsy moth mating with microencapsulated disparlure. *Science* (Wash.) **183**:972-973.

(34)  H.H. Shorey, R.S. Kaae, L.K. Gaston, and J.R. McLaughlin. 1972. Sex pheromones of Lepidoptera. XXX. Disruption of sex pheromone communication in *Trichoplusia ni* as a possible means of mating control. *Environ. Entomol.* **1**:641-645.

(35)  R.T. Cardé, K. Trammel, and W.L. Roelofs. 1975. Disruption of sex attraction of the redbanded leafroller *(Argyrotaenia velutinana)* with microencapsulated pheromone components. *Environ. Entomol.* **4**:448-450.

(36)  M. Beroza and E.F. Knipling. 1972. Gypsy moth control with the sex attractant pheromone. *Science* (Wash.). **177**:19-27.

(37)  J.H.H. Phillips. 1973. Monitoring for oriental fruit moth with synthetic sex pheromone. *Environ. Entomol.* **2**:1039-1042.

Chapter 24

# Advancing toward Operational Behavior-Modifying Chemicals

C. S. Koehler
*University of California*
*Berkeley, California*


J. J. McKelvey, Jr.
*The Rockefeller Foundation*
*New York, New York*


W. L. Roelofs
*New York State Agricultural Experiment Station*
*Geneva, New York*


H. H. Shorey
*Department of Entomology*
*University of California*
*Riverside, California*


R. M. Silverstein
*SUNY College of Environmental Science and Forestry*
*Syracuse, New York*


D. L. Wood
*Department of Entomological Sciences*
*University of California*
*Berkeley, California*

The study of chemicals that modify insect behavior is a dynamic, emerging field of science. These chemicals offer considerable promise that they will become important components of pest-management systems. During an intensive conference at Bellagio, Italy the contributors to this book isolated and identified certain issues that are likely to influence the future growth and direction of this field. This analysis led to a number of conclusions and recommendations which are set forth here for the information not only of those who are interested in the basic or applied research in this field, but also of those who sponsor or supervise such research, and of those who might develop commercial applications of the research results.

## I. STATUS AND PROSPECTS OF BEHAVIOR-MODIFYING CHEMICALS IN PEST MANAGEMENT

Certain types of behavior-modifying chemicals have already been put to practical use. For example, some of the chemicals are naturally occurring, deterrent compounds found in strains of plants that have been bred for their insect-resistant properties (Chapter 17). In the case of tephritid fruit flies, chemical attractants have been used extensively for both survey and detection of infestations, and as components of toxic baits that have been used effectively to eradicate new fruit- fly infestations from certain areas (Chapter 19). Behavior-modifying chemicals have also been used for some time in conventional pest-control practices to attract a number of other agricultural pests to toxicants in the field (Chapter 21). The use of repellents to ward off mosquitoes and other blood-sucking insects (Chapter 7 and 18) is familiar to everyone. The recent identification of the pheromones of a number of pest insects has provided a new, especially powerful tool for monitoring the spread of insects into previously uninfested areas, or the build-up of infestations to densities that require suppression (Chapters 19-23). Pheromones, although not yet fully operational as direct control agents in pest-management systems, offer early promise as safe, selective, and effective control agents against a number of important pests (Chapters 20-22).

Relative to some of the other new approaches to pest management—that is, the mass release of sterile male insects, the use of hormones or other insect growth regulators, and the production of genetically inferior strains of insects—behavior-modifying chemicals have already achieved a high level of operational success. During the past decade, much of the enthusiasm directed toward some of these new approaches has either disappeared or been tempered by the failure of the approaches to perform as expected. Together with the manipulation and dissemination of classical biological control agents (parasites, predators, and pathogens) and the implementation of cultural techniques that are disadvantageous to pest insects, behavior-modifying chemicals are potentially a viable alternative or supplement to the use of conventional chemical insecticides. Some observers have evidenced impatience with the various programs that have been directed toward the development of behavior-modifying chemicals (especially the pheromones) for insect pest management. However, it must be recalled that almost all of the practical research directed toward operational control programs has been initiated only during the past 5 years, and by a small number of investigators. Additionally, only limited resources have been made available for this research. Thus, on any reasonable basis of expectation in science, progress should be considered quite remarkable.

## II. PHEROMONES: ONLY ONE OF MANY BEHAVIOR-MODIFYING CHEMICAL GROUPS

With the recent advent of intensive chemical, biologic, and pest-management research centered around insect pheromones, some workers have concluded that they have the ultimate behavior-modifying tool. These workers have single-mindedly emphasized pheromone research without considering the great variety of other chemicals and stimuli that influence insects and that might be used in conjunction with—or instead of—pheromones in pest-management systems. This tunnel vision is perhaps advantageous during the pioneering stages of the research; it allows the investigator to understand the biology and chemistry of the pheromone-communication system before attempting to integrate this knowledge with other pest-suppression methods and theory. But now it is time to take a broader view—initially of integrative systems based on pheromones plus other behavior-modifying chemicals, and then of behavior-modifying chemicals plus all other available pest-management tools.

There are several reasons why a researcher working with pheromones should think of integrated systems. First, although pheromones are extremely potent biologic agents, they are not panaceas, anymore than was DDT. Pheromones are being used widely in survey traps to monitor the distribution or abundance of pests, but their use as direct pest-control agents will probably be limited to certain key pests that have great economic importance (see Section IV). Second, pheromone components released from insects represent only one of the important types of chemical stimuli that influence the behavior of pest species. Other chemical messengers, especially those released from favored plant or animal hosts, can act independent of or in conjunction with pheromones to optimize the responses of the insects. This important phenomenon has been known for many years by investigators working with forest pests (Chapters 11 and 22), although it is not yet as apparent with agricultural or stored-products insects. Third, many dramatic advances in concepts and methodology, especially in the area of microanalytic chemistry, have been achieved during the last few years in laboratories pioneering pheromone research. However, these advanced techniques have not been widely employed on other behavior-modifying- chemical systems.

## III. AREAS NEEDING SPECIAL RESEARCH ATTENTION

The rapid progress that has been made during recent years in the identification of behavior-modifying chemicals, especially pheromones, has exceeded the capacity of biologists and pest-management specialists to develop useful applications of these compounds. The key to effective pest-management programs is the acquisition of an intimate knowledge of the life processes of the pest species, so that vital aspects of these processes can be attacked by man. Research areas that should receive high priority by biologists investigating behavior-modifying chemicals include the following:

1. Basic studies of insect behavior are essential. After all, it is the behavior of the pest that will be manipulated by behavior-modifying chemicals. Detailed research is needed on the mechanisms by which insects orient to chemical sources, and the influence of environmental variables such as wind, temperature, and nonchemical stimuli emanating from host plants and animals on this orientation.
2. Sensory-physiology research should emphasize the mechanisms by which stimulatory and inhibitory chemicals are received and integrated in the nervous systems of the insects.
3. Population biology investigations are needed to determine the influence of the chemicals on the pest population and on nontarget organisms—both plant and animal—as well as the effect of a large-scale suppression of a target species on other species in the pest ecosystem.
4. Short-range dispersal as well as long-range migration behavior of the target species must be determined. Also needed are studies of the influence of environmental factors on dispersion, both of the target species and of the behavior-modifying chemical when it is released into the atmosphere.

The above discussion of the problems facing biologic investigators should not be taken as an inference that the chemists have finished their work. The field has moved rapidly from the prevailing view of a little more than a decade ago that a single, unique pheromone chemical was used by each insect species. It now appears that single-component pheromones may be more the exception than the rule. Recent investigations indicate that many insect species use pheromones consisting of multiple components, that some components are "shared" by related species, that some components have little or no potency if they are displayed alone, and that the species specificity of some pheromones is a factor of the ratios in which the components are combined (Chapters 14-16). Many pheromone-isolation techniques of a decade ago seem primitive, as will our present techniques at some future date. Most early identifications of pheromones, as well as all identifications that yielded chemicals having only low levels of biologic activity, should be reviewed.

Specialized strategies must be developed for deploying behavior-modifying chemicals at the correct time and place in the environment so that thay can be maximally effective in pest-management programs. Evaporative substrates need to be designed in order to achieve proper emission rates of the chemicals over large areas and for long periods. Many advances have been made in formulations technology in other chemical fields, and this information needs to be applied to the behavior-modifying chemicals.

A selection process is now necessary as we move toward the operational pest-management phase. The time is ripe for augmented financial support for large-scale studies in a few experimental programs that are directed toward the suppression of major pest species. Such programs could establish principles, methods, and an awareness of the limitations of the approach that utilizes behavior-modifying chemicals. At the same time, we must avoid the risk of financing a narrow, elite group to the exclusion of other research workers. This potential problem can be avoided by also supporting numerous, vigorous, small-scale studies by imaginative investigators.

## IV. SELECTION OF TARGET PESTS

Careful attention should be given to the selection of the most appropriate insect species as candidates for suppression programs based on behavior-modifying chemicals. One must consider the entire pest spectrum before the decision is reached. Many insects have achieved pest status through induction. Insecticides that have been initially directed toward primary or key pests often have served to upset biotic balances by destroying natural biological control agents of other insects, and thus enabled those insects to multiply to the extent that they achieved pest status. These induced, secondary pests may not be the most practical targets of behavior-modifying chemical systems because frequent insecticide applications might still need to be applied for control of the key pests. The induced pests might cease to be a problem if the key pests were controlled by selective, ecologically sound methods. Clearly, the key pests should be the main targets of the behavioral control efforts.

## V. MAKING BEHAVIOR-MODIFYING CHEMICALS OPERATIONAL

For behavior-modifying chemicals to become operational, private industry will be needed to develop and market commercial products. Likely, extensive technical service commitments will be required also. We fully recognize that the synthesis of many behavior-modifying chemicals is not simple, that the amount of the chemicals required will be much smaller than the amount of conventional insecticides currently sold, and that procedures for registering with regulatory agencies the chemicals for use on edible commodities appear formidable at the present time. Because the public at large will be the ultimate benefactor of behavior-modifying chemical usage, governments need to play a major role in subsidizing some phases of the research and development activities.

# INDEX